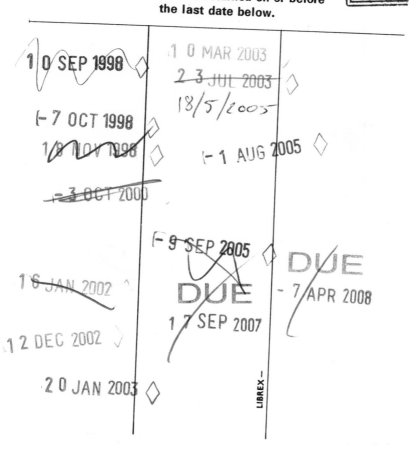

Micro-optics

Micro-optics

Elements, systems and applications

Edited by
HANS PETER HERZIG
Institute of Microtechnology, University of Neuchâtel, Switzerland

Taylor & Francis
Publishers since 1798

UK   Taylor & Francis Ltd, 1 Gunpowder Square, London EC4A 3DE
USA   Taylor & Francis Inc., 1900 Frost Road, Suite 101, Bristol, PA 19007

British Library Cataloguing in Publication Data

A catalogue record for this book is available from the British Library.
ISBN 0-7484-0481-3 (cloth)

Library of Congress Cataloging Publication Data are available

Cover design by Jim Wilkie

The book cover shows an array of refractive micro-lenses (diameter = 190 μm, focal length = 330 μm). The array has been fabricated by Ph. Nussbaum and photographed by R. Völkel.

Typeset in Times 10/12pt by Keyset Composition, Colchester, Essex

Printed in Great Britain by T. J. International Ltd, Cornwall

Contents

Contents

Preface

This book presents the properties and the potential of micro-optical elements, which are key components for building compact optoelectronic systems. Besides the theoretical background and an overview of current fabrication technology, the book covers recent applications and future trends.

The book concentrates on diffractive and refractive micro-optical elements, such as lenses, fan-out gratings, optimized phase elements, and polarizers. These elements are fabricated using modern technology to produce refractive index modulation or surface-relief modulation. Almost any structure shape, including asymmetric aspherics, can be manufactured, which provides all degrees of freedom for the design.

Chapter 1 introduces the design fundamentals of diffractive optical elements (DOEs) and compares their optical properties with those of refractive elements. The design is based on scalar diffraction theory and geometrical ray-tracing. Chapter 2 presents the rigorous diffraction theory of surface-relief gratings, and outlines approximate methods which are applicable when the grating period $d \gg \lambda$ or $d \ll \lambda$. The design data can be transformed into an efficient optical element with the aid of modern micro- and nanotechnology, including e-beam writing, laser beam writing, lithography, etching, and thin film deposition. Attention is also given to replication techniques for low-cost mass-production. Chapter 3 explores key aspects of binary optics technology, focusing on the crucial methodology of device fabrication. Chapter 4 concentrates on the two major technologies for direct writing of continuous-relief microstructures – laser beam writing and electron beam writing. Alternative technologies such as diamond turning and laser ablation are also included. A variety of techniques for the manufacture of refractive microlenses is summarized in Chapter 5. The most important methods are: ion diffusion to produce graded-index lenses, 'thermal reflow', and photolithographic techniques using photosensitive glass. Chapter 6 gives an overview of the main replication technologies applicable to micro-optics and DOE production, i.e. hot embossing, injection moulding, and casting.

New packaging techniques are required for the realization of compact, stable systems with a high alignment precision of a few microns or less between different optoelectronic components. Planar optics and stacked optics are two basic concepts. Chapter 7 describes in detail the planar optics approach to building integrated free-space optical systems. Several potential applications are discussed. Chapter 8 presents the concept of stacked optics, which is a real three-dimensional integration of optical systems.

Furthermore, the book describes selected applications to demonstrate the potential of micro-optical elements. The unique properties of lasers have made them valuable tools in many applications, ranging from medicine to metrology. In many cases, however, the spatial characteristics of a particular laser beam or arrays of beams are not ideally suited to the application, and it is desirable to modify these characteristics. Chapter 9 explores several optical techniques for laser beam shaping. In order to combine the advantages of refractive optics (low dispersion) and diffractive optics (arbitrary shape), considerable work has been invested in the development of hybrid elements. Chapter 10 investigates hybrid optical elements for the most common tasks, i.e. achromatization, athermalization, and the reduction of spherical aberration. Chapter 11 considers the design of elements that are capable of producing an array of point sources with illumination from a single point source. Finally, Chapter 12 investigates the polarization properties of high spatial frequency surface-relief gratings. Emphasis is placed on the design and application of lamellar gratings in the zero-order diffraction regime.

As editor, I wish to express my special thanks to the authors who contributed a chapter to this book. In addition, I am grateful to many friends and colleagues for reading portions of the text and providing helpful comments. I would especially like to mention Peter Blattner, Prof. René Dändliker, Dr Peter Ehbets, Dr Markku Kuittinen, Dr Wolfgang Singer, Dr Reinhard Völkel and Dr Ken Weible.

Hans Peter Herzig
Neuchâtel, Switzerland

Contributors

Gregory P. Behrmann
US Army Research Laboratory, AMSRL-SE-EO, 2800 Powder Mill Road, Adelphi, Maryland 20783, USA

Karl-Heinz Brenner
Lehrstuhl für Optoelektronik, Technische Informatik, Universität Mannheim, B6, 26 Z.320, D-68131 Mannheim, Germany

Mike T. Gale
Paul Scherrer Institute, Badenerstrasse 569, CH-8048 Zurich, Switzerland

Charles W. Haggans
3M Storage Diskette and Opt. Techn. Division, Building 544-2N-01, 1185 Wolters Boulevard, Vadnais Heights, MN 55110-5128, USA

Hans Peter Herzig
Institute of Microtechnology, University of Neuchâtel, Rue A.-L. Breguet 2, CH-2000 Neuchâtel, Switzerland

Michael C. Hutley
National Physical Laboratory, Queen's Road, Teddington, Middlesex TW11 0LW, UK

Jürgen Jahns
Fernuniversität-GH Hagen, Lehrstuhl für Optische Nachrichtentechnik, Feithstrasse 140, D-58084 Hagen, Germany

Raymond K. Kostuk
Electrical and Computer Engineering, University of Arizona, Tucson, AZ 85721, USA

Contributors

James R. Leger
Department of Electrical Engineering, University of Minnesota, 200 Union Street SE, Minneapolis, MN 55455, USA

Joseph N. Mait
US Army Research Laboratory, AMSRL-SS-IA, 2800 Powder Mill Road, Adelphi, MD 20783-1145, USA

Wolfgang Singer
Institut für Technische Optik, Universität Stuttgart, Pfaffenwaldring 9, D-70569 Stuttgart, Germany

Margaret B. Stern
Lincoln Laboratory, Massachusetts Institute of Technology, 244 Wood Street L-237, Lexington, Massachusetts, 02173-9108, USA

Jari Turunen
Department of Physics, University of Joensuu, P.O. Box 111, 80101 Joensuu, Finland

1

Design of Refractive and Diffractive Micro-optics

H. P. HERZIG

1.1 Introduction

Passive optical components, shown in Fig. 1.1, are used in optical systems to collect, distribute or modify optical radiation. Refractive and reflective components, such as lenses, prisms and mirrors, are well-known. Following the trend of miniaturization, novel technologies have been developed to shrink the size of these elements. Refractive microlenses with diameters of 1 mm down to a few microns can now be fabricated with high quality (see Chapter 5).

In parallel, diffractive optics has emerged from holography (Swanson and Veldkamp, 1989; Herzig and Dandliker, 1993; Herzig *et al.*, 1993). Typical diffractive optical elements (DOEs) have multilevel microreliefs ('binary optics') or continuous microreliefs, with features ranging from submicron to millimetre dimensions and relief amplitudes of a few microns. Novel structures can be realized, complementing and exceeding the possibilities of traditional lenses, prisms and mirrors. Almost any structure shape, including asymmetric aspherics, can be manufactured, which provides all degrees of freedom for the design.

Refractive elements consisting of macroscopic surface-relief structures are designed using the laws of geometrical optics, treating light by the refraction and reflection of geometrical rays at optical interfaces. The eikonal equation, as the basis of geometrical optics, can be derived from Maxwell's equations in the limit where the wavelength tends to zero (Born and Wolf, 1980). Therefore, no wavelength-dependent properties are observed, apart from those due to material dispersion. In contrast, DOEs are planar elements consisting of zones, which retard the incident wave by a modulation of the refractive index or by a modulation of the surface profile. The light emitted from the different zones interferes and forms the desired wavefront. Since these phenomena are strongly dependent on the wavelength of light, DOEs are restricted to monochromatic applications. In order to combine the advantages of refractive optics (low dispersion) and diffractive optics (arbitrary shape), considerable work has been invested in the development of hybrid elements (see Chapter 10).

The design of refractive and reflective optics has been extensively discussed in

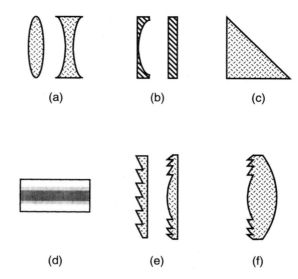

Figure 1.1 Optical elements: (a) refractive lenses, (b) mirrors, (c) prisms, (d) GRIN lenses, (e) diffractive optical elements (DOEs), (f) hybrid elements.

the literature (Kingslake, 1978; Welford, 1986). This chapter presents the design fundamentals of diffractive optical elements (DOEs) and compares their optical properties with those of refractive elements.

The following section 1.2 describes the design of diffractive elements and systems. The transfer of waves through a diffractive element is modelled by the thin element approximation and the propagation through free space by scalar diffraction theory. Section 1.2.1 presents the conversion of the design data into an efficient diffractive element. We concentrate on multilevel and continuous surface-relief structures.

In general, the optical function of the DOE, or its surface-relief profile, has to be found by optimization. The most successful schemes are summarized in section 1.2.2. Ray-tracing is the standard method of designing optical systems consisting of refractive and reflective elements. We show how to apply ray-tracing to the design of diffractive elements. In section 1.2.3, we present another concept for the system design, which is based on wave optics.

Most diffractive elements also have refractive properties. However, the diffractive properties are dominant and determine the optical function. This is different if only a few zones of the element are illuminated. In section 1.3, we discuss the refractive properties of DOEs using the examples of blazed gratings and fan-out elements.

Finally, section 1.4 compares the basic properties of refractive and diffractive microlenses, i.e. the focal length, the dispersion and the aberrations.

1.2 Design of Diffractive Optical Elements and Systems

In scalar diffraction theory, the diffractive optical element with phase profile $\Psi(x, y)$ is modelled as a thin phase screen with a complex amplitude transmittance of

$$t(x, y) = \exp[i\Psi(x, y)]. \tag{1.1}$$

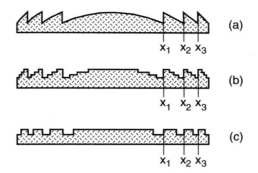

Figure 1.2 Diffractive lens with (a) continuous profile, (b) multilevel profile and (c) binary profile.

The DOE retards the incident wavefront and propagation of the new wavefront is modelled by the appropriate scalar formulation, e.g. angular spectrum, Fresnel diffraction, Fraunhofer diffraction (see Goodmann, 1968).

Note that there is a difference between the phase profile $\Psi(x, y)$ of a DOE and the phase $\Phi(x, y)$, which is generated in the first (or another) diffraction order. Figure 1.2 shows diffractive lenses with different phase profiles $\Psi(x, y)$, which generate all the same phase function $\Phi(x, y)$ in the first order. The elements perform the same wavefront conversion, but with different diffraction efficiency. The diffraction efficiency is defined as the amount of light intensity that goes into a particular diffraction order. The propagation of the first diffraction order can be modelled by replacing $\Psi(x, y)$ with $\Phi(x, y)$ in equation (1.1).

In this section, we explain the design of diffractive elements and systems. First, we start with the calculation of the phase function $\Phi(x, y)$ for first-order DOEs and the implementation of $\Phi(x, y)$ as continuous or multilevel surface-relief elements.

1.2.1 *Phase Function: Description and Realization*

The Phase Function

A thin phase element that is illuminated by an incident wave $\Phi_{in}(x, y)$ generates an output wave $\Phi_{out}(x, y)$. The wavefront conversion is described by

$$\Phi_{out}(x, y) = \Phi_{in}(x, y) + \Phi(x, y). \tag{1.2}$$

From equation (1.2) we can easily find the phase function $\Phi(x, y)$ of the phase element for a given pair of waves:

$$\Phi(x, y) = \Phi_{out}(x, y) - \Phi_{in}(x, y). \tag{1.3}$$

For the diffractive lens shown in Fig. 1.3, which has to connect an object point (x_1, y_1, z_1) with an image point (x_2, y_2, z_2), the phases Φ_{out} and Φ_{in} are of the form

$$\Phi_i(x, y) = \frac{2\pi}{\lambda_0} \sqrt{(x - x_i)^2 + (y - y_i)^2 + (z_i)^2}$$

where λ_0 is the design wavelength and $i = 1, 2$.

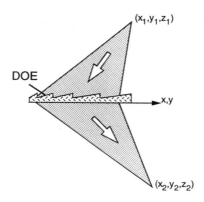

Figure 1.3 Diffractive lens which connects an object point (x_1, y_1, z_1) with an image point (x_2, y_2, z_2).

In general, the optical task is more complex, e.g. if an extended object has to be imaged. In this case, the DOE phase function $\Phi(x, y)$ is typically described by a polynomial:

$$\Phi(x, y) = \frac{2\pi}{\lambda_0} \sum_m \sum_n a_{mn} x^m y^n. \tag{1.4}$$

The DOE is then optimized by optimizing the polynomial coefficients a_{mn} (see section 1.2.2).

The Implementation of the Phase Function as Diffractive Element

The phase function $\Phi(x, y)$ can be implemented as a refractive or as a diffractive element. A refractive element generates the phase distribution Φ by varying the optical path length through a phase plate. In the case of a diffractive element, the phase function is mainly generated by the position and the grating period of a local grating. The shape of the grating period determines the efficiency of the element, which is the amount of light that goes into a particular diffraction order.

In order to realize a diffractive element, the phase function Φ is wrapped to an interval between 0 and an integer multiple of 2π. In the following, without loss of generality, we restrict the discussion to the case of maximum modulation depth equal to 2π. The phase profile Ψ of the DOE is then given by

$$\Psi(x, y) = [\Phi(x, y) + \varphi_0] \mod 2\pi \tag{1.5}$$

where φ_0 is a constant phase offset. The surface-relief profile $h(x, y)$ for a thin DOE in transmission is related to the phase profile $\Psi(x, y)$ by

$$h(x, y) = \frac{\lambda_0}{n(\lambda_0) - 1} \frac{\Psi(x, y)}{2\pi} \tag{1.6}$$

where n is the refractive index of the grating material and λ_0 is the design wavelength. In the case of large diffraction angles, equations (1.5) and (1.6) have to be calculated with a more accurate method (Rossi *et al.*, 1995, 1996).

The consequence of the phase wrapping in equation (1.5) is discrete steps of 2π in the phase profile. Using the phase offset φ_0, the position of the transition points can be shifted laterally. This has no effect on the quality of an ideal diffractive element with the calculated phase function $\Psi(x, y)$, but it can influence the behaviour of the element in the case of shrinkage errors, wavelength errors, or if only a few periods are illuminated (see section 1.3). A well-known example is the binary zone plate. The size of the central ring of a zone plate can be varied by adding a phase offset, which considerably reduces the amount of light in the zero order (Singer *et al.*, 1996).

As already mentioned, we can distinguish between the continuous first-order phase function $\Phi(x, y)$, which is determined by the position x_i of the phase transitions (Fig. 1.2), and the wrapped phase profile $\Psi(x, y)$ of the diffractive element. In the ideal case of a continuous surface-relief element (Fig. 1.2(a)), only one diffraction order is generated and then both functions are identical.

If the calculated phase function Φ is in the order of a few times 2π or even less than 2π, then the phase wrapping described by equation (1.5) yields only a few or no phase transitions. Such an element does not behave in a purely diffractive way (see section 1.3). In this case, we can add a linear phase term or a weak focusing power to increase the number of transitions.

Binarization and Diffraction Efficiency

Scalar theory predicts an ideal diffraction efficiency of 100% for DOEs with continuous surface-relief profile. The technology for fabricating these elements (Chapter 4) has made significant progress; however, experimental results are still 5–15% lower than the theoretical prediction. The standard method of fabricating DOEs in rigid material, such as glass and quartz, is based on multiple mask projection and subsequent etching (Chapter 3). These elements have staircase-like phase profiles as shown in Figs 1.2(b) and (c). (For further reading on the fabrication methods, see Chapters 3 and 4 of this book.) The diffraction efficiency of staircase gratings depends on the number of phase levels M. For linear gratings the first-order diffraction efficiency η is given by (Swanson and Veldkamp, 1989)

$$\eta = \left(\frac{\sin(\pi/M)}{\pi/M} \right)^2. \tag{1.7}$$

For eight phase levels the diffraction efficiency is already 95%. The diffraction efficiency for binary gratings (two phase levels) is only 40.5%. However, they are interesting for applications such as fan-out elements or diffusers, where several diffraction orders are used.

Equation (1.7) can also be applied to estimate the efficiency of diffractive lenses. The lens structure is considered as local gratings with varying periods. In the central part of the lens, the grating periods are large and therefore the quantization of the phase profile into eight phase levels is simple and not critical. The grating periods become smaller towards the rim. The number of phase levels is then limited by the resolution of the lithographic fabrication process. The diffraction efficiency decreases rapidly for small grating periods if only four or two phase levels are feasible. In this case, the quantization of the continuous lens function becomes important. Successful encoding schemes include the RISIDO method presented by Welch *et al.* (1993), which is based on optimization, and the somewhat simpler Direct Sampling (DS) method presented by Kuittinen and Herzig (1995).

The relation between the maximum number M of phase levels and the first-order diffraction angle θ is given by

$$M = \frac{\lambda}{\sin \theta \, m_f} \qquad (1.8)$$

where m_f is the minimum feature size that can be fabricated with the available technology. The diffraction efficiency η for a given number M of phase levels is known from equation (1.7). If we now substitute equation (1.8) in equation (1.7) we get an expression for the maximum diffraction efficiency profile as a function of the diffraction angle:

$$\eta(\theta) = \left(\frac{\sin[(\pi m_f \sin \theta)/\lambda]}{(\pi m_f \sin \theta)/\lambda} \right)^2 . \qquad (1.9)$$

For a diffractive lens, the angle θ in equation (1.9) varies from the centre ($\theta = 0$) up to the rim ($\theta = \theta_{max}$). The maximum angle θ_{max} is determined by the numerical aperture of the lens (NA $= \sin \theta_{max}$). The efficiency η versus the diffraction angle θ is shown in Fig. 1.4 by a solid line. We stress here that equation (1.9) gives the highest possible efficiency for a constant sample space ($= m_f$). In practice the efficiency is lower, because the phase values are quantized. We can now estimate the maximum efficiency of a diffractive lens by integrating $\eta(\theta)$ over the angle. For cylindrical and spherical lenses we can write (Kuittinen and Herzig, 1995)

$$\eta = \int_0^{\theta_{max}} s(\theta) \, \eta(\theta) \, d\theta \qquad (1.10)$$

where $s(\theta) = 2f/(d \cos^2 \theta)$ for cylindrical lenses and $s(\theta) = 8f^2 \tan \theta/(d^2 \cos^2 \theta)$ for spherical lenses. d is the diameter of the lens.

Assuming a minimum feature size of $1 \, \mu m$ and a wavelength of $\lambda = 488 \, nm$, equation (1.10) predicts a maximum diffraction efficiency of 93.3% for a diffractive lens with NA $= 0.1$ and 75.8% for a lens with NA $= 0.2$. The estimation is based on scalar diffraction theory. For lenses with higher NA, rigorous diffraction theory is required (see Chapter 2).

1.2.2 *Optimum Design*

Figure 1.5 shows two basically different design problems: (a) imaging and (b) beam shaping. In the case of imaging a set of continuous waves emitted by the object has to be converted into another continuous set of output waves forming the image. The ideal input and output waves are known, but the optical element that best images all input waves is not yet known. The term 'imaging' includes here not only the classical image formation, but also the design of Fourier lenses (Kedmi and Friesem, 1986), laser scanners (Herzig and Dändliker, 1987), or other design problems, where we have to minimize the geometrical aberrations. In the case of beam shaping (Fig. 1.5(b)), one input wave illuminates the DOE, which should generate the desired intensity distribution in another plane. Now, the output wave is unknown. Typical applications are the conversion of a Gaussian beam into a uniform beam with rectangular shape, or fan-out elements for array generation. Different design methods have been investigated; we present here some of the most common.

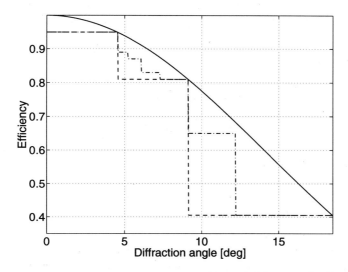

Figure 1.4 Optimum diffraction efficiency profile (solid line) and efficiency profiles when 8, 4 and 2 phase levels are used (dashed line) and when all (8, 7, 6, . . .) phase levels are used (dashed-dotted line). The fabrication limit is assumed to be 1 μm (m_f) and the operating wavelength is 632.8 nm.

Ray-tracing

Ray-tracing is the standard method of designing optical elements (Smith, 1990). In ray-tracing through lens systems the path of the light is determined, with the help of elementary geometry, by successive application of the law of refraction (or reflection). In diffractive optics, the law of refraction has to be replaced by the grating diffraction equation.

In a ray-tracing program, the diffractive phase element is described by the first-order phase function $\Phi(x, y)$ (see section 1.2). The reconstruction process is essentially governed by the condition of phase matching in the DOE plane (x, y), which is given by equation (1.2). The phase matching condition yields relations for the normal projection $k_{x,i}, k_{y,i}$ of the wavevectors $\mathbf{k}_i$, onto the (x, y) plane. The vectors $(k_{x,i}, k_{y,i})$ and the phase functions Φ_i are related by $k_{x,i} = \partial\Phi_i/\partial x$ and $k_{y,i} = \partial\Phi_i/\partial y$. By derivation of equation (1.2), we obtain

$$k_{x,\text{out}} = k_{x,\text{in}} + m \frac{\partial\Phi}{\partial x} \tag{1.11a}$$

$$k_{y,\text{out}} = k_{y,\text{in}} + m \frac{\partial\Phi}{\partial y} \tag{1.11b}$$

where m is the diffraction order. In general, the elements are designed for the first diffraction order ($m = 1$). The gradient $(\partial\Phi/\partial x, \partial\Phi/\partial y)$ in equation (1.11) describes the local grating vector of the diffractive structure. The length of the wavevectors at the reconstructing wavelength is given by $|\mathbf{k}| = |\mathbf{k}_{\text{in}}| = 2\pi/\lambda_{\text{in}}$. For a transmission element, the component $k_{z,\text{out}}$ of the outgoing wave normal to the x, y-plane is determined by

$$k_{z,\text{out}} = \text{sign}(k_{z,\text{in}}) \sqrt{(2\pi/\lambda_{\text{in}})^2 - (k_{x,\text{out}})^2 - (k_{y,\text{out}})^2} \tag{1.12}$$

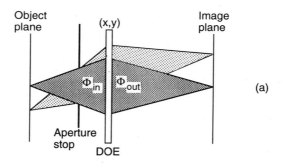

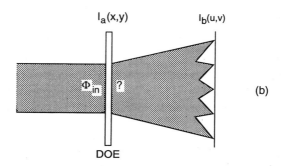

Figure 1.5 (a) Imaging, (b) beam shaping.

where sign($k_{z,\text{in}}$) denotes the sign of $k_{z,\text{in}}$. For reflective elements, sign($k_{z,\text{in}}$) has to be replaced by $-$sign($k_{z,\text{in}}$).

Equations (1.11) and (1.12) describe the grating diffraction, which allows the tracing of a bundle of finite rays through a diffractive component. The results are for example presented as spot diagrams, which are the points of intersection of the calculated rays with the image plane.

For numerical optimization, the DOE phase function $\Phi(x, y)$ is typically described by coefficients of a polynomial (equation (1.4)) and the quality of the system is described by a merit function. Various methods, such as the damped least squares method, are applied to minimize the merit function. A summary of the schemes which have been developed for lens design optimization is presented in the book by Malacara and Malacara (1994).

Most of the modern ray-tracing programs include the design of diffractive lenses. If this feature is not available, it is still possible to apply the Sweatt model. Sweatt (1977) and Kleinhans (1977) showed that a diffractive lens is mathematically equivalent to a thin refractive lens with an infinite refractive index.

Analytic Optimum Design for Imaging

For imaging through a single DOE, there also exist analytic solutions. The basic problem is to convert a continuous set of input waves $\Phi_{\text{in}}(x, y, t)$ into another continuous set of output waves $\Phi_{\text{out}}(x, y, t)$, as shown in Fig. 1.5(a). The ideal waves are spherical waves emitted by points in the object plane which are focused onto the image plane. The parameter t describes for example the direction of the wave, or the position of the focus. Thus, for each pair of input and output waves the desired phase function $\Phi_d(x, y, t)$ of the optical element is known, and is the difference between the phases of the input and output waves (equation (1.3)), i.e.

$$\Phi_d(x, y, t) = \Phi_{\text{out}}(x, y, t) - \Phi_{\text{in}}(x, y, t). \tag{1.13}$$

The phase $\Phi_d(x, y, t)$ varies with the parameter t. We are now looking for a continuous DOE phase function $\Phi(x, y)$, which is as close as possible to $\Phi_d(x, y, t)$ for all t. The performance of the optical element is considered to be optimum when the value of the mean-squared difference between the desired local functions $\Phi_d(x, y, t)$ and the required DOE phase $\Phi(x, y)$ is minimum (Kedmi and Friesem, 1986; Cederquist and Fienup, 1987):

$$\int W(t) P(x, y, t) [\Phi_d(x, y, t) - \Phi(x, y) + \xi(t)]^2 \, dt \, dx \, dy \rightarrow \min. \tag{1.14}$$

The absolute phase of the desired function $\Phi_d(x, y, t)$ is not significant, therefore an arbitrary phase $\xi(t)$ can be added. $W(t)$ is a weighting function and $P(x, y, t)$ is the pupil function. The t-dependence of the pupil function indicates that the readout wave may illuminate different parts of the DOE for different values of the parameter t.

Another approach is based on analytic ray-tracing and relies on propagation vectors and grating vectors rather than on the phase functions (Hasman and Friesem, 1989). This means that the phases Φ_i in equations (1.13) and (1.14) are replaced by the projection of the wavevectors onto the x, y-plane, $k_{iH}(x, y) = \text{grad}[\Phi_i(x, y)]$. As a consequence, the function $\xi(t)$ disappears.

For either approach, variational methods are applied to find the optimum phase function $\Phi(x, y)$ for the given functions, $W(t)$, $P(x, y, t)$ and $\Phi_d(x, y, t)$. In many cases of practical interest, the desired phase function is not precisely known, for example if distortions or field curvature can be tolerated. Alternatively, the optimum position of the aperture stop has first to be found in order to determine the pupil function $P(x, y, t)$. A good insight into these problems is obtained by an alternative method (Herzig and Dändliker, 1987). Instead of minimizing the difference between the desired DOE phase function Φ_d and the optimum phase function Φ, the method requires that the first and second derivatives of Φ_d and Φ are identical. The method has been developed initially for holographic scanners, but also works well for the design of imaging elements. Once the phase function $\Phi(x, y)$ is found, it can be implemented in the form of a diffractive element, as explained in section 1.2.1.

Optimum Design for Beam Shaping

Up to now, we have only considered the optimization of the first-order phase function Φ, because we were mainly interested in an aberration-free imaging. In principle, a beam-shaping element can also be implemented as a first-order DOE, which generates the desired wave in the first diffraction order. However, in many cases several orders are involved in the beam-shaping process (fan-out elements, diffusers), therefore we consider here the direct optimization of the DOE phase profile $\Psi(x, y)$.

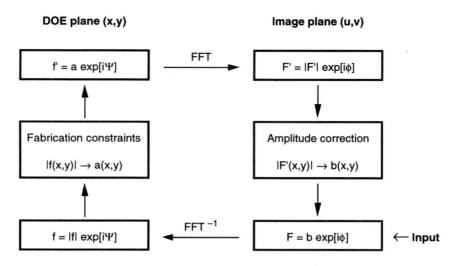

Figure 1.6 Schematic representation of the iterative Fourier transform algorithm.

In general, the beam-shaping problem has no analytical solution. Many different numerical methods have been investigated. Here, we concentrate on the two most successful methods: the iterative Fourier transform (IFT) algorithm and simulated annealing. For further reading, we recommend the articles of Wyrowski (1993), Johnson *et al.* (1993), Mait (1995), and Romero and Dickey (1996).

For this discussion, we consider a wave with an intensity distribution $I_a(u, v) = a^2(x, y)$ and a uniform phase. It is desired to add a DOE with phase $\Psi(x, y)$ in the x, y-plane such that a given intensity distribution $I_b(u, v) = b^2(u, v)$ results in another plane, which we call here the image plane (Fig. 1.5(b)). The problem is now to find the phase profile $\Psi(x, y)$ of the DOE. The phase distribution $\phi(u, v)$ in the image plane is usually a free parameter. In the following, we assume that the complex amplitudes $a(x, y)$ and $b(u, v)$ in the DOE plane and in the image plane, respectively, are related by a simple Fourier transform. However, the algorithms are not restricted to this case.

The iterative Fourier transform algorithm has been developed by Gerchberg and Saxton in 1972. This algorithm has since been modified and improved by a number of authors, e.g. Fienup (1980), Wyrowski and Bryngdahl (1988) and Eismann *et al.* (1989). The basic principle is shown in Fig. 1.6. The algorithm is started in the image plane with the desired amplitude distribution $b(u, v)$ and a guess for the phase distribution $\phi(u, v)$. Then, the field in the DOE plane (x, y) and in the image plane (u, v) are iteratively calculated by using the fast Fourier transform (FFT) and projected in each domain onto a set of specific constraints until it converges. In the DOE plane, the amplitude modulation is clipped to the amplitude $a(x, y)$ of the illumination beam, in order to calculate a phase-only element. In addition, fabrication constraints have to be satisfied. For multilevel surface-relief elements, the fabrication constraints include the discretization of the phase function to N phase levels (Wyrowski, 1990). In the case of fabrication by direct writing, the constraints include the compensation of point spread function (PSF) effects of the writing spot (Bengtson, 1994; Ehbets, 1996).

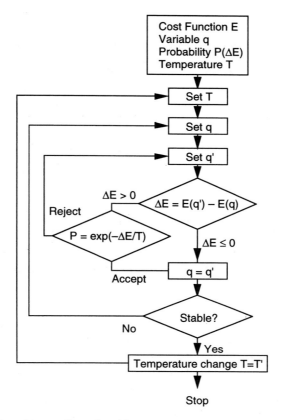

Figure 1.7 Simulated annealing algorithm.

The IFT algorithm is very efficient and can handle large amounts of data. On the other hand, the algorithm is sensitive to the starting parameters. It is difficult to avoid the process stagnating in a local minimum.

Global optimization methods do not depend on the starting point and are very effective in avoiding local minima. The main problem with these methods is that the computing time may become extremely large. An example of global optimization is the simulated annealing algorithm (Kirkpatrick *et al.*, 1983). The basic scheme is shown in Fig. 1.7. The DOE is characterized by a parameter set q. The set q stands, for example, for the pixels of a binary structure, or the coefficients of a polynomial. Then, a cost function has to be defined which describes the difference between the desired pattern ($I_b(u, v)$ in Fig. 1.5(b)) and the calculated pattern, as a function of the parameter set q. During the process the parameter set is altered, which yields a change of ΔE in the cost function. If E is reduced ($\Delta E \leq 0$), the change is accepted. If E increases ($\Delta E > 0$), the change is accepted with probability $\exp(-\Delta E/T)$. T is a control parameter analogous to the temperature in thermodynamics. After a number of cycles, the temperature T is lowered gradually and the process continues. As a result, the probability of a change being accepted, with $\Delta E > 0$, is reduced. The temperature associated with the annealing decision process is continuously reduced until the cost function reaches a minimum value and shows no further signs of decreasing. At this point,

11

the simulated annealing algorithm is terminated. The success of simulated annealing depends strongly on the annealing schedule and on the strategy of the move from one parameter set to the other.

1.2.3 System Design

Ray-tracing, presented in the previous section, is the standard method of designing optical systems. The method is optimum for calculating the propagation of wavefronts through a system and to minimize geometrical aberrations. On the other hand, ray-tracing is not ideal for calculating diffraction phenomena, such as the propagation of a Gaussian beam, or the diffraction at an aperture stop. We would like to introduce here a complementary method to calculate optical systems. The method is a paraxial diffraction theory which is developed in the book of Siegman (1986). The method is suitable for calculating beams with arbitrary amplitude distributions (flat-top beams, Gaussian beams, multiple apertures) through systems including refractive lenses, filters, aperture stops and diffractive elements. The approach describes paraxial propagation entirely in terms of complex $ABCD$ matrices. First, we describe briefly the transfer of an optical ray through a system described by an $ABCD$ matrix.

A ray of light that is travelling approximately in the z-direction is defined by a transverse displacement $r(z)$ from the axis and a small slope of α in a medium with refractive index n, as shown in Fig. 1.8:

$$\mathbf{r}(z) = \begin{pmatrix} r(z) \\ s(z) \end{pmatrix} = \begin{pmatrix} r(z) \\ n(z)\alpha(z) \end{pmatrix}. \tag{1.15}$$

The paraxial ray vectors at the input and output plane of an optical system consisting of simple optical components (free space, refractive or reflective surfaces, lenses and mirrors) are related by a linear transformation, namely

$$\mathbf{r}_2 = \begin{pmatrix} r_2 \\ s_2 \end{pmatrix} = \begin{pmatrix} A & B \\ C & D \end{pmatrix}\begin{pmatrix} r_1 \\ s_1 \end{pmatrix} = \mathbf{Mr}_1 \tag{1.16}$$

where $\mathbf{r}_1$ is the input ray and $\mathbf{r}_2$ the output ray.

Examples are (Siegman, 1986): for free-space propagation (distance d)

$$\mathbf{M} = \begin{pmatrix} 1 & d/n \\ 0 & 1 \end{pmatrix}$$

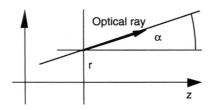

Figure 1.8 Optical ray defined by the transverse displacement $r(z)$ and the angle α within a medium with refractive index n.

and for a thin lens with focal length f

$$\mathbf{M} = \begin{pmatrix} 1 & 0 \\ -1/f & 1 \end{pmatrix}.$$

Note that for a lossless system $AD - BC = 1$.

An important point is that the Fresnel integral for propagation through a cascaded series of conventional elements can be accomplished in one step, using nothing more than the overall $ABCD$ matrix elements for that system. For rotationally symmetric optical systems, the output and input complex fields $U_2(x_2, y_2)$ and $U_1(x_1, y_1)$, respectively, are related by

$$U_2(x_2, y_2) = \frac{\exp(ikL)}{i\lambda B} \int\int_{-\infty}^{\infty} U_1(x_1, y_1) \exp\left[\frac{ik}{2B} (A(x_1^2 + y_1^2) \right.$$

$$\left. - 2(x_1 x_2 + y_1 y_2) + D(x_2^2 + y_2^2)) \right] dx_1 \, dy_1 \tag{1.17}$$

where $k = 2\pi/\lambda$ is the wavenumber in free space. L is the on-axis optical path length through the system:

$$L = \sum_i n_i L_i$$

where each individual element inside the system has a physical thickness L_i and an index of refraction n_i. Equation (1.17) can also be rewritten as a Fourier transform (FT):

$$U_2(x_2, y_2) = \frac{\exp(ikL)}{i\lambda B} \exp\left[\frac{ik}{2B} D(x_2^2 + y_2^2) \right] \mathrm{FT}\left\{ U_1(x_1, y_1) \exp\left[\frac{ik}{2B} A(x_1^2 + y_1^2) \right] \right\}. \tag{1.18}$$

If the system is astigmatic, we use different $ABCD$ matrices in the x and y direction. For the case where the object distribution $U(r_1)$ and the optical system are rotationally symmetric, equation (1.17) becomes

$$U_2(r_2) = \frac{k}{iB} \exp\left[ik\left(L + \frac{D}{2B} r_2^2 \right) \right] \int_{-\infty}^{\infty} r_1 U_1(r_1) J_0\left(\frac{k}{B} r_1 r_2 \right) \exp\left[\frac{ikA}{2B} r_1^2 \right] dr_1 \tag{1.19}$$

where J_0 is the zero-order Bessel function.

We can now calculate a general system consisting of conventional elements (lenses, mirrors), which are described by $ABCD$ matrices and other elements (DOEs, apertures, filters), which are described by a transmittance function (equation (1.1)). The system has to be divided into several cascaded subsystems, as illustrated in Fig. 1.9. Each subsystem is described by a matrix $\mathbf{M}_i$. The propagation through $\mathbf{M}_i$ can therefore be calculated according to equation (1.17). The result is then multiplied with the transmittance of the subsequent element, which is for example a DOE or an aperture stop.

Note that if the matrix describes an imaging system without an aperture stop then B vanishes ($B = 0$). Yariv (1994) has shown that in the limit $B \to 0$, equation (1.17) can be written as

$$U_2(x_2, y_2) = \frac{\exp(ikL)}{-A} \exp\left[\frac{ikC}{2A} (x_2^2 + y_2^2) \right] U_1\left(\frac{x_2}{A}, \frac{y_2}{A} \right). \tag{1.20}$$

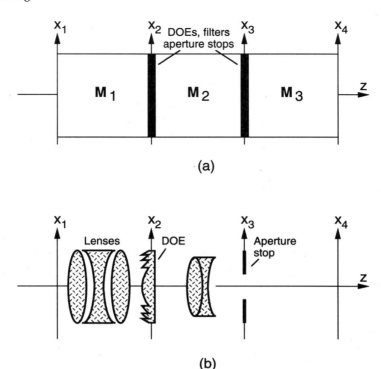

Figure 1.9 Optical system with conventional elements, such as lenses, mirrors, which can be described by *ABCD* matrices (M_i) and other elements such as DOEs, apertures, filters, which can be modelled by a transmittance function: (a) schematic view, (b) example.

This shows that when $B = 0$ the output is an exact, scaled replica of the input field except for a quadratic phase factor. The image magnification is A. In the case of an aperture stop within the imaging system, we first calculate the propagation until the stop, according to equation (1.17), and then from the stop to the image plane ($B \neq 0$ for both matrices). As a result the image is slightly blurred due to diffraction at the stop.

1.3 Refractive Properties of Diffractive Elements

A diffractive element consists of zones which diffract the incoming light. The light diffracted from the different zones interferes and forms the desired wavefront. We can calculate this wavefront, e.g. by ray-tracing, according to equation (1.11). If only a few zones of the element are illuminated, the element does not behave like a pure diffractive element. Consequently, the results are not accurately modelled using equation (1.11). Critical elements are, for example, diffractive field lenses, which are close to an intermediate image plane.

   In this section, we first discuss the refractive properties of blazed gratings as a function of the number of illuminated grating periods. We then consider the

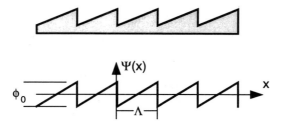

Figure 1.10 Blazed grating with periodicity Λ and phase depth ϕ_0.

refractive properties of continuous-relief fan-out elements. By adding focusing power, or an off-axis angle, the elements can be made diffractive. As a result, the elements show reduced sensitivity to vertical profile scaling errors and are therefore easier to fabricate. Complementary information about refractive and diffractive properties of DOEs can be found in the literature (Rossi *et al.*, 1995; Sinzinger and Testorf, 1995).

1.3.1 *Blazed Gratings*

The difference between a diffractive and a refractive element can be easily understood by considering a blazed grating. A blazed grating consists of small prisms which are repeated with a periodicity Λ, as shown in Fig. 1.10. The transmission $t(x)$ of this grating is given by

$$t(x) = \exp\left[i\,\frac{\phi_0 x}{\Lambda}\right] \text{rect}\left[\frac{x}{\Lambda}\right] \circ \text{comb}\left[\frac{x}{\Lambda}\right] \tag{1.21}$$

where $\circ$ stands for a convolution and the comb function is an infinite series of Dirac functions spaced by Λ. The maximum phase shift introduced by the grating is ϕ_0. Illuminated with a plane wave, the grating diffracts the incident light into the different diffraction orders of the grating. The light distribution is described by the Fourier transform $\hat{U}(p)$ of the transmission function $t(x)$. From equation (1.21), we get

$$\hat{U}(p) = \text{FT}\{t(x)\} = \text{sinc}\left(\Lambda p - \frac{\phi_0}{2\pi}\right)\text{comb}(\Lambda p). \tag{1.22}$$

The diffraction orders appear at the spatial frequencies $p = N/\Lambda$ ($N = 0, \pm 1, \pm 2,$ ...). The amplitudes of the diffraction orders depend on the position $p = \phi_0/2\pi\Lambda$ of the sinc function (Fig. 1.11(a)). For an ideally blazed grating ϕ_0 is equal to 2π and all the light is diffracted into the first order, as shown in Fig. 1.11(b).

The periodic grating function is a diffractive element that generates diffraction orders. The shape of the envelope (sinc function) is generated by diffraction at a single period, which is a prism. On the other hand, the position of the envelope is determined by refraction at the prism.

If both refraction and diffraction coincide, then the efficiency is highest. If many grating periods are illuminated then diffraction determines the direction of the outgoing beam. A change in the phase depths of the grating results in a shift of the envelope. The diffraction efficiency in the first order is reduced, but the

15

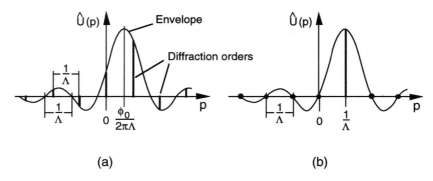

Figure 1.11 Far field of a blazed grating: (a) $\phi_0 \neq 2\pi$, (b) $\phi_0 = 2\pi$.

position remains unchanged (Fig. 1.11(a)). If only a few grating periods are illuminated, then the width of the diffraction orders is broadened. As a result, the outgoing light can now shift within the enlarged diffraction order. In the case where only one grating period is illuminated, the diffraction orders are no longer resolved, and the direction of the light propagation is completely determined by the envelope, i.e. by refraction.

Figure 1.12 shows the refractive and diffractive properties of a blazed grating, assuming a shrinkage error of 10%. Q is a normalized value for the variation of the deflection angle due to shrinkage error, i.e. Q = variation of the deflection angle/variation of the prism angle. Q equal to 1 means that the deflection angle changes according to the laws of refraction at a prism with thickness change of 10%. Q equal to zero means that the deflection angle stays constant. The element behaves refractively if only one zone is illuminated and it behaves diffractively if many zones are illuminated.

In conclusion, the optical function of an ideal diffractive element is encoded as the lateral position of zones. If the phase profile of the zones shrinks, the lateral position stays at the same place. Therefore, diffractive elements are insensitive to shrinkage errors. Refractive elements are encoded as the optical path difference. If the profile of the element shrinks, then the optical path length changes also.

1.3.2 *Fan-out Elements*

A fan-out element is an element that splits an incoming beam into an array of diffraction orders of equal power, as shown in Fig. 1.13. The diffraction orders are generated by the grating, whereas the shape of each grating period equalizes the power in the different orders. Well-known are binary Dammann gratings (Dammann and Görtler, 1971), shown in Fig. 1.13(a), and continuous-relief fan-out elements (Herzig *et al.*, 1990), shown in Fig. 1.13(b). Highest theoretical efficiency, over 95%, is achieved with continuous or multilevel surface-relief grating structures. Unfortunately, very accurate fabrication of the surface-relief profile is required to obtain small uniformity errors. Binary Dammann gratings have lower efficiency (typically 80%, Vasara *et al.*, 1992), but are easier to fabricate; not only because of the simpler binary structure, but also because each grating period is diffractive, i.e. the optical function is determined by the lateral

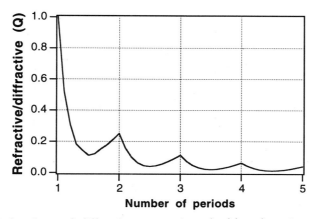

Figure 1.12 Refractive and diffractive properties of a blazed grating, assuming a shrinkage error of 10%. Q is a normalized value for the variation of the deflection angle due to shrinkage error. The element behaves refractively if only one zone is illuminated and it behaves diffractively if many zones are illuminated.

position of the phase transitions. An error in the etch depth only affects the zero order, while the intensity distribution of the remaining diffraction orders is unaffected. In the following, we show that continuous-relief fan-out elements may also become less sensitive to fabrication error if we introduce additional phase transitions, i.e. if we make them more diffractive.

The quality of fan-out elements is given by the diffraction efficiency η and the uniformity error u, which are defined as

$$\eta = \sum_{p=1}^{N} I_p \bigg/ \sum_{p=-\infty}^{\infty} I_p \quad \text{and} \quad u = \frac{I_{\max} - I_{\min}}{I_{\max} + I_{\min}} \tag{1.23}$$

where $I_{\max}$ and $I_{\min}$ represent the maximum and minimum intensities of the N fan-out diffraction orders I_p.

Let us consider an optimized one-dimensional fan-out element, which generates $N = 9$ equal spots (Prongué *et al.*, 1992). Figure 1.14 shows one period of the continuous surface-relief grating which produces the fan-out pattern with a diffraction efficiency of $\eta = 99.3\%$ and perfect uniformity. The ideal phase function Φ is implemented as continuous phase profile $\Psi(x,y) = \Phi(x,y)$ (dashed line), or with 0–2π transitions $\Psi(x,y) = [\Phi(x,y) + \varphi_0] \bmod 2\pi$ (solid line). We assume that the element has been realized as a surface-relief profile $h(x,y)$ according to equation (1.6). Furthermore, we assume a linear scaling error of the phase function $\Psi(x,y)$, which can result from either a wavelength change $\lambda \neq \lambda_0$ or a linear surface-relief error. The scaling results in a transmittance function

$$t_s(x,y) = P(x,y) \exp[i\alpha \Psi(x,y)] \tag{1.24}$$

where α represents the linear scaling factor, and $P(x,y)$ is the pupil function describing the finite size of the element. For a surface-relief DOE in transmission, the scaling factor α is given by

$$\alpha = \frac{\lambda_0[n(\lambda) - 1]}{\lambda[n(\lambda_0) - 1]} \beta.$$

17

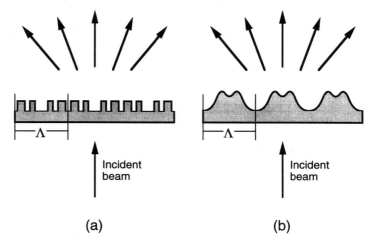

Figure 1.13 (a) Dammann grating, (b) continuous surface-relief fan-out element.

The first factor takes into account the wavelength and material dispersion (Buralli *et al.*, 1989), and β represents the relief depth scaling. Note that the scaled function $\alpha\Psi(x, y)$ has the same periodicity as the ideally wrapped function $\Psi(x, y)$. As a consequence, the transmittance function of the scaled element $t_s(x, y)$ can be expressed by a generalized Fourier series expansion of the form (Ehbets *et al.*, 1995)

$$t_s(x, y) = P(x, y) \sum_{q=-\infty}^{\infty} \exp[iq(\varphi_0 - \pi)] \, \mathrm{sinc}(\alpha - q) \exp[iq\Phi(x, y)]. \tag{1.25}$$

The first diffraction order ($q = 1$) corresponds to the desired optical function. Scaling errors do not influence the phase function $\Phi(x, y)$, but they reduce the power in the desired order by a factor of $\mathrm{sinc}^2(\alpha - q)$ and produce additional noise orders ($q \neq 1$). The constant phase offset φ_0 changes the relative phase of the different orders q in equation (1.25) and can be used to minimize the interference effects in the far field. It has been shown that the phase offset improves fabrication error tolerances of continuous-relief fan-out elements (Ehbets *et al.*, 1995).

In order to define a criterion for the sensitivity of the fan-out function to scaling errors, we calculate the average diffraction efficiency (efficiency sensitivity)

$$\eta_s = 0.5[\eta(\alpha = 0.9) + \eta(\alpha = 1.1)] \tag{1.26}$$

and the average uniformity error (uniformity sensitivity)

$$u_s = 0.5[u(\alpha = 0.9) + u(\alpha = 1.1)] \tag{1.27}$$

resulting from scaling factors $\alpha = 0.9$ and $\alpha = 1.1$. This criterion corresponds to a tolerable scaling error of $\pm 10\%$, which can be achieved even in an industrial fabrication environment.

Using direct write technology (see Chapter 4), the continuous phase function $\Phi(x, y)$ of fan-out elements can either be directly realized or be wrapped to the interval $[0, 2\pi]$ for the fabrication. Both possibilities are shown in Fig. 1.14. In the ideal case, both approaches yield the same efficiency and uniformity. However, they have fundamentally different responses to scaling errors.

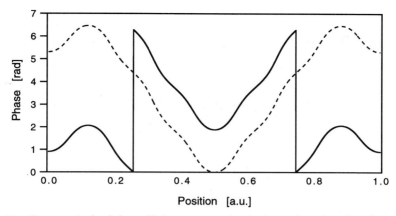

Figure 1.14 One period of the efficiency optimized phase function for the 9×1 fan-out element. The ideal phase function Φ can be implemented as continuous phase profile $\Psi(x,y) = \Phi(x,y)$ (dashed line), or with 0–2π transitions $\Psi(x,y) = [\Phi(x,y) + \varphi_0] \bmod 2\pi$ (solid line).

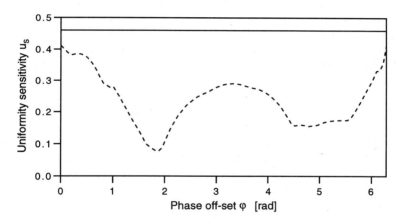

Figure 1.15 Uniformity error sensitivity u_s for the 9×1 fan-out element as a function of the phase offset parameter φ_0 for a phase function with 0–2π transitions (dashed curve). The value of μ_s for the continuous phase function (without discrete transitions) is indicated for comparison (solid curve).

In the case of continuous-phase DOEs without discrete phase transitions, scaling errors modify the fan-out phase function $\Phi(x)$ as if it were a refractive element and therefore directly affect the diffraction orders. Consequently there is a strong dependence of the uniformity error on scaling errors. The sensitivity criterion (equation (1.27)) gives a uniformity error of $u_s = 46\%$ (Fig. 1.15, solid line). In the absence of scaling errors ($\alpha = 0$), the optimized fan-out generates a uniformity error u below 0.1%.

By wrapping the phase function, two phase transitions are introduced (Fig. 1.14) and the function becomes partially diffractive. The position of the phase transitions depends on the phase offset φ_0. The dashed curve in Fig. 1.15 shows the sensitivity u_s of the uniformity error for the fan-out elements as a function of φ_0. The solid

19

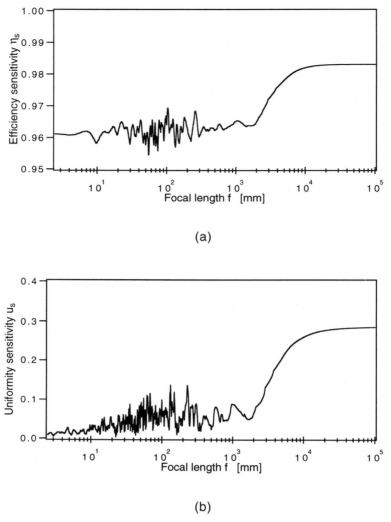

Figure 1.16 (a) Efficiency sensitivity η_s and (b) uniformity sensitivity u_s for the focusing 9×1 fan-out element as a function of the focal length f. The curves are calculated for a fixed value of the phase offset parameter φ_0.

curve is the value u_s for the continuous phase function without discrete transitions. It can be seen that by choosing the optimum value for φ_0, the sensitivity is reduced from over 40% to 8%.

The element becomes even less sensitive to scaling errors if more phase transitions are introduced. This can be done either by adding a linear phase term (Turunen *et al.*, 1990), which leads to an off-axis element, or by adding focusing power (Ehbets *et al.*, 1995), which leads to a focusing fan-out element. Figure 1.16 presents (a) the efficiency sensitivity η_S and (b) the uniformity sensitivity u_s for the focusing 9×1 fan-out element as a function of the focal length f. The curves were calculated for a fixed value of the phase offset parameter φ_0. The other parameters are: wavelength $\lambda = \lambda_0 = 632.8$ nm, diameter of the element

$D = 2.56$ mm, fan-out period $\Lambda = 160\,\mu$m. The transition between refractive behaviour and diffractive behaviour can clearly be seen in both curves, for the efficiency sensitivity η_s and for the uniformity sensitivity u_s. Similarly to the non-focusing fan-out element (Fig. 1.15), we can additionally optimize the phase offset φ_0 in order to obtain a minimum for each focal length f in Fig. 1.16.

It is important to note that even a weak focusing power adds sufficient supplementary transition points to the phase function to create fan-out elements that are almost independent of scaling errors. If too many transitions are introduced, then the losses due to fabrication errors increase and will be more important than the scaling behaviour. This concept has been proven experimentally by Ehbets *et al.* (1995).

1.4 Diffractive and Refractive Lenses

Lenses are probably the most widely used optical elements. In this section, we compare the basic properties of refractive and diffractive microlenses, i.e. the focal length, the dispersion and the aberration.

1.4.1 *Paraxial Properties of Lenses*

Figure 1.17 shows examples of refractive and diffractive lenses. A refractive lens is described by a refractive index $n(\lambda)$ and two curvatures c_1 and c_2. A diffractive lens on a planar substrate is described by a phase function (section 1.2.1). The phase function of a rotationally-symmetric diffractive lens with an arbitrary profile can be of the form

$$\Phi(r) = 2\pi(a_2 r^2 + a_4 r^4 + \ldots) \tag{1.28}$$

where r is the radial coordinate in the plane of the diffractive lens. The optical power of the diffractive lens in the mth diffraction order is then given by

$$1/f_0 = -2a_2\lambda_0 m \tag{1.29}$$

where λ_0 is the design wavelength and f_0 is the design focal length.

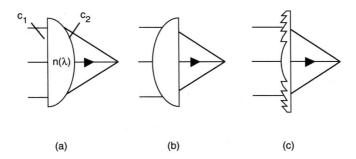

(a)   (b)   (c)

Figure 1.17 Refractive and diffractive lenses.

We now compare the focal length f_r of a refractive lens with the focal length f_d of a diffractive lens, which are given by

$$f_r(\lambda) = \frac{1}{n(\lambda) - 1} \frac{1}{c_1 - c_2} \tag{1.30}$$

and

$$f_d(\lambda) = f_0 \frac{\lambda_0}{\lambda}. \tag{1.31}$$

For a refractive lens the variation of the focal length with respect to the wavelength is small and depends on the factor $n(\lambda) - 1$. The dispersion of such a lens is described by the Abbe number ν_r, defined as

$$\nu_r = \frac{n(\lambda_1) - 1}{n(\lambda_2) - n(\lambda_3)}. \tag{1.32}$$

For a diffractive lens the focal length is linearly proportional to the wavelength λ. Therefore, the Abbe number becomes

$$\nu_d = \frac{\lambda_1}{\lambda_2 - \lambda_3} \tag{1.33}$$

where $\lambda_3 > \lambda_1 > \lambda_2$. Since the refractive index of glass decreases with increasing wavelength, ν_r is always positive, whereas ν_d becomes negative. The Abbe number is normally given for the standard wavelengths $\lambda_1 = 587.6$ nm, $\lambda_2 = 486.1$ nm and $\lambda_3 = 656.3$ nm. For the various optical glasses, the Abbe number ν_r ranges from about 80 to 20, whereas ν_d becomes -3.45, independent of the material. Note that a strong dispersion corresponds to a small Abbe number.

The strong negative dispersion of diffractive elements is used in hybrid (refractive/diffractive) elements in order to compensate for the chromatic aberration of the refractive component. This enables relatively thin and lightweight achromatic systems made of only one material. More information about hybrid micro-optics can be found in Chapter 10.

The dispersion can be calculated for any wavelength range. Figure 1.18 shows a comparison of the dispersion between refractive and diffractive elements versus the wavelength λ. The dispersion ratio is defined as $D_r = -\nu_{\text{refractive}}/\nu_{\text{diffractive}}$. Two different optical glasses have been chosen with high (SF6) and small (BK7) dispersion. The Abbe numbers ν have been calculated for the three wavelengths λ, $\lambda - 5$ nm, and $\lambda + 5$ nm. For longer wavelengths ($\lambda = 600$–1000 nm), the dispersion of diffractive elements is about 20 times higher than the dispersion of refractive elements. The difference becomes significantly less for shorter wavelengths.

1.4.2 *Aberrations of Refractive and Diffractive Optical Elements*

In this section, the aberrations of refractive and diffractive lenses are discussed. Various authors have investigated the aberrations of diffractive, or rather holographic, optical elements (Meier, 1965; Dändliker *et al.*, 1980). In this chapter we present the equations for Seidel aberrations derived by Buralli and Morris

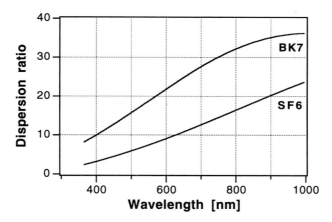

Figure 1.18 Comparison of the dispersion between refractive and diffractive elements versus the wavelength λ. The dispersion ratio is defined as $D_r = -\nu_{\text{refractive}}/\nu_{\text{diffractive}}$. The Abbe numbers ν have been calculated for the three wavelengths λ, $\lambda + 5\,\text{nm}$, and $\lambda - 5\,\text{nm}$.

(1991). These equations are more general and enable a direct comparison between refractive and diffractive lenses. The derivation is based on the results of Sweatt (1977) and Kleinhans (1977) who showed that a diffractive lens is mathematically equivalent to a thin refractive lens with an infinite refractive index.

To third order, the wavefront aberration polynomial W, as a function of normalized object coordinate h and normalized polar pupil coordinates ρ and ϕ_P, is (Welford, 1986)

$$W(h, \rho, \phi_P) = \tfrac{1}{8}S_I\rho^4 + \tfrac{1}{2}S_{II}h\rho^3\cos\phi_P + \tfrac{1}{2}S_{III}h^2\rho^2\cos^2\phi_P$$
$$+ \tfrac{1}{4}(S_{III} + S_{IV})h^2\rho^2 + \tfrac{1}{2}S_V h^3\rho\cos\phi_P. \tag{1.34}$$

The Seidel sums S_I–S_V refer to spherical aberration, coma, astigmatism, Petzval curvature and distortion, respectively. The fourth term in equation (1.34), including astigmatism and Petzval curvature, describes the field curvature. One form of the Seidel sums for a thin lens with aperture stop in contact is conveniently given in terms of dimensionless bending and conjugate parameters V and T, defined as

$$V = \frac{c_1 + c_2}{c_1 - c_2} \quad \text{and} \quad T = \frac{u + u'}{u - u'}. \tag{1.35}$$

In equation (1.35), c_1, c_2 are the curvatures of the lens surfaces and u, u' are the angles of the paraxial marginal ray before and after passing through the thin lens. The basic configuration is shown in Fig. 1.19. Using these parameters, the Seidel sums for a thin refractive lens with refractive index n and focal length f the forms (Welford, 1986):

$$\text{spherical aberration,} \quad S_I = \frac{y^4}{4f^3}\left[\left(\frac{n}{n-1}\right)^2 + \frac{n+2}{n(n-1)^2}V^2 + \frac{4(n+1)}{n(n-1)}VT \right.$$
$$\left. + \frac{3n+2}{T}T^2\right] \tag{1.36a}$$

23

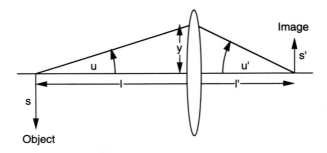

Figure 1.19 Imaging configuration: l and u' are negative as shown, while the other quantities are positive as shown.

coma,
$$S_{II} = \frac{-y^2 H}{2f}\left[\frac{n+1}{n(n-1)}V + \frac{2n+1}{n}T\right] \tag{1.36b}$$

astigmatism,
$$S_{III} = H^2/f \tag{1.36c}$$

Petzval curvature,
$$S_{IV} = H^2/fn \tag{1.36d}$$

distortion,
$$S_V = 0. \tag{1.36e}$$

We have denoted the height of the paraxial marginal ray at the lens by y and the Lagrange invariant by H, where $H = su = s'u'$ (in air). The distortion S_V vanishes, because the principal ray meets the thin lens at the axis and is transmitted undeviated.

We now consider the Seidel sums of diffractive lenses. We assume a rotationally-symmetric diffractive lens with an arbitrary profile given by equation (1.28). The substrate of the diffractive lens can be bent. The bending parameter B is defined by

$$B = 2c_s f \tag{1.37}$$

where c_s is the curvature of the diffractive lens substrate and $f = f(\lambda)$ the focal length (equation (1.31)).

The aberration coefficients for the diffractive lens are given by (Buralli and Morris, 1991):

spherical aberration,
$$S_I = \frac{y^4}{4f^3}(1 + B^2 + 4BT + 3T^2) - 8\lambda a_4 m y^4 \tag{1.38a}$$

coma,
$$S_{II} = \frac{-y^2 H}{2f^2}(B + 2T) \tag{1.38b}$$

astigmatism,
$$S_{III} = H^2/f \tag{1.38c}$$

Petzval curvature,
$$S_{IV} = 0 \tag{1.38d}$$

distortion,
$$S_V = 0. \tag{1.38e}$$

In both cases (refractive and diffractive lenses), the coefficients S_I–S_V change if the aperture stop is moved to a location distant from the lens. After the stop-shift, the aberration coefficients (denoted by S_I^*–S_V^*) are given by (Welford, 1986)

$$S_I^* = S_I \tag{1.39a}$$

$$S_{II}^* = S_{II} + \varepsilon S_I \tag{1.39b}$$

$$S_{III}^* = S_{III} + 2\varepsilon S_{II} + \varepsilon^2 S_I \tag{1.39c}$$

$$S_{IV}^* = S_{IV} \tag{1.39d}$$

$$S_V^* = S_V + \varepsilon(3S_{III} + S_{IV}) + 3\varepsilon^2 S_{II} + S_I. \tag{1.39e}$$

In equations (1.39), ε is given by $\varepsilon = \delta\bar{y}/y$, where $\delta\bar{y}$ is the change in paraxial chief ray height (denoted by $\bar{y}$) caused by the stop-shift, and y is the height of the paraxial marginal ray at the lens, which is not affected by the stop-shift.

1.4.3 Discussion of Aberration

A diffractive lens can make a perfect image of a single point for a single wavelength. Aberrations are due to oblique illumination with respect to the design axis or a change in the wavelength. In this section, the discussion is limited to rotationally symmetric elements. In that case, only spherical aberration can occur on-axis, i.e. $h = 0$ in equation (1.34). The spherical aberration S_I of diffractive lenses can be eliminated by a proper choice of the coefficient a_4 (equation (1.38a)). This is not possible for refractive lenses with spherical surfaces (equation (1.36a)). The spherical aberration can be minimized by varying the shape V, i.e. by bending the lens. S_I depends quadratically on V. The minimum of S_I occurs at (Welford, 1986)

$$V = -\frac{2(n^2 - 1)}{n + 2} T. \tag{1.40}$$

Equation (1.40) gives the best shape lens for given conjugates T. For equal conjugates ($T = 0$) the best shape is $V = 0$, or equiconvex. For $T = \pm 1$, object or image at infinity, the best shape is approximately plano-convex with the convex side towards the infinity conjugate. The exact shape depends on the refractive index.

Figure 1.20 shows spherical aberration S_I and coma S_{II} versus the bending parameter V for a refractive focusing lens ($T = -1$) with two different refractive indices ($n = 1.5$ and 1.7). A higher refractive index requires a lower curvature of the lens surface, therefore the aberrations are smaller. Equation (1.36b) shows that coma (S_{II}) is a linear function of bending when the stop is at the lens. The shape where $S_{II} = 0$ is given by

$$V = \frac{(n - 1)(2n + 1)}{n + 1} T. \tag{1.41}$$

Numerically, this is close to the shape for minimum spherical aberration.

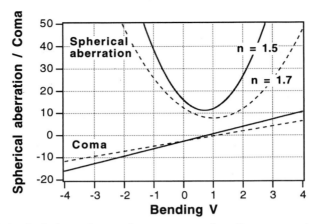

Figure 1.20 Spherical aberration and coma versus bending parameter V. The spherical aberration is given by $4S_{\mathrm{I}}f^3/y^4$ and coma by $2S_{\mathrm{II}}f^2/(-y^2H)$. The refractive indices are $n = 1.5$ (solid line) and $n = 1.7$ (dashed line).

Coma for diffractive lenses is described by equation (1.38b). For a planar diffractive lens coma disappears only for 1:1 imaging ($T = 0$). Otherwise the substrate has to be bent. In this case coma vanishes if $B = -2T$. The curvature c_{s} of the substrate is then given by

$$c_{\mathrm{s}} = \frac{-T}{f}. \tag{1.42}$$

Astigmatism cannot be eliminated as long as the stop is at the lens.

The effects of moving the aperture stop to a location different from the lens are described by equations (1.39a–e). We can now eliminate the Seidel coma and astigmatism by shifting the stop and by introducing spherical aberration. For a focusing lens with object at infinity, the aperture stop has to be placed in the front focal plane of the lens (Buralli and Morris, 1989). Since the Petzval term is also zero, both tangential and sagital fields are flat. This is in contrast to a refractive lens, which has flat field in only one meridian. The remaining spherical aberration limits the diffractive lens to operating at relatively modest apertures.

A summary of the main differences between the aberrations of refractive and diffractive lenses is shown in Table 1.1. The spherical aberration has been deduced from equations (1.36) and (1.38) for a focusing lens ($T = -1$). Furthermore, we have used the focal length f and the numerical aperture $\mathrm{NA} \approx y/f$ to write the equations. The refractive lenses are plano-convex lenses as shown in Fig. 1.17. For these lenses, the spherical aberration depends on the direction of the light propagation through the lens ($C_a \approx 4C_b$). Diffractive lenses have no spherical aberration for a single wavelength λ_0. For other wavelengths, spherical aberration is present, but still relatively small. The dominant error is the change in the focal length, which varies proportionally to the wavelength λ. An interesting consequence of this wavelength dependence is that the Airy disk radius generated by diffractive lenses does not vary with λ. For a fixed aperture, the spherical

Table 1.1 Comparison between the aberrations of refractive and diffractive lenses

	Refractive lens (plano-convex)		Diffractive lens
Focal length	$f_{a,b} = \dfrac{R}{n(\lambda) - 1}$		$f_c = \dfrac{\lambda_0}{\lambda} f_0$
Airy disk radius	$r(\lambda) = 1.22 \dfrac{\lambda f}{D}$		$r(\lambda) = 1.22 \dfrac{\lambda_0 f_0}{D} \approx$ constant
Spherical aberration $S_I = (NA)^4 f C_i$	$C_a = \dfrac{n^2}{(n-1)^2}$	$C_b = \dfrac{n^3 - 2n^2 + 2}{n(n-1)^2}$	$C_c = 1 - \left(\dfrac{\lambda_0}{\lambda}\right)^2$
Coma	can be eliminated using an aspheric lens shape can be eliminated using an optimized lens shape		arbitrary phase functions are standard can be eliminated using a curved substrate
Astigmatism	can be eliminated by shifting the aperture stop		can be eliminated by shifting the aperture stop
Petzval curvature	$S_{IV} = \dfrac{H^2}{nf}$ cannot be eliminated		0

Indices a, b, c refer to Fig. 1.17: Focal length f, Numerical aperture $NA \approx y/f$, Refractive index n, Lagrange invariant H.

aberration scales with the focal length. Interesting also is that the Petzval curvature disappears for diffractive lenses.

1.5   Conclusions

Micro-optical elements can be refractive, diffractive or hybrid (refractive/ diffractive). The decision to use diffractive or refractive optics for a specific optical problem depends on many parameters, e.g. the spectrum of the light source, the optical task (beam shaping, imaging), the efficiency required, the acceptable straylight, etc. Full design freedom is provided by diffractive optics. Arbitrary wavefronts can be generated very accurately. A drawback for many applications is the strong wavelength dependence. Diffractive optics are therefore mostly used with laser light and for non-conventional imaging tasks, like beam shaping, diffusers, fan-in/fan-out elements, filters and detectors. For broadband applications DOEs are combined with refractive optics to correct for the chromatic aberration. This combination allows systems with low weight, or which consist of only one material (Chapter 10). Refractive optical elements have in general higher efficiency and less straylight. On the other hand it is more difficult to make refractive lenses with precise focal lengths or aspheric shapes. Note that in many cases, the question is not **if** refractive or diffractive elements are more suitable because of their optical properties, it is rather a question of which technology provides the possibility of making segmented elements, aperture filling and compact systems.

This work was supported by the Swiss Priority Program OPTIQUE.

References

BENGTSTON, J. (1994) Direct inclusion of the proximity effect in the calculation of kinoforms. *Appl. Opt.* **33**, 4993–4996.

BORN, M and WOLF, E. (1980) *Principles of Optics*, 6th edn, Chapter 3. Oxford: Pergamon Press.

BURALLI, D. A. and Morris, G. M. (1989) Design of a wide field diffractive landscape lens. *Appl. Opt.* **28**, 3950–3959.

(1991) Design of diffractive singlets for monochromatic imaging. *Appl. Opt.* **30**, 2151–2158.

BURALLI, D. A., MORRIS, G. M. and ROGERS, J. R. (1989) Optical performance of holographic kinoforms. *Appl. Opt.* **28**, 976–983.

CEDERQUIST, J. N. and FIENUP, J. R. (1987) Analytic design of optimum holographic optical elements. *J. Opt. Soc. Am.* A, **4**, 699–705.

DAMMANN, H. and GÖRTLER, K. (1971) High-efficiency in-line multiple imaging by means of multiple phase holograms. *Opt. Commun.* **3**, 312–315.

DÄNDLIKER, R., HESS, K. and SIDLER, T. (1980) Astigmatic pencils of rays reconstructed from holograms. *Israel J. Technol.* **18**, 240–246.

EHBETS, P. (1996) *Efficient Phase Gratings for Beam Splitting and Beam Shaping*, PhD. Thesis, University of Neuchâtel.

EHBETS, P., ROSSI, M. and HERZIG, H. P. (1995) Continuous-relief fan-out elements with optimized fabrication tolerances. *Opt. Eng.* **34**, 3456–3464.

EISMANN, M. T., TAI, A. M. and CEDERQUIST, J. N. (1989) Iterative design of a holographic beamformer. *Appl. Opt.* **28**, 2641–2650.

FIENUP, J. R. (1980) Iterative method applied to image reconstruction and to computer-generated holograms. *Opt. Eng.* **19**, 297–305.

GERCHBERG, R. W. and SAXTON, W. O. (1972) A practical algorithm for the determination of phase from image and diffraction plane pictures. *Optik* **35**, 237–246.

GOODMAN, J. W. (1968) *Introduction to Fourier Optics*. San Francisco: McGraw-Hill.

HASMAN, E. and FRIESEM, A. A. (1989) Analytic optimization for holographic optical elements. *J. Opt. Soc. Am.* A, **6**, 62–72.

HERZIG, H. P. and DÄNDLIKER, R. (1987) Holographic optical scanning elements: Analytical method for determining the phase function. *J. Opt. Soc. Am.* A, **4**, 1063–1070.

(1993) Diffractive components: Holographic optical elements. In Lalanne, Ph. and Chavel, P. (eds) *Perspectives for Parallel Interconnects*, pp. 43–69. Berlin: Springer.

HERZIG, H. P., PRONGUÉ, D. and DÄNDLIKER, R. (1990) Design and fabrication of highly efficient fan-out elements. *Jpn. J. Appl. Phys.* **29**, L1307–L1309.

HERZIG, H. P., GALE, M. T., LEHMANN, H. W. and MORF, R. (1993) Diffractive components: Computer generated elements. In Lalanne, Ph. and Chavel, P. (eds) *Perspectives for Parallel Interconnects*, pp. 71–107. Berlin: Springer.

JOHNSON, D. E. G., KATHMAN, A. D., HOCHMUTH, D. H., COOK, A. L., BROWN, D. R. and DELANEY, B. (1993) Advantages of genetic algorithm optimization methods in diffractive optic design. In Lee, S.-H. (ed.) *Diffractive and Miniaturized Optics*, Vol. CR49, pp. 54–74. Bellingham: SPIE.

KEDMI, J. and FRIESEM, A. A. (1986) Optimized holographic optical elements, *J. Opt. Soc. Am.* A, **3**, 2011–2018.

KINGSLAKE, R. (1978) *Lens Design Fundamentals*. Boston: Academic Press.

KIRKPATRICK, S., GELATT, C. D. and VECCHI, M. P. (1983) Optimization by simulated annealing. *Science* **220**, 671–680.

KLEINHANS, W. A. (1977) Aberrations of curved zone plates and Fresnel lenses. *Appl. Opt.* **16**, 1701–1704.

KUITTINEN, M. and HERZIG, H. P. (1995) Encoding of efficient diffractive microlenses. *Opt. Lett.* **20**, 2156–2158.

MAIT, J. N. (1995) Understanding diffractive optic design in the scalar domain. *J. Opt. Soc. Am.* A, **12**, 2145–2158.

MALACARA, D. and MALACARA, Z. (1994) *Handbook of lens design*. New York: Marcel Dekker.

MEIER, R. W. (1965) Magnification and third-order aberrations in holography. *J. Opt. Soc. Am.* **55**, 987–992.

PRONGUÉ, D., HERZIG, H. P., DÄNDLIKER, R. and GALE, M. T. (1992) Optimized kinoform structures for highly efficient fan-out elements. *Appl. Opt.* **31**, 5706–5711.

ROMERO, L. A. and DICKEY, F. M. (1996) Lossless laser beam shaping. *J. Opt. Soc. Am.* A, **13**, 751–760.

ROSSI, M., KUNZ, R. E. and HERZIG, H. P. (1995) Refractive and diffractive properties of planar micro-optical elements. *Appl. Opt.* **34**, 5996–6007.

ROSSI, M., BLOUGH, C. G., RAGUIN, D. H., POPOV, E. K. and MAYSTRE, D. (1996) Diffraction efficiency of high-NA continuous-relief diffractive lenses. In *Diffractive Optics and Micro-optics*, OSA Technical Digest Series, Vol. 5, pp. 233–236, Washington, D.C.: OSA.

SIEGMAN, A. E. (1986) *Lasers*. Mill Valley, California: University Science Books.

SINGER, W., HERZIG, H. P., KUITTINEN, M., PIPER, E. and WANGLER, J. (1996) Diffractive beamshaping elements at the fabrication limit, *Opt. Engin.* **35**, 2779–2787.

SINZINGER, S. and TESTORF, M. (1995) Transition between diffractive and refractive micro-optical components, *Appl. Opt.* **34**, 5970–5976.

SMITH, W. J. (1990) *Modern Optical Engineering*. New York: McGraw-Hill.

SWANSON, G. J. and VELDKAMP, W. B. (1989) Diffractive optical elements for use in infrared systems. *Opt. Eng.* **28**, 605–608.

SWEATT, W. C. (1977) Describing holographic optical elements as lenses. *J. Opt. Soc. Am.* **67**, 803–808.

TURUNEN, J., FAGERHOLM, J., VASARA, A. and TAGHIZADEH, M. R. (1990) Detour-phase kinoform interconnects: the concept and fabrication considerations. *J. Opt. Soc. Am.* A, **7**, 1202–1208.

VASARA, A., TAGHIZADEH, M. R., TURUNEN, J., WESTERHOLM, J., NOPONEN, E., ICHIKAWA, H., MILLER, J. M., JAAKKOLA, T. and KUISMA, S. (1992) Binary surface-relief gratings for array illumination in digital optics. *Appl. Opt.* **31**, 3320–3336.

WELCH, W. H., MORRIS, J. E. and FELDMAN, M. R. (1993) Iterative discrete on-axis encoding of radially symmetric computer-generated holograms. *J. Opt. Soc. Am.* A, **10**, 1729–1738.

WELFORD, W. T. (1986) *Aberrations of Optical Systems*. Bristol: Adam Hilger.

WYROWSKI, F. (1990) Diffractive optical elements: iterative calculation of quantized, blazed phase structures. *J. Opt. Soc. Am.* A, **7**, 961–969.

(1993) Design theory of diffractive elements in the paraxial domain. *J. Opt. Soc. Am.* A, **10**, 1553–1561.

WYROWSKI, F. and BRYNGDAHL, O. (1988) Iterative Fourier-transform algorithm applied to computer holography. *J. Opt. Soc. Am.* A, **5**, 1058–1065.

YARIV, A. (1994) Imaging of coherent fields through lenslike systems. *Opt. Lett.* **19**, 1607–1608.

2

Diffraction Theory of Microrelief Gratings

J. TURUNEN

2.1 Introduction

A significant part of micro-optical technology is based on the use of diffractive structures to control the amplitude, the phase, or the state of polarization of an optical field. In diffractive micro-optics, periodic structures (gratings) play a central role, because most diffractive elements may be treated as gratings at least in some sense. The modulated structure may either be globally periodic, perhaps with a complicated surface-relief or index-modulation profile, or locally periodic such as a diffractive lens. In both cases, grating-diffraction theory can be used to synthesize the modulation profile and to analyze its performance.

Both scalar and electromagnetic grating theories are well-established in optics. Loosely speaking, approximate scalar theories are applicable if the variations of the surface-relief or index-modulation profile are slow compared to the wavelength of light λ, provided that the thickness of the modulated region is comparable to λ. In such circumstances the magnitude of the diffracted spatial-frequency spectrum decreases rapidly as the diffraction angle increases. Rigorous electromagnetic theories are generally applicable. Unfortunately, few rigorous analytical solutions of diffraction problems are known, and the numerical methods tend to be prohibitively time-consuming if the grating period d is large, in particular for so-called crossed gratings with three-dimensional modulation profiles.

Books and review papers exist, which deal with approximate (Beckmann and Spizzichino, 1963) and exact (Petit, 1980; Maystre, 1984; Gaylord and Moharam, 1985) grating theories. The present discussion differs from the previous texts primarily because of the emphasis on micro-optical instrumentation rather than, for example, spectroscopy or scattering by rough surfaces. Our treatment encompasses a selection of exact and approximate theories. We concentrate on surface-relief gratings, which are rapidly gaining ground in micro-optics, as thorough treatments of index-modulated gratings already exist (Solymar and Cooke, 1981; Korpel, 1988).

This chapter is organized as follows. In section 2.2, we present the basic concepts of grating theory and treat the exact electromagnetic approach. Various numerical methods exist, of which a certain eigenmode formulation is presented in detail and compared to other approaches. Approximate methods, which are applicable when $d \gg \lambda$ or $d \ll \lambda$ are outlined in section 2.3. In section 2.4, numerical results are presented for some important grating profiles when d/λ is reduced. Section 2.5 deals with the principles of constructing modulated gratings for diffractive micro-optics.

2.2 Electromagnetic Grating Theory

We begin with a rigorous solution of the grating-diffraction problem by means of electromagnetic diffraction theory. To make the treatment reasonably self-contained, some concepts of electromagnetic wave theory are first reviewed.

2.2.1 *Principles of Electromagnetic Theory*

We consider throughout this chapter a time-harmonic electromagnetic field of frequency ω, characterized by electric and magnetic field vectors of the form

$$E(r, t) = \mathcal{R}\{E(r)\exp(-i\omega t)\}$$

$$H(r, t) = \mathcal{R}\{H(r)\exp(-i\omega t)\}$$

where $r = (x, y, z)$ and $\mathcal{R}$ denotes the real part. In a continuous medium, the time-harmonic field satisfies Maxwell's equations

$$\nabla \times E(r) = i\omega B(r) \tag{2.1}$$

$$\nabla \times H(r) = J(r) - i\omega D(r) \tag{2.2}$$

$$\nabla \cdot D(r) = \rho(r) \tag{2.3}$$

$$\nabla \cdot B(r) = 0 \tag{2.4}$$

where $D(r)$ and $B(r)$ are the spatial parts of the electric displacement and the magnetic induction, respectively, $J(r)$ denotes the electric current density, and $\rho(r)$ represents the electric charge density. These equations are supplemented (in non-relativistic optics of linear isotropic media) by constitutive relations

$$D(r) = \epsilon(r)E(r) \tag{2.5}$$

$$B(r) = \mu(r)H(r) \tag{2.6}$$

$$J(r) = \sigma(r)E(r) \tag{2.7}$$

where the quantities ϵ, μ and σ are known as the dielectric constant (permittivity), the magnetic permeability and the specific conductivity, respectively.

Diffractive micro-optical elements contain abrupt boundaries between continuous media. The transformations of the field vectors across such boundaries are covered by boundary conditions. We denote by n_{12} a unit normal vector of the boundary, which points from medium 1 to medium 2, and indicate the boundary values of the field vectors in media 1 and 2 by corresponding subscripts. Then

$$n_{12} \cdot (B_2 - B_1) = 0 \tag{2.8}$$

$$\boldsymbol{n}_{12} \cdot (\boldsymbol{D}_2 - \boldsymbol{D}_1) = \rho_S \tag{2.9}$$

$$\boldsymbol{n}_{12} \times (\boldsymbol{E}_2 - \boldsymbol{E}_1) = 0 \tag{2.10}$$

$$\boldsymbol{n}_{12} \times (\boldsymbol{H}_2 - \boldsymbol{H}_1) = \boldsymbol{J}_S \tag{2.11}$$

where ρ_S and $\boldsymbol{J}_S$ denote the surface charge and current density, respectively. If medium 2 is perfectly conducting ($\sigma_2 = \infty$), the field cannot penetrate into it, and $\boldsymbol{B}_2 = \boldsymbol{D}_2 = \boldsymbol{E}_2 = \boldsymbol{H}_2 = 0$. If the conductivities are finite (or zero), $\boldsymbol{J}_S = \rho_S = 0$.

We will need measures for the energy properties of the electromagnetic field. The quantity of primary interest in grating theory is the Poynting vector $\boldsymbol{S} = \boldsymbol{E} \times \boldsymbol{H}$. Owing to the large values of ω in the optical region, only its time-average

$$\langle \boldsymbol{S}(\boldsymbol{r}) \rangle = \lim_{T \to \infty} \frac{1}{2T} \int_{-T}^{T} \boldsymbol{E}(\boldsymbol{r}, t) \times \boldsymbol{H}(\boldsymbol{r}, t) \, \mathrm{d}t = \frac{1}{2} \mathcal{R} \{ \boldsymbol{E}(\boldsymbol{r}) \times \boldsymbol{H}^*(\boldsymbol{r}) \}$$

is observable.

2.2.2 *The Grating-diffraction Problem*

It is a well-known physical fact that a periodic scatterer, illuminated by an infinite plane wave, generates a discrete set of propagating plane waves. The basic aim of grating theory is to predict the complex amplitudes of these plane waves, provided that the grating structure is known. There is also considerable interest in the inverse problem: finding the grating structure from measurements of the diffracted field. In diffractive optics, we are often interested in solving a synthesis problem: constructing a grating structure that generates a diffracted field with some specified properties.

Since rigorous analytical solutions of diffraction problems are rare (see Born and Wolf, 1980, Chapter 11 and section 13.5, and Bouwkamp, 1954), we must resort to a numerical solution of Maxwell's equations and the appropriate boundary conditions. We assume that an electromagnetic plane wave, which propagates in the xz-plane, is incident from a homogeneous dielectric medium I ($z < 0$) upon a modulated region II, which occupies the volume $0 \le z < h$, and is periodic in the x-direction with a period d. Region III ($z > h$) is again homogeneous, either dielectric or metallic. Region II may contain variations of ϵ, μ and σ, which may be of volume or surface-relief type, or both. However, they are assumed y-invariant.

An exact solution of the grating-diffraction problem is obtained if we find an electromagnetic field that satisfies Maxwell's equations and the associated boundary conditions everywhere, and if the diffracted part of this field satisfies certain physically obvious radiation conditions: in region I, the reflected field must be an outgoing bounded wave as $z \to -\infty$, and the transmitted field must behave similarly in region III when $z \to \infty$. We consider dielectric or finitely conducting materials, with $\boldsymbol{J}_S = \rho_S = 0$.

All partial y-derivatives in equations (2.1) and (2.2) vanish in the two-dimensional geometry considered here. In component form, we therefore have

$$i\omega B_x(x, z) = -\frac{\partial}{\partial z} E_y(x, z) \tag{2.12}$$

$$i\omega B_z(x, z) = \frac{\partial}{\partial x} E_y(x, z) \tag{2.13}$$

$$\frac{\partial}{\partial z} H_x(x, z) - \frac{\partial}{\partial x} H_z(x, z) = J_y(x, z) - i\omega D_y(x, z) \tag{2.14}$$

$$J_x(x, z) - i\omega D_x(x, z) = -\frac{\partial}{z} H_y(x, z) \tag{2.15}$$

$$J_z(x, z) - i\omega D_z(x, z) = \frac{\partial}{\partial x} H_y(x, z) \tag{2.16}$$

$$\frac{\partial}{\partial z} E_x(x, z) - \frac{\partial}{\partial x} E_z(x, z) = i\omega B_y(x, z). \tag{2.17}$$

According to equations (2.5)–(2.7), $E \| D \| J$ and $B \| H$. This implies that the groups (2.12)–(2.14) and (2.15)–(2.17) are independent. By the use of equations (2.5)–(2.7), and a substitution of equations (2.12) and (2.13) into equation (2.14), we obtain

$$\frac{\partial}{\partial x}\left[\frac{1}{\mu(x, z)}\frac{\partial}{\partial x}E_y(x, z)\right] + \frac{\partial}{\partial z}\left[\frac{1}{\mu(x, z)}\frac{\partial}{\partial z}E_y(x, z)\right]$$
$$+ \omega^2[\epsilon(x, z) + i\sigma(x, z)/\omega]E_y(x, z) = 0. \tag{2.18}$$

Once E_y is known, the magnetic field is given by

$$H(x, z) = \frac{1}{i\omega\mu(x, z)}\left[-x\frac{\partial}{\partial z}E_y(x, z) + z\frac{\partial}{\partial x}E_y(x, z)\right].$$

Hence E_y specifies the complete electromagnetic solution, and we speak of the TE mode of polarization. We are primarily interested in non-magnetic media, with $\mu = \mu_0 =$ constant. If we define a complex relative permittivity $\hat{\epsilon}_r(x, z)$ and a complex refractive index $\hat{n}(x, z)$ by the relations

$$\hat{\epsilon}_r(x, z) = [\hat{n}(x, z)]^2 = \epsilon(x, z)/\epsilon_0 + i\sigma(x, z)/\omega\epsilon_0$$

then equation (2.18) reduces to a Helmholtz equation

$$\frac{\partial^2}{\partial x^2}E_y(x, z) + \frac{\partial^2}{\partial z^2}E_y(x, z) + k^2\hat{\epsilon}_r(x, z)E_y(x, z) = 0 \tag{2.19}$$

where $k = \omega/c$ is the wave number and $c = (\epsilon_0\mu_0)^{-1/2}$ is the speed of light in a vacuum.

Similarly, equations (2.15)–(2.17) yield a propagation equation for H_y. If we use equations (2.5)–(2.7) and insert equations (2.15) and (2.16) into equation (2.17), we obtain

$$\frac{\partial}{\partial x}\left[\frac{1}{\hat{\epsilon}_r(x, z)}\frac{\partial}{\partial x}H_y(x, z)\right] + \frac{\partial}{\partial z}\left[\frac{1}{\hat{\epsilon}_r(x, z)}\frac{\partial}{\partial z}H_y(x, z)\right] + k^2 H_y(x, z) = 0. \tag{2.20}$$

Once H_y is known, the electric field is obtained from Maxwell's equations. This is the TM mode of polarization.

Let us denote by U the field component that appears in the appropriate propagation equation, i.e. $U = E_y$ for TE polarization and $U = H_y$ for TM polarization. The total field may be written in the form $U = U_P + U_D$, where the incident field

$$U_P(x, z) = \exp[ikn_I(x \sin \theta + z \cos \theta)] \tag{2.21}$$

is non-zero only in region I, U_D denotes the diffracted field and n_I is the refractive index in region I. Clearly, the function $U_P(x, z) \exp(-i\alpha_0 x)$, where $\alpha_0 = kn_I \sin \theta$, is periodic with the period d of the grating. It follows from the Floquet–Bloch theorem and the boundary conditions that U_D has the same property, i.e. $U_D(x + d, z) = \exp(i\alpha_0 d) U_D(x, z)$ in all regions. Such fields are called pseudo-periodic. If we express the periodic function

$$G(x, z) = U_D(x, z) \exp(-i\alpha_0 x) \tag{2.22}$$

in the form of a Fourier series in x, denote the Fourier coefficients by $G_m(z)$, and define

$$\alpha_m = \alpha_0 + 2\pi m/d \tag{2.23}$$

which will lead to the grating equations, we obtain an expression

$$U_D(x, z) = \sum_{m=-\infty}^{\infty} G_m(z) \exp(i\alpha_m x) \tag{2.24}$$

for the diffracted field.

2.2.3 *Fields Outside the Modulated Region*

In the homogeneous regions I and III, the propagation equations (2.19) and (2.20) both reduce to a constant-coefficient Helmholtz equation

$$\frac{\partial^2}{\partial x^2} U(x, z) + \frac{\partial^2}{\partial z^2} U(x, z) + k^2 \hat{n}_j^2 U(x, z) = 0 \tag{2.25}$$

where $j = I$ or $j = III$. Let us first consider region III and denote the transmitted part of the diffracted field U_D by U_T. If we assume that the material in region III is dielectric, i.e. $\hat{n}_{III} = n_{III}$ is real, and insert equation (2.24) into equation (2.25), we obtain

$$\frac{\partial^2}{\partial z^2} G_m(z) + t_m^2 G_m(z) = 0 \tag{2.26}$$

where $t_m^2 = (kn_{III})^2 - \alpha_m^2$. The solution of equation (2.26) is

$$G_m(z) = T_m \exp[it_m(z - h)] + R_m \exp[-it_m(z - h)] \tag{2.27}$$

with the sign convention

$$t_m = \begin{cases} [(kn_{III})^2 - \alpha_m^2]^{1/2} & \text{if } |\alpha_m| \le kn_{III} \\ i[\alpha_m^2 - (kn_{III})^2]^{1/2} & \text{if } |\alpha_m| > kn_{III} \end{cases} \tag{2.28}$$

The solution for $U_T(x, z)$ is obtained by inserting equation (2.27) into equation (2.24). The resulting expression is a superposition of plane waves with wavevectors $\mathbf{k} = \alpha_m \mathbf{x} + t_m \mathbf{z}$. For the latter term in equation (2.27), the case $|\alpha_m| \le kn_{III}$ corresponds to plane waves that arrive from $z = +\infty$, whereas $|\alpha_m| > kn_{III}$ implies exponentially growing waves when $z \to \infty$. Both must vanish ($R_m = 0$) because there are neither sources nor scatterers in the half-space $z > h$. Hence

$$U_{\mathrm{T}}(x, z) = \sum_{m=-\infty}^{\infty} T_m \exp\{i[\alpha_m x + t_m(z - h)]\}. \tag{2.29}$$

Here the real values of t_m correspond to homogeneous plane waves, i.e. propagating diffraction orders of the grating. If we write $\alpha_m = k n_{\mathrm{III}} \sin \theta_m$, where θ_m is the diffraction angle of the mth-order diffracted wave, equation (2.23) gives the grating equation

$$n_{\mathrm{III}} \sin \theta_m = n_{\mathrm{I}} \sin \theta + m\lambda/d.$$

The imaginary values of t_m correspond to evanescent waves, which decay exponentially as $z \to \infty$. If the material in region III is a metal ($\hat{n}_{\mathrm{III}}$ is complex-valued), there are no homogeneous plane waves in region III.

In a similar fashion, one can derive a plane-wave representation for the diffracted field in region I:

$$U_{\mathrm{R}}(x, z) = \sum_{m=-\infty}^{\infty} R_m \exp[i(\alpha_m x - r_m z)] \tag{2.30}$$

(the subscript D is replaced by R to emphasize reflection). The expression for r_m is identical to equation (2.28), with the exception that n_{III} is replaced by n_{I}. The grating equation for reflected waves takes the form

$$n_{\mathrm{I}} \sin \theta_m = n_{\mathrm{I}} \sin \theta + m\lambda/d.$$

The plane-wave expansions (2.29) and (2.30) are commonly known as Rayleigh expansions (Lord Rayleigh, 1907).

We are most often interested in the energy distribution amongst the diffracted orders. The diffraction efficiency of the mth reflected order is defined as the z-component of $\langle S \rangle$, normalized to that of the incident wave:

$$\eta_{\mathrm{R},m} = \mathscr{R}\{r_m/r_0\}|R_m|^2.$$

Similarly, for the mth transmitted order,

$$\eta_{\mathrm{T},m} = C\mathscr{R}\{t_m/r_0\}|T_m|^2$$

where $C = 1$ for TE polarization and $C = (n_1/n_{\mathrm{III}})^2$ for TM polarization. For evanescent waves $\langle S \rangle_z = 0$, i.e. they carry no energy in the z-direction.

2.2.4 *Field Inside the Modulated Region*

Some of the grating profiles commonly encountered in diffractive optics are illustrated in Fig. 2.1. The lamellar profile of Fig. 2.1(a), with

$$\hat{\epsilon}_r(x, z) = \hat{n}_l^2 \quad \text{when } x_{l-1} \leq x < x_l, \ l = 1, \ldots, L \tag{2.31}$$

is particularly popular: if $\hat{n}_l$ has only two permitted values, the profile of equation (2.31) can be generated by a single-step microlithographic process. Multilevel lamellar profiles of the type shown in Fig. 2.1(b), i.e.

$$\hat{\epsilon}_r(x, z) = \hat{n}_{lq}^2 \quad \text{when } x_{l-1} \leq x < x_l \text{ and } (q-1)h/Q \leq z < qh/Q \tag{2.32}$$

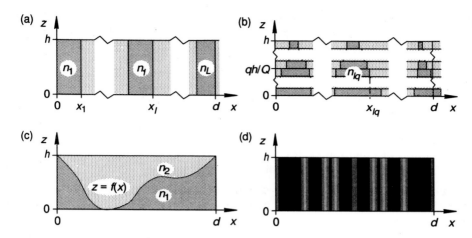

Figure 2.1 (a) Lamellar profile, (b) multilevel lamellar profile, (c) continuous surface-relief profile, (d) index-modulated profile.

may be produced by an *N*-stage lithographic process when $Q = 2^N$. Arbitrary surface-relief profiles of the form of Fig. 2.1(c) can be modelled, at any desired level of accuracy, with multilevel profiles of equation (2.32) by increasing Q. Smoothly varying index-modulation profiles illustrated in Fig. 2.1(d) are obtained, for example, in acousto-optics and volume holography. If such a profile is not *z*-invariant within the entire modulated region, it can again be divided into Q approximately *z*-invariant slabs. We therefore first consider the *z*-invariant case, using an eigenmode theory developed by Burckhardt (1966), Kaspar (1973), Peng *et al.* (1975), and Knop (1978).

Let us first assume TE-mode of polarization. By a separation of variables $E_y(x, z) = X(x)Z(z)$ in equation (2.19), we arrive at two ordinary differential equations

$$\frac{d^2}{dx^2} X(x) + [k^2 \hat{\epsilon}_r(x) - \gamma^2] X(x) = 0 \tag{2.33}$$

and

$$\frac{d^2}{dz^2} Z(z) + \gamma^2 Z(z) = 0 \tag{2.34}$$

where γ^2 is the separation constant. The solution of equation (2.34) can be expressed as

$$Z(z) = a \exp(i\gamma z) + b \exp[-i\gamma(z - h)] \tag{2.35}$$

where a and b are as-yet undetermined constants. To solve equation (2.33), we express the periodic function $\hat{\epsilon}_r$ in the form of a Fourier series

$$\hat{\epsilon}_r(x) = \sum_{p=-\infty}^{\infty} \varepsilon_p \exp(i2\pi px/d) \tag{2.36}$$

37

and search for a pseudo-periodic solution of the form

$$X(x) = \sum_{m=-\infty}^{\infty} P_m \exp(i\alpha_m x). \tag{2.37}$$

If we insert equations (2.36) and (2.37) into equation (2.33), we obtain

$$\sum_{m=-\infty}^{\infty} \sum_{p=-\infty}^{\infty} [(\alpha_m^2 + \gamma^2) \exp(i\alpha_p x)\delta_{mp} - k^2 \varepsilon_p \exp(i\alpha_{m+p}x)]P_m = 0. \tag{2.38}$$

Multiplication of equation (2.38) by $\exp(-i\alpha_l x)$ and integration over the period d gives a system of equations (numbered by the index l)

$$\sum_{m=-\infty}^{\infty} (k^2 \varepsilon_{l-m} - \alpha_m^2 \delta_{lm})P_m = \gamma^2 P_l. \tag{2.39}$$

This linear set of equations may be expressed in matrix form $\mathbf{MP} = \gamma^2 \mathbf{P}$, where the elements of the matrix $\mathbf{M}$ and the vector $\mathbf{P}$ are $M_{lm} = k^2 \varepsilon_{l-m} - \alpha_m^2 \delta_{lm}$ and P_l, respectively. The eigenvalues γ_n^2 and the elements of the corresponding eigenvectors P_{ln} (with $n = 1, \ldots, \infty$) can be solved from this matrix eigenvalue problem by standard numerical techniques. The complete solution of equation (2.19) is then obtained by combining the solutions of equations (2.33) and (2.34):

$$E_y(x, z) = \sum_{l=-\infty}^{\infty} \sum_{n=1}^{\infty} P_{ln} \exp(i\alpha_l x)\{a_n \exp(i\gamma_n z) + b_n \exp[-i\gamma_n(z - h)]\} \tag{2.40}$$

where the sign convention for γ_n is $\mathcal{R}\{\gamma_n\} + \mathcal{I}\{\gamma_n\} > 0$.

In TM polarization, we again attempt a separable pseudo-periodic solution

$$H_y(x, z) = \{a \exp(i\gamma z) + b \exp[-i\gamma(z - h)]\} \sum_{m=-\infty}^{\infty} P_m \exp(i\alpha_m x). \tag{2.41}$$

An eigenvalue equation corresponding to equation (2.39) can be derived by substituting equation (2.41) and the expansion

$$\frac{1}{\hat{\varepsilon}_r(x)} = \sum_{p=-\infty}^{\infty} \xi_p \exp(i2\pi p x/d) \tag{2.42}$$

into equation (2.20). However, the eigenvalue equation can be cast into a more convenient (but numerically equivalent) form if we introduce another function

$$Q(x, z) = \frac{1}{\hat{\varepsilon}_r(x)} \frac{\partial}{\partial z} H_y(x, z) \tag{2.43}$$

and its pseudo-periodic expansion

$$Q(x, z) = i\gamma\{a \exp(i\gamma z) - b \exp[-i\gamma(z - h)]\} \sum_{m=-\infty}^{\infty} Q_m \exp(i\alpha_m x). \tag{2.44}$$

Multiplication of equation (2.43) by $\hat{\varepsilon}_r(x)$, followed by the insertion of equations (2.36), (2.41) and (2.44), gives

$$P_l = \sum_{m=-\infty}^{\infty} \varepsilon_{l-m} Q_m. \tag{2.45}$$

By combining equations (2.20) and (2.43), we obtain

$$\frac{\partial}{\partial x}\left[\frac{1}{\hat{\epsilon}_r(x)}\frac{\partial}{\partial x}H_y(x,z)\right] + \frac{\partial}{\partial z}Q(x,z) + k^2 H_y(x,z) = 0. \tag{2.46}$$

If we finally insert equations (2.41), (2.42) and (2.44) into equation (2.46), we arrive at

$$\sum_{m=-\infty}^{\infty}(k^2\delta_{lm} - \alpha_l\xi_{l-m}\alpha_m)P_m = \gamma^2 Q_l. \tag{2.47}$$

In matrix form, equations (2.45) and (2.47) read as $\mathbf{NQ} = \mathbf{P}$ and $\mathbf{MP} = \gamma^2\mathbf{Q}$, where the elements of $\mathbf{N}$ and $\mathbf{M}$ are $N_{lm} = \varepsilon_{l-m}$ and $M_{lm} = k^2\delta_{lm} - \alpha_l\xi_{l-m}\alpha_m$, respectively. We therefore obtain $\mathbf{MNQ} = \gamma^2\mathbf{Q}$, which gives the eigenvalues γ_n and the eigenvector-elements Q_{ln}. The eigenvector-elements P_{ln} are given by $\mathbf{P} = \mathbf{NQ}$.

The TM eigenvalue problem can also be cast in forms that are not numerically equivalent with equations (2.45) and (2.47), and can be numerically inferior or superior (Li, 1996). For example, a somewhat different derivation gives

$$\mathbf{L}^{-1}\mathbf{MD} = \gamma^2\mathbf{D} \tag{2.48}$$

where the elements of matrix $\mathbf{L}$ are $L_{lm} = \xi_{l-m}$, the elements of matrix $\mathbf{M}$ remain the same, and the elements of vector $\mathbf{D}$ are the expansion coefficients of D_x, the x-component of the electric displacement, which turn out to be equal to the expansion coefficients of H_y. The form (2.48) of the eigenvalue equation is attractive because, unlike the previous formulation, it provides the correct form-birefringence limit (see section 2.3.1) when only the lowest-order term is retained in each summation.

2.2.5 *Solution of the Boundary-value Problem*

We now have exact representations of the electromagnetic field in all regions I–III, but the parameters $\{R_m\}$ and $\{T_m\}$ in equations (2.29) and (2.30), and $\{a_n\}$ and $\{b_n\}$ in equation (2.40) are still unspecified. They are solved from electromagnetic boundary conditions, equations (2.8)–(2.11), at $z = 0$ and $z = h$.

Let us first consider TE polarization, i.e. $U = E_y$. At the boundaries $z = 0$ and $z = h$ the surface normal $\mathbf{n}_{12}\|z$. Hence, in view of equation (2.10), E_y must be continuous across these boundaries. At $z = 0$, it follows from equations (2.21), (2.30) and (2.40) that

$$\delta_{l0} + R_l = \sum_{m=1}^{\infty}[a_m + b_m\exp(i\gamma_m h)]P_{lm} \tag{2.49}$$

where $l = -\infty, \ldots, \infty$. At $z = h$, equations (2.29) and (2.40) give

$$T_l = \sum_{m=1}^{\infty}[a_m\exp(i\gamma_m h) + b_m]P_{lm}. \tag{2.50}$$

In view of equation (2.11), H_x and therefore $\partial E_y(x,z)/\partial z$ must also be continuous across the two boundaries. At $z = 0$ we have

$$r_l(\delta_{l0} - R_l) = \sum_{m=1}^{\infty}\gamma_m[a_m - b_m\exp(i\gamma_m h)]P_{lm} \tag{2.51}$$

39

and at $z = h$

$$t_l T_l = \sum_{m=1}^{\infty} \gamma_m [a_m \exp(i\gamma_m h) - b_m] P_{lm}. \tag{2.52}$$

If we insert R_l and T_l from equations (2.49) and (2.50) into equations (2.51) and (2.52), we obtain a system of simultaneous equations

$$\sum_{m=1}^{\infty} (r_l + \gamma_m) P_{lm} a_m + \sum_{m=1}^{\infty} (r_l - \gamma_m) \exp(i\gamma_m h) P_{lm} b_m = 2r_l \delta_{l0}$$

$$\sum_{m=1}^{\infty} (t_l - \gamma_m) \exp(i\gamma_m h) P_{lm} a_m + \sum_{m=1}^{\infty} (t_l + \gamma_m) P_{lm} b_m = 0$$

which can be solved for a_m and b_m, e.g. by Gauss elimination. Consequently, T_l and R_l are solved from equations (2.49) and (2.50).

In TM polarization $U = H_y$. In view of equation (2.11), H_y is continuous across $z = 0$ and $z = h$. On the other hand, equation (2.10) implies the continuity of E_x, i.e. the function $Q(x, z)$ in equation (2.43). These conditions lead to

$$\sum_{m=1}^{\infty} (n_{\mathrm{I}}^{-2} r_l P_{lm} + \gamma_m Q_{lm}) a_m + \sum_{m=1}^{\infty} (n_{\mathrm{I}}^{-2} r_l P_{lm} - \gamma_m Q_{lm}) \exp(i\gamma_m h) b_m = 2n_{\mathrm{I}}^{-2} r_l \delta_{l0}$$

and

$$\sum_{m=1}^{\infty} (n_{\mathrm{III}}^{-2} t_l P_{lm} - \gamma_m Q_{lm}) \exp(i\gamma_m h) a_m + \sum_{m=1}^{\infty} (n_{\mathrm{III}}^{-2} t_l P_{lm} + \gamma_m Q_{lm}) b_m = 0.$$

In the multilevel case, we solve the eigenvalue problem separately in each slab (Nyyssonen and Kirk, 1988). Hence we obtain, in slab q, quantities $\gamma_{m,q}$ and $P_{lm,q}$ in TE polarization, and also $Q_{lm,q}$ in TM polarization. We need additional relations at the boundaries $z = z_q$ to determine the new unknowns $\{a_{m,q}\}$ and $\{b_{m,q}\}$. In TE polarization, the boundary conditions give

$$\sum_{m=1}^{\infty} \{a_{m,q} \exp[i\gamma_{m,q}(z_q - z_{q-1})] + b_{m,q}\} P_{lm,q}$$

$$= \sum_{m=1}^{\infty} \{a_{m,q+1} + b_{m,q+1} \exp[i\gamma_{m,q+1}(z_{q+1} - z_q)]\} P_{lm,q+1} \tag{2.53}$$

and

$$\sum_{m=1}^{\infty} \{a_{m,q} \exp[i\gamma_{m,q}(z_q - z_{q-1})] - b_{m,q}\} \gamma_{m,q} P_{lm,q}$$

$$= \sum_{m=1}^{\infty} \{a_{m,q+1} - b_{m,q+1} \exp[i\gamma_{m,q+1}(z_{q+1} - z_q)]\} \gamma_{m,q+1} P_{lm,q+1} \tag{2.54}$$

In TM polarization, $P_{lm,q}$ are replaced by $Q_{lm,q}$ in equation (2.54), but not in equation (2.53).

2.2.6 Numerical Considerations

The method presented above is applicable to all grating profiles illustrated in Fig. 2.1. The infinite matrices and vectors, which appear in the eigenvalue problem and in the boundary-value problem, must be truncated in numerical calculations. If we retain N Rayleigh orders and, consequently, N lowest-order eigenmodes of the modulated structure, the eigenvalue matrix has dimensions $N \times N$. The dimensions of the boundary-value matrix are $2N \times 2N$ in the single-slab case, and $2N(Q + 1) \times 2N(Q + 1)$ in the Q-slab case. In the latter case, the matrix is sparse, because there is coupling between subsequent slabs only (see Gaylord and Moharam (1985) for details). This implies considerable savings in terms of computer time and memory. In general, we must retain all propagating Rayleigh orders and a sufficient number of evanescent orders to achieve accurate results; the recommended method of investigating the accuracy of the numerical solution is to increase N and observe the convergence of the efficiencies $\eta_{T,m}$ and $\eta_{R,m}$. The theoretical formulation presented above (in particular the choice of exponentials in equation (2.35)) is such that no numerical instability should be encountered when N is increased, even if the modulated region is arbitrarily thick.

Because of the necessity of retaining a sufficient number of evanescent orders, the method scales poorly as the period d increases. This is, however, also true for alternative methods. The computation time is proportional to the third power of N, and thus of d/λ. The computer memory requirements increase quadratically with N. The convergence is rapid for dielectric gratings (the efficiencies can be computed at an accuracy of several decimal places), but slower for metallic gratings, in particular if the conductivity is high. Serious problems have been encountered in TM polarization if the profile contains abrupt boundaries, i.e. in all cases except for (d) in Fig. 2.1. These problems are common to all methods that employ a Fourier expansion of the permittivity profile (Nèviere and Vincent, 1988; Li and Haggans, 1993), but they can be avoided (Li, 1996).

The rigorous analysis methods most commonly employed in grating theory may be divided into integral, differential, modal and coupled-wave approaches. The coupled-wave method (Moharam and Gaylord, 1981), which has been extended to anisotropic gratings (Rokushima and Yamakita, 1983), is basically equivalent with the eigenmode formulation presented above, although the representation of the field inside the grating is different.

In rigorous modal methods (Maystre and Petit, 1971; Botten *et al.*, 1981; Sheng *et al.*, 1982), the representation of the field inside the grating is a superposition of exact modal functions of the periodic waveguide, in contrast to the truncated pseudo-periodic expansions employed above. Such exact eigenmodes can only be found for some specific grating geometries. Fortunately these include the binary and multilevel (Surrateau *et al.*, 1983) structures of Figs 2.1(b) and (c). If the grating contains features with dimensions considerably larger than the wavelength, or if the conductivity is high, care has to be exercised in the numerical implementation (Roberts and McPhedran, 1987). With this precaution, the modal methods converge well even for metallic gratings in TM polarization. However, the search for eigenvalues can be quite an arduous task, although approximate values can be obtained by the numerically efficient method presented in detail above.

The differential method (Petit, 1980, Chapter 4), in which the propagation equations are integrated directly, also employs a Fourier-series expansion of the permittivity profile and thus suffers from convergence problems in the case of TM polarization and highly conducting gratings. This method is particularly efficient for continuous dielectric surface-relief gratings. Integral methods (Petit, 1980,

Chapter 3) are also good for highly conducting gratings, but they are theoretically rather complicated. Of the more recent methods, we cite the Legendre polynomial expansion method of Morf (1993), which overcomes the convergence problems of highly conducting gratings in TM polarization, and a finite-element approach of Delort and Maystre (1993).

Thus far, the incident wave has been assumed to lie in the xz-plane. If this is not the case, one speaks about conical incidence. This situation can only be dealt with easily in the case of perfectly conducting gratings, where a certain superposition of TE- and TM-solutions provides an exact solution (Petit, 1980). Otherwise the TE/TM decomposition is not valid, and the rigorous analysis methods have to be modified accordingly: see, for example, Moharam and Gaylord (1983) and Li (1993). If the grating structure is three-dimensional, or 'doubly periodic', consisting of, for example, pillars, holes or pyramidal structures on a plane substrate, the rigorous analysis is significantly more complicated both theoretically and numerically (Vincent, 1978; Bräuer and Bryngdahl, 1993). Here we must retain $N \times N$ Rayleigh orders in the analysis, including a sufficient number of evanescent waves. This implies a high computational cost.

2.3 Approximate Grating Theories

Approximate methods to describe the grating's response can be applicable if either $d \gg \lambda$ or $d \ll \lambda$.

2.3.1 *Subwavelength-period Gratings*

If $d \ll \lambda$, the grating does not diffract light in the classical sense, since all orders other than the transmitted and the reflected zeroth order are evanescent. The grating can, nevertheless, significantly modify the properties of the incident wave, primarily its phase and its state of polarization. It can also reduce the reflectance of the material boundary, and if the substrate is covered by a dielectric layer of higher refractive index, it can excite leaky guided modes in this layer.

The simplest model for subwavelength-period gratings is the classical theory of form birefringence (Born and Wolf, 1980, pp. 705–708). Let us consider a dielectric grating with $\epsilon(x, z) = \epsilon_1$ in the interval $0 \leq x < c$ and $\epsilon(x, z) = \epsilon_2$ if $c \leq x < d$. In TE polarization, equation (2.10) enforces the only non-vanishing component of the electric field, E_y, to be continuous across the boundaries at $x = 0$ and $x = c$. If $d \ll \lambda$, we may assume E_y to be constant in sections $0 < x < c$ and $c < x < d$; hence it is constant inside the entire grating. In view of equation (2.5), $D_y = \epsilon_1 E_y$ when $0 < x < c$ and $D_y = \epsilon_2 E_y$ when $c < x < d$. The mean value of D_y is then $\bar{D}_y = [\epsilon_1 c/d + \epsilon_2(1 - c/d)]E_y$, and we may define an effective dielectric constant

$$\epsilon_{TE} = \bar{D}_y / E_y = \epsilon_1 c/d + \epsilon_2(1 - c/d). \tag{2.55}$$

In TM polarization, at normal incidence, the only non-vanishing components of $\boldsymbol{E}$ and $\boldsymbol{D}$ are E_x and D_x. Equation (2.9) requires that D_x be continuous across the boundaries and hence constant inside the grating. The mean value of E_x is $\bar{E}_x = [\epsilon_1^{-1} c/d + \epsilon_2^{-1}(1 - c/d)]D_x$, and the effective dielectric constant is given by

$$\epsilon_{TM} = D_x / \bar{E}_x = \frac{\epsilon_1 \epsilon_2}{\epsilon_2 c/d + \epsilon_1(1 - c/d)}. \tag{2.56}$$

The difference $\epsilon_{TE} - \epsilon_{TM} \geq 0$, which means that the grating acts like a negative uniaxial crystal; hence the term form birefringence.

The expressions (2.55) and (2.56) can also be obtained from the rigorous theory presented in section 2.2.4 if we retain only the lowest-order terms in equations (2.39) and (2.48) for TE and TM modes, respectively, and define the effective dielectric constants as $(\gamma/k)^2$, in view of equation (2.35).

Because of the form birefringence, a single-slab subwavelength-period grating with a suitably chosen thickness can obviously act as a wave plate. If each slab in the grating structure is thought to be replaced by an effective medium with the permittivity given by equation (2.55) or (2.56), the theory of multilayer films (Born and Wolf, 1980, pp. 51–70) is applicable to the calculation of the efficiencies of the transmitted and reflected zeroth-order beams. If, for example, $\epsilon_{TE} = n_I n_{III}$ and $h = \lambda/4\epsilon_{TE}^{1/2}$, the grating acts as an antireflection layer for TE-polarized light at normal incidence.

More refined theories of effective media can be derived, e.g. by retaining only one mode and the zeroth Rayleigh orders in the modal theory of lamellar gratings (Botten *et al.*, 1981; McPhedran *et al.*, 1982). The determination of the effective dielectric constant reduces to a numerical solution of a transcendental equation, which is derived in a somewhat different manner by Yeh *et al.* (1977).

2.3.2 *Complex-amplitude Transmittance Method*

We next assume that $d \gg \lambda$, and that the grating profile is coarse. In the geometry of Fig. 2.1(a), this means that $x_l - x_{l-1} \gg \lambda$ for all l, and in the case of Fig. 2.1(b) $x_{l,q} - x_{l-1,q} \gg \lambda$ for all l, q. In the geometry of Fig. 2.1(c), we require that $df(x)/dx \ll 1$, and in the case of Fig. 2.1(d) $\partial\hat{n}(x, z)/\partial x \ll 1$. These assumptions imply that the angular spectrum of diffracted plane waves concentrates around the z-axis, although abrupt boundaries in the x-direction always generate finite amplitudes for arbitrarily high orders. Hence the longitudinal components of the electromagnetic field are negligible and the use of scalar diffraction theory is justified (it should be noted, however, that scalar theory may give useful results even if its use cannot be formally justified).

In the simplest approximation, we may assume that light rays pass the modulated region without deflection. The effect of a transmission grating may then be described as $U_T(x, h) = t(x) U_P(x, 0)$, where the so-called complex-amplitude transmission function $t(x)$ takes the form

$$t(x) = \exp\left[ik \int_0^h \hat{n}(x, z)\, dz\right].$$

Here we have a pointwise relationship between the incident field and the transmitted part of the diffracted field (in fact, the reflected part of the diffracted field is neglected). The complex amplitudes of the transmitted orders are given by

$$T_m = \frac{1}{d} \int_0^d U_T(x, h) \exp(-i\alpha_m x)\, dx \tag{2.57}$$

which is obtained by Fourier inversion of equation (2.29) at $z = h$. In scalar theory the efficiencies are the same for both states of polarization: $\eta_m = |T_m|^2$.

The complex-amplitude transmittance method often permits an analytical solution of the grating's response. Let us consider a Q-level dielectric surface-relief grating with $n(x, z) = n$ if $z \leq h(x)$, $n(x, z) = 1$ elsewhere, and $h(x) = h(Q - q + 1)/Q$ when $d(q - 1)/Q \leq x < dq/Q$. A straightforward calculation gives

$$\eta_m = \text{sinc}^2(m/Q) \left| \frac{1}{Q} \frac{1 - \exp\{-i2\pi[h(n - 1)/\lambda + m]\}}{\exp\{i2\pi[h(n - 1)/\lambda + m]/Q\} - 1} \right|^2$$

where $\text{sinc}(x) = \sin(\pi x)/(\pi x)$. When $2\pi[h(n - 1)/\lambda + m] \to 0$

$$\eta_m \to \text{sinc}^2(1/Q).$$

Hence the maximum (minus) first-order efficiency of a Q-level staircase grating is $\eta_{-1} = \text{sinc}^2(1/Q)$. For a binary grating we have $\eta_{-1} \approx 40.5\%$, for a four-level grating $\eta_{-1} \approx 81\%$, for an eight-level grating $\eta_{-1} \approx 95\%$, etc. When $Q \to \infty$, we obtain a triangular profile. In this case $\eta_{-1} \to 1$.

2.3.3 *Thin-grating Decomposition Method*

As the analysis methods of volume gratings are well documented, we only present a natural extension of the amplitude transmittance theory not covered in the standard works of Solymar and Cook (1981), and of Gaylord and Moharam (1985).

Let us consider a weakly index-modulated, thick ($h/\lambda \gg 1$) dielectric grating with $n_{II}(x, z) = \bar{n} + \Delta n(x, z)$, where $\bar{n}$ is the mean refractive index and $|\Delta n(x, z)| \ll \bar{n}$. The modulated region is divided into Q slabs with $h(q - 1)/Q < z < hq/Q$, where $q = 1, \ldots, Q$. In each slab the material is assumed homogeneous with a refractive index $\bar{n}$. The true index modulation is modelled by assuming that infinitely thin gratings exist at the slab boundaries $z_q = hq/Q$, and that the complex-amplitude transmission functions of these gratings are of the form $t_q(x) = \exp[i\Phi_q(x)]$, with

$$\Phi_q(x) = k\bar{n}h/Q + k \int_{h(q-1)/Q}^{hq/Q} \Delta n(x, z)\, dz \approx k[\bar{n} + \Delta n(x, hq/Q)]h/Q.$$

The approximation is good when Q is sufficiently large.

We consider the periodic function $G(x, z)$ of equation (2.22) and denote its values immediately before and after a thin grating by superscripts $-$ and $+$, respectively. Starting from an initial value $G^+(x, 0) = 1$, the field is propagated through the qth slab with the aid of the paraxial form of the Rayleigh expansion:

$$G^-(x, z_q) = \exp(ik\bar{n}h/Q) \sum_{m=-\infty}^{\infty} T_m(z_{q-1}) \exp[-i(\alpha_m^2/2k\bar{n})h/Q] \exp(i2\pi mx/d) \quad (2.58)$$

where

$$T_m(z_{q-1}) = \frac{1}{d} \int_0^d G^+(x, z_{q-1}) \exp(-i2\pi mx/d)\, dx \quad (2.59)$$

and the response of the qth thin grating is $G^+(x, z_q) = t_q(x)G^-(x, z_q)$.

The numerical efficiency of this thin-grating decomposition method is based on the fast Fourier transform algorithm, which is applicable if the integral in equation (2.59) is sampled at N points and N Rayleigh orders are retained in equation (2.58). If N and Q are large enough and $|\Delta n(x, z)|$ is always less than ~ 0.1, the method gives accurate results for gratings with periods as small as a few λ. It can be useful even when $d \approx \lambda$, in particular if the exact Rayleigh expansion is used in equation (2.58), but in such a case the results should be compared to rigorous theory. Several versions of the thin-grating decomposition method are widely used in, for example, fibre and integrated optics (Lagasse and Baets, 1987).

2.3.4 Local Plane-wave Method

In the case of surface-relief gratings of the type shown in Fig. 2.1(c), approximate results may be obtained even if we cannot assume that the rays propagate through the grating without deflection. The considerations that follow are based on the classic ideas of Beckmann and Spizzichino (1963), and of Beckmann (1967).

Let us consider a surface-relief grating with a continuous profile, as shown in Fig. 2.1(c). The goal is to predict the transmitted field in the plane $z = h$, and then to calculate the complex amplitudes of the diffraction orders from equation (2.57). The key assumption is that the profile $f(x)$ acts locally as if it were an infinite plane boundary between two dielectric media. We may thus employ the Fresnel reflection and transmission coefficients to evaluate the local transmittance and reflectance. Ray-tracing, Snell's law, the law of reflection, and optical path calculations along the rays are used to predict $U_T(x, h)$. Multiple-scattering effects within the modulated structure can be taken into account, but the results are not rigorously valid because of the local plane-wave approximation. They may be expected to apply whenever all locally straight grating-profile sections are large compared to λ, but the method can be useful even if this assumption is not valid, as will be illustrated below.

2.4 Diffraction Characteristics of Gratings

We will now examine the validity of the amplitude transmittance theory as d/λ is decreased. Exact efficiency curves are presented for some of the most commonly used surface-relief profiles. Parametric optimization of the permittivity structure in the resonance domain ($d \approx \lambda$) is considered, and some results are given.

2.4.1 Validity of Scalar Theory

Let us first consider a binary dielectric surface-relief grating at normal incidence. We assume an aspect ratio $c/d = \frac{1}{2}$, a relief depth $h/\lambda = 1$, and a refractive index $n = 1.5$. The amplitude transmittance theory gives $\eta_{\pm 1} = 4/\pi^2$ when $d/\lambda \gg 1$. Figure 2.2(a) shows the rigorously calculated first-order and zeroth-order efficiency curves for both states of polarization. Apart from the fluctuations at $d/\lambda < 3$, the results are in good agreement with the amplitude transmittance theory, provided that the latter is corrected by the Fresnel transmission factor of

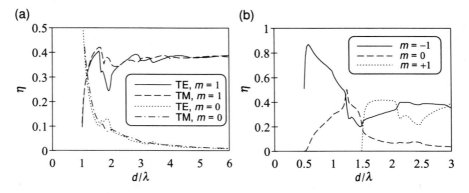

Figure 2.2 (a) Efficiencies $\eta_{\pm 1}$ and η_0 of a binary grating at $\theta = 0$ for both states of polarization, (b) efficiencies η_{-1}, η_0 and η_1 at Bragg incidence and TE polarization.

0.96. Figure 2.2(b) shows the efficiency curves of the three central orders at Bragg incidence $\sin \theta = \lambda/(2nd)$. The amplitude transmission theory still predicts $\eta_{\pm 1} = 0.96 \times 4/\pi^2 \approx 0.389$, but in the rigorous curves the symmetry is broken: η_{-1} reaches a high value, 87%, at $d/\lambda \approx 0.585$, shortly above the cut-off at $d/\lambda = \frac{1}{2}$.

Figure 2.3(a) shows the first-order efficiency curves of multilevel gratings with $Q = 4$ (solid line) and $Q = 16$ (dashed line). The efficiencies are significantly below the asymptotic limits in the entire range $1 < d/\lambda < 6$ shown in Fig. 2.3(a), and fall below 20% before reaching a peak at $d/\lambda < 2$. The dotted line in Fig. 2.3(a) shows the result of the local plane wave theory for a triangular grating: this theory is obviously far more accurate than the amplitude-transmittance approach. The steadily falling portion of this curve is calculated in the geometry of Fig. 2.3(b). The peak is predicted by assuming total internal reflection and the double-scattering model of Fig. 2.3(c), where all other reflections and refractions are neglected. This model gives some physical insight into the diffraction characteristics of triangular gratings, because the peak position coincides with $\beta = \theta_{-1}$, where θ_{-1} represents the diffraction angle of the order $m = -1$ (Noponen *et al.*, 1993).

2.4.2 *Parametric Optimization of Grating Profiles*

The examples of grating-diffraction characteristics shown in Figs 2.2 and 2.3 raise the question: is it possible to increase the efficiency η_{-1} by optimization of the parameters that define the grating structure? The answer is, in general, affirmative.

There is little room for improvement in the case of binary gratings at $\theta = 0$, since rigorously $\eta_{-1} = \eta_1 \leq 50\%$ for a symmetric profile. Significant gains can be achieved at Bragg incidence if $0.5 < d/\lambda < 1.5$, since the only propagating transmitted orders are then $m = -1$ and $m = 0$. If we optimize the aspect ratio c/d and the depth h/λ, it is often possible to achieve $\eta_{-1} > 90\%$, at least for a single polarization state. For example, a solution $d/\lambda = 1$, $c/d = \frac{1}{2}$, $h/\lambda = 1.63$ gives $\eta_{-1} = 97.7\%$ in TE polarization. All such transmission-type solutions possess a rather severe angular selectivity. In the reflective mode of operation, perfectly

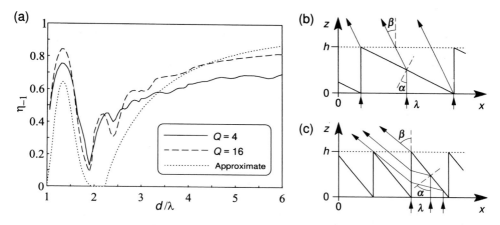

Figure 2.3 (a) Analysis of multilevel gratings. Solid line: four-level grating; dashed line: 16-level grating; dotted line: triangular grating by local plane wave theory (courtesy of E. Noponen). (b) Local plane wave model when α is below the critical angle α_C of total internal reflection. (c) The model when $\alpha > \alpha_C$.

conducting binary gratings can achieve $\eta_{-1} = 100\%$ over wide parameter ranges (Hessel *et al.*, 1975; Jull *et al.*, 1977; Loewen *et al.*, 1979).

At normal incidence, we are forced to employ multilevel gratings to achieve high efficiencies η_{-1}. As the regular Q-level profile fails in view of Fig. 2.3(a), we must look for a better combination of surface-relief transition points in each layer. Such combinations can indeed be found (Noponen *et al.*, 1992), and these solutions permit us to avoid the drop in the η_{-1} curves shown in Fig. 2.3(a).

2.5 Modulated Gratings

In diffractive micro-optics, gratings are frequently used as carriers, into which complicated wavefront transformations may be encoded once the desired output wave is known. Sometimes this wave may be obtained analytically by optical path considerations or by ray-tracing. In other cases, sophisticated numerical synthesis techniques may be employed (Bryngdahl and Wyrowski, 1990; Wyrowski and Bryngdahl, 1991). The wavefront transformation is most often realized by modulating the first diffraction order of the carrier grating, but the zeroth order can also be used. If we consider reconstruction fidelity, phase information is far more important than amplitude information. Moreover, the desired output wave can often be made phase-only by the use of synthesis algorithms. We now consider the physical principles of modulated gratings, or locally periodic diffractive elements. The best-known example of such devices is undoubtedly the diffractive lens (Nishihara and Suhara, 1987).

2.5.1 *Modulation of the First Order*

Let us first consider the case in which the carrier-order $m = -1$ is employed. The phase modulation is, in this case, based on Lohmann's detour-phase principle (Brown and Lohmann, 1966; Lohmann and Paris, 1967).

For an arbitrary grating, the transmitted field in the plane $z = h$ is given by equation (2.29), and the complex amplitudes T_m are in the form of equation (2.57). If we translate the permittivity profile by a distance Δx, the periodic function $G(x, h)$ in equation (2.22) satisfies $G'(x + \Delta x, h) = G(x, h)$ and thus

$$T'_m = \frac{1}{d} \int_0^d G(x - \Delta x, h) \exp(-i2\pi mx/d) \, dx \qquad (2.60)$$

(the primes denote translated quantities). By a change of variables $x' = x - \Delta x$ and use of the periodicity of $G(x, h)$, we obtain

$$T'_m = T_m \exp(-i2\pi m \Delta x/d) \qquad (2.61)$$

which is the rigorous mathematical form of the detour-phase principle. It implies that the phase of order m is modulated as $\phi'_m = \phi_m - 2\pi m \Delta x/d$ when the grating is shifted by a distance Δx.

The detour-phase principle provides the means for phase modulation: the shift Δx is modulated according to the phase of the desired signal wave $U_T(x, h)$. This method is mathematically justifiable if the shift Δx changes only slightly within the period d, but Lohmann's original experiments show that the modulation can be much more rapid. The rigorous nature of the detour-phase principle also permits, for example, the use of binary resonance-domain Bragg-type carrier gratings to achieve a high diffraction efficiency (Turunen *et al.*, 1993), and the application of parametric optimization results in the construction of high numerical aperture diffractive lenses and lens arrays (Noponen *et al.*, 1993).

Amplitude modulation, if required, can be obtained by controlling the local first-order diffraction efficiency of the carrier grating. In the case of a binary carrier grating, this is achieved most conveniently by modulating the aspect ratio c/d and keeping the pulse centres at their original positions to avoid undesired detour-phase modulation. The required mapping between the local aspect ratio c/d and the efficiency η_{-1} is obtained analytically by amplitude-transmittance theory if $d \gg \lambda$ and numerically, by exact electromagnetic theory, in the general case.

2.5.2 *Modulation of the Zeroth Order*

Whilst the modulation of the first diffraction order has been at the root of many milestones in diffractive optics, modulation of the zeroth order is a more recent approach. Gale (1976) and Knop (1976) designed image-plane diffractive elements by controlling the local zeroth-order efficiency of a dielectric surface-relief grating. In their approach, grating-depth variations are used to deflect a controlled amount of incident energy into higher orders, which are filtered out: if we assume an aspect ratio $c/d = \frac{1}{2}$, amplitude-transmittance theory predicts $\eta_0 = \cos^2[\pi(n - 1)h/\lambda]$ for a binary grating. This method permits, for example, the display of pictorial information in the form of diffractive colour slides.

Another approach, which provides the means for phase modulation, is based on the theory of effective media. It was discovered independently by Streibl and his co-workers (Stork *et al.*, 1991; Haidner *et al.*, 1992) and Farn (1992). A binary surface-relief grating with a subwavelength period is employed, and the aspect ratio is again varied. As only the zeroth order propagates, the amplitude modulation is slight, but the phase can be controlled over the entire required range $(0, 2\pi)$.

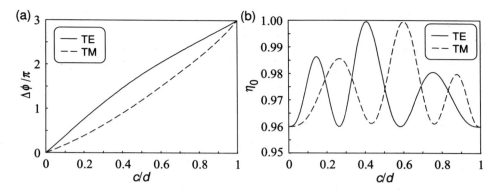

Figure 2.4 (a) The mapping between the phase delay $\Delta\phi$ and the aspect ratio c/d for a subwavelength-period grating with $d/\lambda = 0.4$, (b) zeroth-order efficiency η_0 as a function of c/d (courtesy of E. Noponen).

A first approximation of the mapping between the aspect ratio c/d and the phase delay $\Delta\phi$ is obtained from equation (2.55) or (2.56), depending on the state of polarization: the grating is thought to be replaced by a slab of effective material of refractive index $n_{eff} = \epsilon_{eff}^{1/2}$, and the phase change is $\Delta\phi = khn_{eff}$. For this approximation to be good, we must have $d \ll \lambda$. More accurate results are obtained by more refined theories of effective media or, ultimately, by the use of rigorous diffraction theory. Figure 2.4(a) shows the exact mapping between the phase delay and the aspect ratio for a grating of period $d = 0.4\lambda$ and thickness $h = 3\lambda$ (adapted from Noponen and Turunen, 1994). Here $n = 1.5$ when $0 \leq x < c$ and $n = 1$ when $c \leq x < d$. The zeroth-order efficiency shown in Fig. 2.4(b) varies between 0.96 and unity. The lower limit is equal to the Fresnel transmittance at an interface between materials with indices $n = 1.5$ and $n = 1$, and at one value of c/d the grating acts as an antireflection layer: $\eta_0 \approx 1$.

The drawback of this method is the large ratio (>10) of the relief depth and the minimum transverse feature size. The first demonstrations were carried out using water waves, but the method appears feasible at least in the mid-infrared ($\lambda \approx 10 \,\mu m$), perhaps also in the near-infrared and visible regions.

2.6 Conclusions

It may be stated that the theory of gratings is already rather well understood, although new contributions continue to appear. The choice of the analysis method greatly affects the complexity of solution: rigorous methods are universally valid, but often require considerable numerical computations. However, they are necessary when accurate results are needed and the grating period d is of the order of the wavelength λ. If $d/\lambda > 5$–10, simple scalar theories such as the complex-amplitude transmittance method for surface-relief gratings are typically adequate. The local plane-wave method provides reasonable accuracy even for small grating periods, and methods such as the thin-grating decomposition approach can give good results for volume gratings even when $d/\lambda \approx 1$–3. If, on the other hand, $d/\lambda < \frac{1}{5} - \frac{1}{10}$, the simplest theory of form birefringence may often be applied.

Many diffractive micro-optical elements have complicated periodic structures with $d/\lambda \gg 1$, but the dimensions of the smallest features may nevertheless be close to λ. One example is the Dammann grating, which divides an incident beam into several diffracted orders of equal efficiency (Dammann and Görtler, 1971). Other elements are locally periodic, and the local period can be of the order of λ in some regions such as the edges of a diffractive lens of high numerical aperture. In the former case a global rigorous analysis is needed to predict the performance precisely, which can be an extremely time-consuming numerical task. If the grating is periodic in both x- and y-directions, computer storage presently limits the maximum analyzable period to $\sim 10\lambda \times 10\lambda$. If the element is only locally periodic, rigorous analysis can also be applied locally to evaluate, for example, the efficiency of a lens in regions where the period is small.

Not all diffractive elements are even locally periodic. Approximate theories such as the complex-amplitude transmittance approach can be readily extended to treat such elements. The selection of rigorous methods is, however, much more limited, because the assumption of periodicity greatly simplifies the solution. However, the existing methods typically also permit the treatment of an arbitrary incident wave (Mendez *et al.*, 1983; Roberts, 1987).

References

BECKMANN, P. (1967) Scattering of light by rough surfaces. In Wolf, E. (ed.) *Progress in Optics*, Vol. VI, pp. 53–68. Amsterdam: North-Holland.

BECKMANN, P. and SPIZZICHINO, A. (1963) *The Scattering of Electromagnetic Waves from Rough Surfaces*. Oxford: Pergamon.

BORN, M. and WOLF, E. (1980) *Principles of Optics*, 6th edn. Oxford: Pergamon.

BOTTEN, L. C., CRAIG, M. S., McPHEDRAN, R. C., ADAMS, J. L. and ANDREWARTHA, J. R. (1981) The dielectric lamellar diffration grating. *Opt. Acta* **28**, 413–428.

BOUWKAMP, C. J. (1954) Diffraction theory. *Rep. Progr. Phys.* **17**, 35–100.

BRÄUER, R. and BRYNGDAHL, O. (1993) Electromagnetic diffraction analysis of two-dimensional gratings. *Opt. Commun.* **100**, 1–5.

BROWN, B. R. and LOHMANN, W. (1966) Complex spatial filtering with binary masks. *Appl. Opt.* **5**, 967–969.

BRYNGDAHL, O. and WYROWSKI, F. (1990) Digital holography – Computer-generated holograms. In Wolf, E. (ed.) *Progress in Optics*, Vol. XXVIII, pp. 1–86. Amsterdam: North-Holland.

BURCKHARDT, C. B. (1966) Diffraction of a plane wave at a sinusoidally stratified dielectric grating. *J. Opt. Soc. Am.* **56**, 1502–1509.

DAMMANN, H. and GÖRTLER, K. (1971) High-efficiency in-line multiple imaging by means of multiple phase holograms. *Opt. Commun.* **3**, 312–315.

DELORT, T. and MAYSTRE, D. (1993) Finite-element method for gratings. *J. Opt. Soc. Am.* A **10**, 2592–2601.

FARN, M. W. (1992) Binary gratings with increased efficiency. *Appl. Opt.* **31**, 4453–4458.

GALE, M. T. (1976) Sinusoidal relief gratings for zero-order reconstruction of black-and-white images. *Opt. Commun.* **18**, 292–297.

GAYLORD, T. K. and MOHARAM, M. G. (1985) Analysis and applications of optical diffraction by gratings. *Proc. IEEE* **73**, 894–937.

HAIDNER, H., KIPFER, P., STORK, W. and STREIBL, N. (1992) Zero-order gratings used as an artificial distributed index medium. *Optik* **89**, 107–112.

HESSEL, A., SCHMOYS, J. and TSENG, D. Y. (1975) Bragg-angle blazing of diffraction gratings. *J. Opt. Soc. Am.* **65**, 380–384.

JULL, E. V., HEATH, J. W. and EBBESON, G. R. (1977) Gratings that diffract all incident energy. *J. Opt. Soc. Am.* **67**, 1686–1688.

KASPAR, F. G. (1973) Diffraction by thick, periodically stratified gratings with complex dielectric constant. *J. Opt. Soc. Am.* **63**, 37–45.

KNOP, K. (1976) Color pictures using the zero diffraction order of phase grating structures. *Opt. Commun.* **18**, 298–303.

—— (1978) Rigorous diffraction theory for transmission phase gratings with deep rectangular grooves. *J. Opt. Soc. Am.* **68**, 1206–1210.

KORPEL, A. (1988) *Acousto-Optics.* New York: Marcel Dekker.

LAGASSE, P. C. and BAETS, R. (1987) Applications of propagating beam methods to electromagnetic and acoustic wave propagation problems. *Radio Sci.* **22**, 1225–1233.

LI, L. (1993) A modal analysis of lamellar diffraction gratings in conical mountings. *J. Mod. Opt.* **40**, 553–573.

LI, L. (1996) Use of Fourier series in the analysis of discontinuous periodic structures, *J. Opt. Soc. Am.* A, **13**, 1870–1876.

LI, L. and HAGGANS, C. E. (1993) Convergence of the coupled-wave method for metallic lamellar diffraction gratings. *J. Opt. Soc. Am.* A **10**, 1184–1189.

LOEWEN, E. G., NEVIÈRE, M. and MAYSTRE, D. (1979) Efficiency optimization of rectangular groove gratings for use in the visible and IR regions (TE). *Appl. Opt.* **18**, 2262–2266.

LOHMANN, A. W. and PARIS, D. P. (1967) Binary Fraunhofer holograms, generated by computer. *Appl. Opt.* **6**, 1739–1748.

LORD RAYLEIGH (1907) On the dynamical theory of gratings. *Proc. R. Soc.* A **79**, 399–416.

MAYSTRE, D. (1984) Rigorous vector theories of diffraction gratings. In Wolf, E. (ed.) *Progress in Optics*, Vol. XXI, pp. 1–67. Amsterdam: North-Holland.

MAYSTRE, D. and PETIT, R. (1971) Diffraction par un reseau lamellaire infinement conducteur. *Opt. Commun.* **28**, 413–428.

MCPHEDRAN, R. C., BOTTEN, L. C., CRAIG, M. S., NEVIÈRE, M. and MAYSTRE, D. (1982) Lossy lamellar gratings in the quasistatic limit. *Opt. Acta* **29**, 289–312.

MENDEZ, O. M., CADILHAC, M. and PETIT, R. (1983) Diffraction of a two-dimensional electromagnetic field by a thick slit pierced in a perfectly conducting screen. *J. Opt. Soc. Am.* **73**, 328–331.

MOHARAM, M. G. and GAYLORD, T. K. (1981) Rigorous coupled-wave analysis of planar-grating diffraction. *J. Opt. Soc. Am.* **71**, 811–818.

—— (1983) Three-dimensional vector coupled-wave analysis of planar-grating diffraction. *J. Opt. Soc. Am.* **73**, 1105–1112.

MORF, R. (1993) Diffraction theory. In Lalanne, Ph. and Chavel, P. (eds) *Perspectives for Parallel Optical Interconnects*, pp. 79–85. Berlin: Springer-Verlag.

NÈVIERE, M. and VINCENT, P. (1988) Differential theory of gratings: answer to an objection on its validity for TM polarization. *J. Opt. Soc. Am.* **5**, 1522–1524.

NISHIHARA, H. and SUHARA. T. (1987) Micro Fresnel lenses. In Wolf, E. (ed.) *Progress in Optics*, Vol. XXIV, pp. 1–40. Amsterdam: North-Holland.

NOPONEN, E. and TURUNEN, J. (1994) Binary high-frequency-carrier diffractive optical elements: electromagnetic theory. *J. Opt. Soc. Am.* A **11**, 1097–1109.

NOPONEN, E., TURUNEN, J. and VASARA, A. (1992) Parametric optimization of multilevel diffractive optical elements by electromagnetic theory. *Appl. Opt.* **31**, 5910–5912.

—— (1993) Electromagnetic theory and design of diffractive-lens arrays. *J. Opt. Soc. Am.* A **10**, 434–443.

NYYSSONEN, D. and KIRK, C. P. (1988) Optical microscope imaging of lines patterned in thick layers with variable edge geometry. *J. Opt. Soc. Am.* A **5**, 1270–1280.

PENG, S. T., TAMIR, T. and BERTONI, H. L. (1975) Theory of periodic dielectric waveguides. *IEEE Trans. Microwave Theory Tech.* **MTT-23**, 123–133.

PETIT, R. (ed.) (1980) *Electromagnetic Theory of Gratings*. Berlin: Springer-Verlag.

ROBERTS, A. (1987) Electromagnetic theory of diffraction by a circular aperture in a thick, perfectly conducting screen. *J. Opt. Soc. Am.* A **4**, 1970–1983.

ROBERTS, A. and McPHEDRAN, R. C. (1987) Power losses in highly conducting lamellar gratings. *J. Mod. Opt.* **34**, 511–538.

ROKUSHIMA, K. and YAMAKITA, J. (1983) Analysis of anisotropic dielectric gratings. *J. Opt. Soc. Am.* **73**, 901–908.

SHENG, P., STEPLEMAN, R. S. and SANDA, P. N. (1982) Exact eigenfunctions for square wave gratings: Application to diffraction and surface-plasmon calculations. *Phys. Rev.* B, **26**, 2907–2916.

SOLYMAR, L. and COOKE, D. J. (1981) *Volume Holography and Volume Gratings*. London: Academic Press.

STORK, W., STREIBL, N., HAIDNER, H. and KIPFER, P. (1991) Artificial distributed-index media fabricated by zero-order gratings. *Opt. Lett.* **16**, 1921–1924.

SURRATEAU, J. Y., CADHILLAC, M. and PETIT, R. (1983) Sur la détermination numérique des efficacités de certains réseaux diélectriques profonds. *J. Opt.* (Paris) **14**, 273–288.

TURUNEN, J., BLAIR, P., MILLER, J. M., TAGHIZADEH, M. R. and NOPONEN, E. (1993) Bragg holograms with binary synthetic surface-relief profile. *Opt. Lett.* **18**, 1022–1024.

VINCENT, P. (1978) A finite-difference method for dielectric and conducting crossed gratings. *Opt. Commun.* **26**, 293–296.

WYROWSKI, F. and BRYNGDAHL, O. (1991) Digital holography as part of diffractive optics. *Rep. Progr. Phys.* **54**, 1481–1571.

YEH, P., YARIV, A. and HONG, C.-S. (1977) Electromagnetic propagation in stratified media. I General theory. *J. Opt. Soc. Am.* **67**, 423–437.

3

Binary Optics Fabrication

M. B. STERN

3.1 Binary Optics Technology – a Universal Method for Synthesizing Diffractive Micro-optics

Planar arrays of diffractive micro-optics lie at the centre of a quiet revolution in optical system design. Microscopic relief structures incised into the surface of flat or curved substrates form the basis of novel optical elements that, individually or collectively, steer, concentrate, multiplex, image, or otherwise transform an incident wavefront (Stone and Thompson, 1991; Cole and Pittman, 1993; and others). These diffractive optical elements (DOEs) modulate the phasefront according to instructions encoded in the continuous or stepped surface-relief profile. The principal technique for manufacturing phase-only computer-generated DOEs with a stepped surface-relief profile has been named binary optics technology. The name reflects the inherent binary coding of both the phase quantization and the process sequence. Binary optics technology exploits the flexibility and precision of VLSI (very large-scale integration) circuit processing techniques and computer-aided design (CAD) tools to fabricate diffractive micro-optics in virtually any dielectric, metallic or semiconductor substrate (Veldkamp and Swanson, 1983; Swanson, 1989). It offers new opportunities in design optimization by providing a universal method to fabricate DOEs such as multifocal or bichromatic lenses, generalized phase plates, and space variant arrays, that would be difficult or even impossible to make by standard holographic methods (Leger *et al.*, 1988a, 1988b; Goltsos and Holz, 1990; Stern *et al.*, 1991b; Leger and Goltsos, 1992). To obtain high-quality optics, however, the lithography and substrate patterning techniques developed for semiconductor processing must be adapted to meet the specific and distinct demands of optical structure fabrication (Stern *et al.*, 1991a, 1992). Coupling these processing techniques with recent advances in the replication of microstructures holds great promise for the low-cost production of diffractive micro-optics (Gale *et al.*, 1993; Shvartsmann, 1993).

The present chapter explores key aspects of binary optics technology, focusing on the crucial methodology of device fabrication. A brief history of blazed DOE fabrication and a survey of diffractive optical components designed and built at

MIT Lincoln Laboratory (since the beginning of this effort in 1978) are presented in section 3.2. Section 3.3 offers a general review of fabrication procedures and outlines the steps necessary to transform a diffractive phase profile into a multilevel surface-relief structure. A systematic evaluation of the optical performance of diffractive microlenses, including the effects of fabrication-related errors, as described in section 3.4, underlines the challenges of fabricating high-quality diffractive microlenses and of measuring their optical efficiency. Unique processes developed to fabricate high-sag refractive micro-optical elements are summarized in section 3.5. Future applications that depend on the integration of micro-optics with other microcomponents are considered in section 3.6.

3.2. Background and Applications of Binary Optics at MIT Lincoln Laboratory

3.2.1 *Phase Relief Structures*

Significant efforts to increase the efficiency of diffractive structures can be traced to the prediction that blazed phase-only elements could diffract light into a single order with 100% efficiency (Sluisarev, 1957; Miyamoto, 1961; Lesem *et al.*, 1969). The thickness modulation of the optical surface could be achieved by controlling the intensity modulation of the source used to expose a radiation-sensitive film, generally by amplitude modulation techniques. Early researchers experimented with bleaching techniques using dichromated gelatin or photographic emulsions as the recording medium and computer-generated grey-scale masks to modulate the amplitude to form both discrete and continuous (kinoform) phase-only structures (Lohmann and Paris, 1967; Lesem *et al.*, 1969; Lee, 1978). Others explored direct-write variable exposure dose lithographic techniques such as laser beam lithography (LBL) (Gale and Knop, 1983; Koronkevich *et al.*, 1984) and low-resolution electron beam lithography (EBL) to fabricate the kinoform profile (Fujita *et al.*, 1982; Suhara *et al.*, 1982; Nishihara and Suhara, 1987). However, poor control over the profile shape and the phase modulation depth limited the diffraction efficiencies of these early DOEs. (In fact, although recent improvements in resist and pattern-generation technology have resulted in increased efficiency, accurate profile control is still a limiting factor for continuous phase profile DOEs (Gale *et al.*, 1992; Suleski and O'Shea, 1995).) The suggestion that the blaze could be approximated by a stepped or quantized phase profile, made independently by Dammann (1970) and Goodman and Silvestri (1970), inaugurated a new direction in the manufacture of DOEs. Their calculations showed that the first-order diffraction efficiency of a multilevel structure increased with the number of quantization levels (Dammann, 1970, 1979; Goodman and Silvestri, 1970). The stepped approximation to the phase profile would circumvent the difficulties encountered in fabricating a continuous phase shape and could be achieved, for example, by a series of binary amplitude mask exposure and pattern transfer sequences. In one of the first technology demonstrations, d'Auria *et al.* (1972) fabricated a four-level Fresnel phase microlens using three lithographic mask iterations. The plastic replicas had low efficiencies; higher efficiencies would require many more phase levels – an expensive, fabrication-intensive process when each additional lithographic sequence yields only one additional phase step.

3.2.2 *Binary Optics*

VLSI technology was not fully exploited for micro-optics fabrication until the 1980s, when advances in CAD tools, lens design software, and integrated circuit (IC) manufacturing technology coincided with systems needs for diffractive elements, to create the climate for the birth and development of binary optics technology. Researchers at MIT Lincoln Laboratory uniquely demonstrated that with computer-generated design data and IC fabrication techniques, practical, robust and efficient diffractive optical elements could be built. In particular, the binary coding of the phase quantization and fabrication sequence employed by the Lincoln researchers increased the number of phase levels formed in M process steps from $M + 1$ to 2^M, greatly reducing the number of process iterations and processing costs needed to fabricate DOEs with high diffraction efficiency (Swanson, 1989). Lincoln researchers developed a test bed for the design and fabrication of diffractive optics, being the first to evaluate the efficacy of DOEs in infrared systems applications. As the technology continued to evolve to encompass monochromatic and wideband imaging optics, monolithically integrated diffractive systems, and refractive micro-optic arrays, the optical performance – particularly high diffraction efficiency – has been the overriding concern. Over the last decade and a half the work at MIT Lincoln Laboratory has focused on identifying key roles for diffractive optics, fabricating and demonstrating prototype diffractive optical components, and transferring this technology to industry.

Driven by the practical need to reduce the weight, size and cost of complex optical systems as well as the desire to create optics with novel functionality, researchers fabricated and tested subwavelength structures, hybrid diffractive/refractive optics, and large arrays of diffractive micro-optics (Veldkamp, 1991). A two-level binary structure designed to split a single laser beam into 12 multiplexed beams inaugurated the effort (Veldkamp and Van Allen, 1983), while a diffractive subwavelength telescope became the first element formally labelled by the term binary optics (Veldkamp and Swanson, 1983). Weak diffractive structures with 8 and 16 phase levels have been used to correct primary chromatic and residual monochromatic wavefront aberrations of refractive lenses at wavelengths in the far-infrared, mid-infrared, visible and deep ultraviolet regions of the spectrum as well as to provide a second focal power for bifocal imaging and for dual wavelength focusing (Swanson, 1989; Swanson and Veldkamp, 1989; Veldkamp, 1991). With the development of more sophisticated fabrication techniques, as described in section 3.3, it became possible to build large, uniform arrays of microlenses with apertures ranging from a few tens of microns to several millimetres (Goltsos and Holz, 1990; Stern *et al.*, 1991b). The flexibility inherent to these lithographic techniques easily accommodated rectangular, round or hexagonal close-packed layouts, as well as spatially variant arrays (Goltsos and Holz, 1990; Stern *et al.*, 1991b; Leger and Goltsos, 1992). No longer restricted by fabrication constraints, optimal surface profiles such as spheric, parabolic, aspheric, astigmatic or anamorphic, and also ones that combined multiple optical functions in a single element, could be specified according to the application requirements. Significant applications of microlens arrays demonstrated at Lincoln Laboratory included diffractive optics for laser beam shaping and addition (covered in Chapter 9), wavefront sensing (Shack–Hartmann sensors), agile beam steering, beam multiplexing, wavefront transformation, focal plane imaging and fill factor enhancement, antireflection surfaces, diffusers, and colour discrimination optics, to list a few examples (Veldkamp and Van Allen, 1983; Leger *et al.*, 1988a, 1988b; Swanson and Veldkamp, 1989; Goltsos and Holz, 1990; Stern *et al.*, 1991b; Veldkamp, 1991; Leger and Goltsos, 1992; Farn *et al.*, 1993a, 1993b; Wong and Swanson, 1993). Coherent lenslet arrays of more than 20 000 identical *f*/1 microlenslets with 100% fill factor have been demonstrated (Stern *et al.*, 1991b). Very deep echelon-like gratings have been used to diffract visible colour wavebands into separate diffraction orders (Farn *et al.*, 1993a).

3.3 Basics of Binary Optics Design and Fabrication

3.3.1 *Binary Optics Technology*

To exploit the many promising designs for diffractive optics that are found in the literature (Stone and Thompson, 1991; Cole and Pittman, 1993; and others) requires a flexible and efficient technology free from artificial restrictions imposed by fabrication or materials constraints. By drawing on the disciplines of solid state physics, materials science, chemistry and engineering, as well as optics, binary optics technology provides a generic formula for manufacturing wavefront engineering devices in the broad range of optical materials that are of commercial and government interest. The major components of this formula include optical design of the phase profile, CAD translation of the stepped phase topography into a set of amplitude patterns (photomasks), fabrication of the substrate surface relief via an iterative set of lithographic replication and pattern transfer processes, device evaluation, and, when feasible, replication of the master. With this formula, a mathematical description of almost any arbitrary optical phase profile can be translated into an efficient diffractive element. It is necessary, however, to customize existing VLSI techniques to address the stringent tolerances on step heights, linewidths, and overlay registration imposed by optical performance considerations.

In the optical design stage, the optical function is modelled, depending on the design details, by geometrical optics (simple ray-tracing), scalar diffraction theory (Fourier optics), optimization techniques (simulated annealing and phase retrieval), or electromagnetic vector theory, using numerical methods or commercial software design tools to compute the diffractive phase profile. Optical design methodology is considered in other chapters in this book as well as in numerous review articles (Stone and Thompson, 1991; Cole and Pittman, 1993; and others). The optical design yields a specification of the continuous phase function $\phi(x, y)$, which must be converted into a surface relief height $d(x, y)$ at each point on the optical surface, Fig. 3.1(a). Passing the phase through a 2π delimiter at the design wavelength λ_0 results in a piecewise continuous structure, or kinoform, with a scalar diffraction efficiency of 100% in the first order, Fig. 3.1(b). The maximum surface-relief height of any particular kinoform facet is then given by $d_{\max} = [\lambda_0/2\pi(n_0 - 1)]$, where n_0 is the index of refraction of the substrate at λ_0. To simplify fabrication, the phase topography can be quantized into N discrete phase levels of height $2\pi/N$, which approximate the continuous curved or angled blaze by a staircase profile, as shown in Fig. 3.1(c) (Dammann, 1970; Goodman and Silvestri, 1970). Each of these N steps has a surface-relief height of d_0/N.

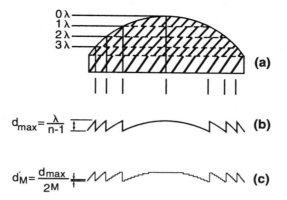

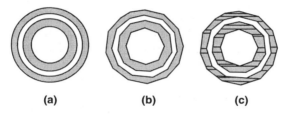

Figure 3.1 Illustration of the conversion of a continuous phase function into a diffractive one: (a) a refractive phase function, (b) the resultant diffractive kinoform and (c) a multilevel approximation to the diffractive surface.

Figure 3.2 Translating data contours into a graphics format for mask generation: (a) constant thickness contours, (b) high-level graphics description and (c) data fractured into trapezoids for MEBES format.

The phase topography is divided into contours of constant surface-relief height by locating the transition points of the phase stair along regions of constant phase. These contours can be determined analytically, by interpolation of contours of constant phase height defined on a regular grid, or individually delineated (Kathman, 1989; Logue and Chisholm, 1989; Swanson, 1989). CAD techniques are used to encode the phase contour data into a graphics format compatible with either direct-write lithography systems (EBL, LBL) or photomask-pattern generators. For example, CIF or GDS II (Mead, 1980; Rubin, 1987) is used for MEBES (moving electron beam exposure system) machines (Sze, 1988). The data is translated into a binary amplitude or intensity code that results in opaque Cr and clear regions on a photomask or modulation of the beam intensity in a direct-write technique. MEBES machines use a raster scan to write the pattern while specialized direct-write instruments often employ a vector scan or variable exposure aperture (JENOPTIK Technologie, Germany). A single microlens pattern, shown in Fig. 3.2(a), can be generated by describing polygons, Fig. 3.2(b), or by inscribing graphical primitives such as rectangles or trapezoids to section a patterned area, Fig. 3.2(c), a procedure known as fracturing the data. The fractured pattern can be repeated and arrayed. The result is a set of N photomasks, each containing a set of amplitude contours corresponding to a particular

Table 3.1 Diffraction efficiency for various numbers of phase levels

Number of levels N	First-order efficiency η_1^N
2	**0.41**
3	0.68
4	**0.81**
5	0.87
6	0.91
8	**0.95**
12	0.98
16	**0.99**

phase-relief height d/N, which provide the template for subsequent substrate fabrication.

The N-level surface-relief structure is built up by N sequential lithographic exposure and pattern transfer steps. To achieve the efficiencies in excess of 90% required by most practical optics applications represents a fabrication-intensive task, since the first-order scalar diffraction efficiency increases only as the square of the number of steps N, $\eta_1^N(\lambda_0) = |\mathrm{sinc}(1/N)|^2$, where $\mathrm{sinc}\,x = \sin(\pi x)/(\pi x)$, Table 3.1 (Dammann, 1979; Swanson, 1989, 1991). By implementing a binary coding scheme in the phase quantization, M mask layers result in $N = 2^M$ phase levels with height $d = \lambda_0/[2^M(n_0 - 1)]$. Values of η_1^N for N equal to M powers of 2 are highlighted in Table 3.1. Only four process iterations now result in 16 phase levels with an efficiency of 99%! Note that this does not restrict the number of phase levels to numbers exactly divisible by 2^M. In general $M + 1$ mask layers are used to create between 2^M and 2^{M+1} phase levels.

Not to be glossed over is the fabrication design step which occurs prior to mask generation. In this process, the optical design is analyzed to determine the fabrication requirements and compatibility with existing process equipment. The quantized phase topography is viewed as a linear grating with a spatially varying frequency, height and orientation; in processing parlance the half-period of each local grating is equated to a linewidth, the phase height to an etch depth, and the number of quantization levels to the number of mask layers and fabrication iterations. Of particular interest are the minimum fringe spacing, or linewidth, on each mask layer which defines the critical dimensions (CDs) and dictates the lithographic and registration precision required to manufacture the optic, the maximum and minimum phase step heights, and the step height tolerance. These parameters must be compatible with existing process tolerances for linewidth resolution, overlay accuracy, and etching or deposition control. Extant materials processing capabilities will influence the selection of the optical material. Feedback from these fabrication design considerations results in modifications to the optical design and vice versa. Most important, lithographic alignment marks used to register sequential mask layers to the substrate and patterns required for lithography tools, for process control, and for process tracking must be specified and included on the mask design. Often, defining these latter patterns may require more time than the optical design!

3.3.2 *Lithography and Substrate Patterning Techniques*

Integral to binary optics technology are the basic fabrication methods of VLSI technology including lithography, masking, etching and coating (Sze, 1988; Moreau, 1989; Vossen and Kern, 1991). The planar, two-dimensional fabrication techniques that are used to build up the multilevel relief structures are shown schematically in Fig. 3.3. Lithographic tools form the intermediate mask pattern, or stencil, by controlling the intensity of a beam of photons, electrons or ions impinging either onto a thin layer of a radiation-sensitive polymeric material coating the substrate, i.e. a resist film, Fig. 3.3(a), or impinging directly onto the substrate surface, Figs 3.3(b) and (c). Maskless direct-write techniques, such as laser-assisted chemical etching or deposition, Figs 3.3(b) and (c), are not in general use and are mentioned only for completeness (Bloomstein and Ehrlich, 1991).

The resist is patterned, Fig. 3.3(a), by exposing it through an amplitude mask comprising opaque and transparent regions (optical imaging lithography), or by modulating the radiation dose to expose the resist directly (non-imaging or direct-write lithography). Upon irradiation, the film composition is altered through either a photochemical reaction or crosslinking of the polymer chains. The excess resist material is then removed by dissolution in a developer solution, yielding a resist stencil, Fig. 3.3(d). Positive resists duplicate the mask tone, while negative resists reverse the mask polarity. Top-surface-imaging (TSI) resists are designed to form a thin etch-resistant film during plasma development of the remaining thick resist layer (Palmateer *et al.*, 1995). Resists developed for the semiconductor industry have been optimized for high contrast, or gamma, although developer chemistries can be varied to impart a more linear response exhibiting less sensitivity to the radiation dose (Moreau, 1989). Scattering of radiation from the substrate can interfere with or cause unwanted resist exposure. For example, substrate reflection

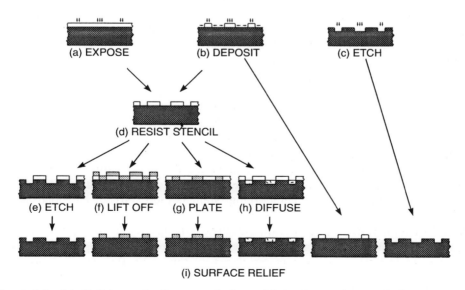

(a) EXPOSE (b) DEPOSIT (c) ETCH

(d) RESIST STENCIL

(e) ETCH (f) LIFT OFF (g) PLATE (h) DIFFUSE

(i) SURFACE RELIEF

Figure 3.3 (a)–(i) Schematic diagram of planar fabrication techniques, described in the text, which are used to create surface-relief structures.

forms standing waves in photoresists patterned optically; back-scattering and secondary electrons cause proximity effect exposure during EBL (Moreau, 1989).

The lithography method chosen for pattern generation depends on the CD, the resolution limit of the lithography tool, the overlay requirement, the exposure field size, the patterning time, and the cost of the tool. Optical photolithography, which images a patterned Cr photomask onto a photoresist film, has the highest throughput. Commercial photomask facilities can currently produce CDs of 0.5 μm annuli and 0.35 μm lines using MEBES III machines, and 0.25 μm CDs with MEBES IV machines (Micro Mask, Inc., Sunnyvale, CA). Optical lithography tools include non-contact projection lithography systems such as 1× large field scanners and 5× or 4× reduction step and repeat systems, and contact/proximity mask aligners. For standard semiconductor wafer sizes, steppers and scanners with advanced positioning stages and alignment algorithms can achieve 50–100 nm placement accuracy across the wafer and linewidth resolution of 350 nm (365 nm Hg i-line, with phase shifting mask technology), 250 nm (248 nm excimer laser), and 180 nm (193 nm excimer laser) (ASM Lithography, Inc., Tempe, AZ). The small stepper field sizes (10–25 mm square) can introduce stitching artifacts in a large-area optical pattern, while a full-field scanner is more sensitive to runout and field distortions. Vacuum contact mask aligners are cheaper than projection systems; can accommodate substrates having the non-standard thicknesses, diameters and geometries encountered in micro-optics fabrication; can achieve overlay registration ≤0.25 μm; and can have high resolution (Stern *et al.*, 1991a). However, particulate contamination can impair mask life and introduce local distortions. Proximity printers are limited to features ≥2 μm, depending on the gap between substrate and mask. EBL, which offers the highest resolution of any lithography technique (≤10 nm) and is essential for fabricating X-ray optics (Kirz, 1974; Anderson and Kern, 1991; Tennant *et al.*, 1991), has limited throughput, small field size and high equipment costs. Laser writers can achieve approximately 1 μm resolution and can be configured in either a centrosymmetric format (compact disc writers) or a raster scan (Koronkevich *et al.*, 1984; Goltsos and Liu, 1990; Gale *et al.*, 1993). Direct-write methods such as EBL, focused ion beam lithography and laser beam lithography are most powerful if used to write an analog structure rather than a discrete stepped profile; however, the continuously varying resist profile requires accurate control over both the exposure dose and the developer chemistry.

The patterned resist serves as a stencil for subsequent pattern transfer, i.e. the building up or removal of layers of material to create the substrate surface relief, Figs 3.3(e)–(i) (Sze, 1988; Vossen and Kern, 1991). Additive methods include blanket deposition of a secondary material onto a patterned surface by evaporation, sputtering, chemical vapour deposition (CVD), laser ablation, or plasma enhanced CVD (PECVD), followed by the subsequent lift-off of the excess material by dissolution of the underlying stencil, Fig. 3.3(f); electroplating, or epitaxial growth of material through a stencil, Fig. 3.3(g); and ion diffusion into unmasked regions of the substrate to cause a change in the refractive index or volume, Fig. 3.3(h). Subtractive methods use wet chemical etching (WCE) or dry (gas phase) etching to remove material not protected by the stencil mask, Fig. 3.3(e). Most WCE is isotropic, with material removed equally in all directions; this results in mask undercutting and linewidth narrowing. Crystallographic WCE can be highly anisotropic along certain crystal planes, but is incapable of exact reproduction of arbitrary or curvilinear features. Dry etching, which offers the greatest amount of profile control and flexibility, is the primary method of pattern transfer used in binary optics technology.

In dry etching, material is removed from the substrate by a combination of physical sputtering and/or chemical reaction mechanisms (Flamm and Donnelly, 1981). Etching tools can be categorized according to the ion-bombardment energy and ion density, whether the plasma results from a glow discharge or is remotely generated, and the gas pressure in the etching chamber (Lehman, 1991). Systems fall into three categories: rf-powered parallel plate plasma etchers (plasma etchers and reactive ion etchers, high-density plasma etchers (magnetically-enhanced reactive ion etchers), electron-cyclotron resonance, helical resonator (helicon), transverse coupled plasmas and inductively coupled plasmas), and ion beam etchers (ion millers, reactive ion beam etchers, and chemically-assisted ion beam etchers). In reactive ion etching (RIE), bombardment of the substrate surface by reactive neutrals and energetic ions generates volatile by-products that are pumped away by the vacuum system. Extrinsic parameters, such as the gas chemistry, chamber pressure, ion energy and cathode material, determine the profile anisotropy (the sidewall angle of the etched cross-section), linewidth acuity, etch rate, mask-to-substrate selectivity, etch uniformity across the wafer, and etched surface texture (Flamm and Donnelly, 1981). Commonly used etch gases for electronic and optical materials are listed in Table 3.2. Inert gases, such as Ar and He, and additives such as O_2 enhance ionization and dissociation processes or scavenge unsaturated species.

3.3.3 *Fabrication of Binary Optics Elements*

Compared to VLSI processing – which may have more then a dozen cycles of photolithography and etching interspersed with ion implantation, polysilicon and metal deposition, and insulator growth steps – lithographic fabrication of multilevel optical structures is simple and inexpensive. The multistep surface-relief structure is built up sequentially by an iterative series of lithographic and etching steps, as shown in Fig. 3.4. Each of the M patterns are replicated individually by optical photolithography onto a substrate coated with a thin layer of positive photoresist. High-fidelity pattern transfer into the substrate to the precise depth d is usually

Table 3.2 Etch gases for optical materials

Material	Gases
SiO_2	CF_4-H_2, CHF_3, C_2F_6, NF_3
Si	CF_4, CF_4-O_2, SF_6-O_2, $CCIF_3$, $SiCl_4$, HBr, Cl_2
Ge	CF_4, CF_4-O_2, SF_6-O_2, $CCIF_3$
Organic materials	O_2, O_2-CF_4, O_2-SF_6
Mo	CF_4, C_2F_6, SF_6
GaAs	CF_3Cl, $SiCl_4$, HBr, Cl_2, BCl_3, BCl_3-Cl_2
InP	HI, $SiCl_4$, HBr, BCl_3-Cl_2
Al	BCl_3, BCl_3-Cl_2, $SlCl_4$

Inert Gases (Ar, He) may be added.

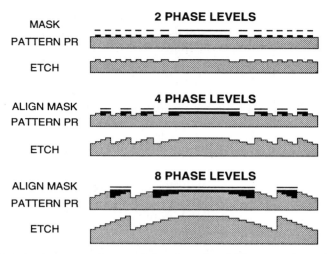

Figure 3.4 Fabrication steps for an eight-phase-level binary optics microlens.

accomplished by RIE. The first mask creates a two-step profile, and each succeeding mask doubles the number of phase levels. The mask with the smallest features is usually replicated first, although the reverse order may also be used. Each successive mask has feature widths twice as large (half the spatial frequency) and requires etching to twice the depth. Each mask must be accurately registered to the patterned substrate; two-mask layers, for example, require one alignment, three-mask layers require two alignments, etc. This efficient fabrication scheme has been demonstrated at a number of laboratories using EBL or LBL to create the binary coded phase profile, in addition to conventional photolithography (Andersson *et al.*, 1990; Stone and Thompson, 1991; Gale *et al.*, 1992; Cole and Pittman, 1993; and others).

All aspects of lithographic pattern definition and replication affect the ultimate optical efficiency of multilevel structures and must be precisely controlled (Cox *et al.*, 1990; Farn and Goodman, 1990; Swanson, 1991). Fabrication-related errors can occur during photomask or substrate fabrication. With current MEBES capabilities, the round-off, pattern conversion and plate positioning inaccuracies generated during the electron beam mask-writing process are minimal (Micro Mask, Inc., Sunnyvale, CA). Process tolerances are much more stringent for micro-optics than microelectronics. As seen from Fig. 3.4, successive mask layers must be registered to previously etched pattern edges with 0% overlay tolerance, compared to conventional VLSI tolerances of 0.1–0.25 μm! Moreover, this must be done over areas which are large compared to the available field sizes in projection lithography and EBL tools. Etching tolerances must often be held to better than 1% of the total phase-relief height, for both shallow (approximately 10 nm) and deep (approximately 10 μm) structures, and achieve a uniformity of 1–2% over large areas, without benefit of end-point detection. Photolithographic patterning must replicate with high fidelity curvilinear linewidths that can vary between 0.5 and 50 μm on a single mask level. Hence, the photolithographic linewidth bias (Moreau, 1989) must be incorporated into the photomask pattern, a computationally intensive task. Photolithographic linewidth errors, pattern

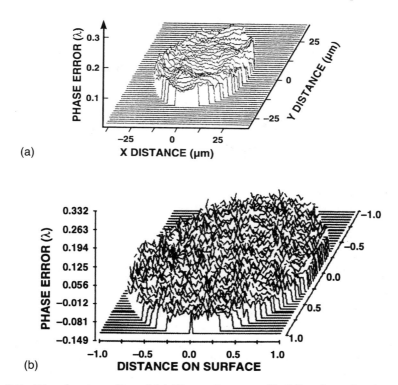

Figure 3.5 Wavefront quality of (a) 50 μm diameter f/2 diffractive microlens with rms phase error λ/50 over the lenslet aperture, and (b) diffractive microlens array with rms error λ/30 over a 6 mm aperture, which contains about 900 f/5 lenslets of 200 μm diameter.

misregistration, etch mask undercutting, etch mask edge erosion, and the angled profile or overlap of materials deposited by lift-off will result in concentric 'ridges' or 'trenches' in the surface-relief structures. In general, fabrication errors should be held to less than 5–10% of the minimum linewidth and minimum phase step height. In practice one can often do better.

3.4 Performance of Binary Optics Microlens Arrays

3.4.1 Fabrication and Optical Performance

Although enormous interest is focused on diffractive optics, scant data on optical performance of DOEs is found in the literature. This may be attributed to the difficulty of fabricating high-quality diffractive optics and of measuring their optical efficiency. Light not routed into the design order, usually the first order, will be scattered into the zero and higher diffraction orders, degrading the optical performance. Measurements of both single lenslets (Leger *et al.*, 1988b) and coherent arrays (Goltsos and Holz, 1990) show that the wavefront quality of diffractive optics is essentially that of the optical blank in which the structures are etched, Fig. 3.5. For example, the rms error of the wavefront profile of 900 lenslets measured within a 6 mm aperture of a macroscopic array is $\lambda/30$, for a Strehl ratio of 0.96 (Goltsos and Holz, 1990), Fig. 3.5(b).

Understanding the intrinsic efficiency limits of diffractive structures is fundamental to binary optics technology, where many applications require high-speed microlenses in the visible and near-infrared. Simple scalar calculations predict, for example, that eight-phase-level microlenses should exhibit efficiencies of 95%, regardless of f-number or fringe spacing. Scalar approximations, however, are only valid when the grating period-to-wavelength ratio $T/\lambda > 10$, i.e. for large features (Swanson, 1989, 1991). For fast lenses where the local T/λ ratio approaches unity, light is increasingly shadowed by – on a wavelength scale – the non-planar relief structure. More rigorous models predict lower efficiency values when the period-to-wavelength ratio drops below 5, due to shadowing and vector electromagnetic effects (Swanson, 1991). In addition, fabrication-related effects can also impact the optical performance. Scalar estimates predict that variations in etch depths of only $\pm 2.5\%$ of d_{min} will reduce the first-order efficiency by about 1% (Swanson, 1989; Cox *et al.*, 1990); however, 0.1 μm linewidth errors and 0.3 μm misalignment (zone position errors) in an $f/3$ microlens cause efficiency losses of 10% (Cox *et al.*, 1990; Farn and Goodman, 1990). To fully realize the potential of binary optics devices, nanometre alignment tolerances and linewidth control may be necessary. Practicality dictates, however, that the expected gain in optical performance be balanced against unreasonable fabrication requirements.

As an example of the fabrication tolerances imposed on optical device fabrication, consider the design parameters of a phase Fresnel microlens for use in the visible. For an aspheric lenslet, the k zone boundaries r_k are given by $r_k = [(k\lambda_0/N)(2f + k\lambda_0/N)]^{1/2}$, where f is the lens focal length and k is an integer. The chirped zonewidths resemble a slowly varying grating and can range from 0.5 μm to tens of microns on a single mask layer depending on the lenslet aperture, D, and focal length. The total number of annuli is determined by setting $r_k = D/2$. The dimension of the smallest zone annulus, Δr, depends on the f-number, f/D, λ_0, and the quantization level $N = 2^M$, $\Delta r \approx 2\lambda_0(f\text{-number})/N$. Alternatively, setting $CD = \Delta r$ determines the smallest f-number (highest numerical aperture) lens that can be fabricated. Fast lenslets in the visible and near-infrared have small grating periods and are inherently more difficult to fabricate. In fact, the 0.5 μm CD on commercial photomasks limits an eight-phase-level lens operating at 633 nm to a speed of $f/3.2$. Assuming that the etch depths should be kept to 5% of the minimum phase height and the alignment accuracy within 10% of the minimum feature size, a fused silica lens would have an overlay tolerance of 50 nm and etch depth precision of 8.8 nm. Fused silica optical elements have been fabricated with an etch depth control of ± 5 nm for a 1.454 μm depth (0.3% precision) (Wong and Swanson, 1991, 1993).

3.4.2 Diffractive Microlens Test Set

To systematically quantify the efficiency limits of diffractive microlenses and to correlate losses in optical efficiency with specific fabrication errors, we designed, built and measured a set of diffractive microlenses with ten different focal lengths, the Best Efficiency Array SeT or BEAST. Microlenses with 200×200 μm square apertures, and focal lengths between 170 μm and 14 mm at 633 nm (lens speed

Table 3.3 Design parameters for diffractive microlens test set

Focal length (μm)	$\mathcal{F}/\#^{\dagger}$	Minimum zone size on mask layer (μm)		
		1	2	3
14 000	60	39.0	17.8	7.6
7000	30	17.8	8.0	4.0
3500	15	7.6	4.0	2.0
2000	9	4.5	2.3	1.1
1400	6	3.2	1.6	0.8
900	4	2.0	1.0	0.5
700	3	1.6	0.8	
425	2	1.0	0.5	
325	1.4	0.8		
170	0.8	0.5		

$^{\dagger}\mathcal{F}/\# = f/D_{\text{eff}}$, where D_{eff} is the diameter of a circle with an effective area equal to that of a square $200 \times 200 \, \mu\text{m}^2$.

from $f/0.8$ to $f/60$) were fabricated both as isolated lenslets and as 10×10 microlens arrays in a 1.2×1.8 cm area (Stern *et al.*, 1991a). Most important, the mask set included 16 vernier alignment patterns, etch evaluation features, and clear apertures for reference calibration. Design parameters are listed in Table 3.3. Three sets of lenslets were fabricated with two, four and eight phase levels, respectively. A 0.5 μm CD limited the two-phase-level lens to $f/0.8$, the four-phase-level lens to $f/2$, and the eight-phase-level lens to $f/4$. The maximum zone width was 130 μm. The 2 in. $\times$ 6 mm Suprasil discs had a $\lambda/10$ surface finish. Substrates were reactive ion etched in a CHF_3 plasma at 10 mTorr and 180 W rf power; etch rates were about 16.5 nm min^{-1}. Etch depths were controlled by etch time. Selectivity between the photoresist mask and the quartz substrate was approximately 2:1. The calculated etch depths in fused silica ($n = 1.4572$ at $\lambda_0 = 633$ nm) were 0.175, 0.349 and 0.699 μm for layers 3, 2 and 1, respectively, for cumulative etch depths of 1.223 μm (eight phase levels) and 1.048 μm (four phase levels). The eight-phase-level $f/6$ microlens, shown in Fig. 3.6, had etch depths measured by stylus profilometry of 0.22, 0.33 and 0.70 μm or a maximum error of 3%. Although the RIE process was optimized to maintain linewidth fidelity, the photolithography bias reduced linewidths by as much as 0.1 μm. Particular effort was devoted to reducing pattern translation errors. The overlay registration accuracy was evaluated by optical microscopy of a background grid of two-dimensional verniers which spanned a ± 0.5 μm range in 0.1 μm increments about the centre, in both x and y, and could be read to ± 0.05 μm precision. Using these verniers, we achieved overlay registration to better than 0.1 μm over the 1.2×1.8 cm field (Stern *et al.*, 1991a).

3.4.3 *Efficiency Measurements and Discussion of Results*

Benchmark optical efficiency measurements were made on these fused silica microlenses at $\lambda = 633$ nm using a dual-beam, single-detector, autoreferencing

Figure 3.6 SEM of an eight-phase-level 200 μm square aperture $f/6$ quartz microlens. The minimum zone width on level 3 is 0.8 μm.

apparatus, shown in Fig. 3.7 (Holz *et al.*, 1991). The measured first-order efficiency was normalized to the relative focal spot power of a diffraction limited lens; $\eta_{norm} = \eta_0/0.815$ for a square aperture. The measurements intrinsically excluded substrate transmission and reflection losses. Relative efficiency could be measured with better than 0.001 precision, and with calibration, absolute efficiency is accurate to better than 0.005. The first-order efficiencies for four- and eight-phase-level microlenses are plotted as a function of lens speed (focal length) in Figs 3.8 and 3.9(b) and compared with predictions of direct electromagnetic calculations from a first-principles diffraction code (Knowlden, 1987) for ideal two-, four- and eight-phase-level fused silica microlenses at $\lambda_0 = 633$ nm. Simple scalar calculations, strictly valid only for large features ($T/\lambda > 10$), are also shown. Because the vector diffraction code is restricted to periodic structures, the diffractive lens was locally modelled as a linear grating – a valid assumption for the slowly varying grating period of these diffractive lenslets. To reduce computation time, TE and TM diffraction efficiencies were calculated and averaged over a set of discrete grating periods and then linearly interpolated to yield a radially varying diffraction efficiency. When integrated, this gave the overall first-order diffraction efficiency for each microlens.

The measured efficiencies of the lenses demonstrate good correspondence with the benchmark calculations, Figs 3.8 and 3.9(b). The single eight-phase-level lenslets on device OB2 (which in the central area had a maximum misalignment in x or y of no more than 0.1 μm between layers 1 and 3, and 0.05 μm between layers 2 and 3) exhibit an excess loss of no more than 4% at all focal lengths compared to the calculated values. For slower lenses ($f/30$ and above) measurements agree with scalar theory predictions. For an eight-phase-level $f/4.5$ microlens having less than 0.1 μm misalignment error we have measured an absolute efficiency of 0.85, a value that is, we believe, the highest efficiency reported to

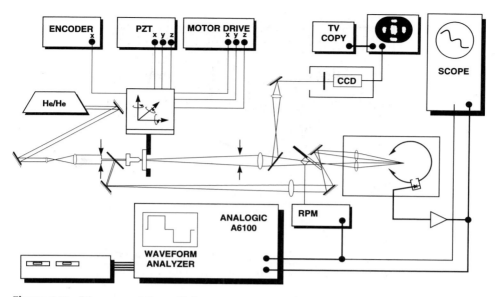

Figure 3.7 Diagram of the efficiency measurement apparatus.

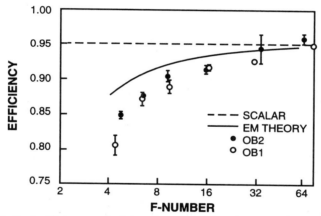

Figure 3.8 Optical efficiency of eight-phase-level lenslets versus *f*-number. Solid line: electromagnetic theory. Dashed line: scalar theory. Symbols: experimental data. The error bars denote the peak-to-peak variation of lenslets in the array, the peak-to-peak spread due to illumination non-uniformity, or the peak-to-peak reference value uncertainty, whichever is largest.

date for such a fast, binary optics lens in the visible. This corresponds to 96% of the predicted value for this lens and implies that net fabrication errors contributed only a 4% efficiency loss.

3.4.4 *Impact of Fabrication Errors*

To explicitly quantify the parametric dependence of optical efficiency on process errors, we have intentionally introduced a 0.35 μm translational error between

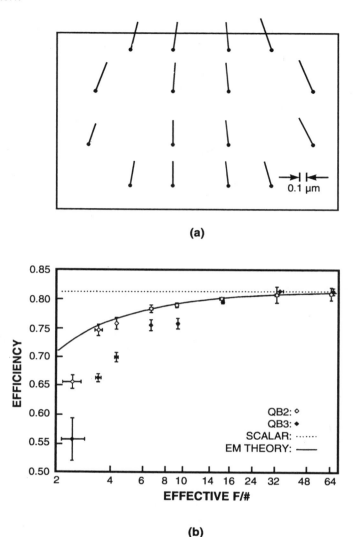

(a)

(b)

Figure 3.9 (a) Alignment results for 'misaligned' BEAST sample QB3 at each of the 16 vernier sites for layers 2→1. The average and standard deviation of these 16 values, weighted equally, are −0.01 ± 0.17 (*x*), 0.38 ± 0.04 (*y*). (b) Optical efficiency of four-phase-level microlenses as a function of *f*-number. Unfilled diamonds: data from QB2 (the 'well-aligned' sample). Solid diamonds: data from QB3 (the intentionally misaligned sample). Dotted line: scalar theory efficiency for four phase levels (81%). Solid line: predicted value from the full electromagnetic vector calculation.

mask layers 1 and 2 in one set of four-phase-level lenslets (Stern *et al.*, 1992). The alignment results are plotted in Fig. 3.9(a) for the misaligned optic, QB3. Optical efficiencies measured for this 'misaligned' set of microlenses are then compared with a nominally identical, four-phase-level, 'well-aligned' set of microlenses, Fig. 3.9(b). It is immediately apparent from the data plotted in Fig. 3.9(b) that pattern registration has a significant effect on the absolute optical efficiencies when measured as a function of the lens speed. The performance of the misaligned optic has fallen below the well-aligned lenslets at $f/15$; at $f/6$ the efficiency of the misaligned microlenses is 5% less than the well-aligned lenslets; and at $f/2$ the misaligned lenslets show a 10% decrease in efficiency below that of the well-aligned microlenses. The sizeable efficiency losses displayed by the fastest microlenses can be correlated with the substantial fraction of the zone width occluded by the intentional translational error. For example, 0.35 μm is 25% of the minimum zone width of an $f/2$ microlens but only 3% of the minimum full zone of an $f/60$ lenslet. Even the nominally well-aligned optic exhibits decreased efficiency for the fastest lenslets ($f/2$). Linewidth errors in the nominally 0.5 μm outer zones of these fast lenslets may contribute to the efficiency losses in both sets of microlenses. The etch depth errors recorded for these microlenses – 1–2% of the total phase step height – have negligible impact on the efficiency.

3.5 Fabrication of Deep Three-dimensional Micro-optics Profile Control in Deep Structures

Deep, high aspect ratio structures needed for fast broadband micro-optics, focal plane microlens arrays, and colour discrimination optics can be fabricated by extending conventional binary optics fabrication techniques (Stern *et al.*, 1991b; Stern and Medeiros, 1992; Farn *et al.*, 1993a). The difficulty of maintaining tight tolerances over pattern registration, submicron feature widths and multimicron etch depths – critical for high optical efficiencies – increases with the depth of the topography. In particular, control over the profile evolution during the etching process is critical to successful fabrication of deep analog and digital structures in infrared and visible materials such as Si, Ge, CdTe and SiO$_2$. New etching processes must be developed to define both analog and digital aspheric surface profiles.

The etched profile is dependent on the sidewall anisotropy $A = (1 - \text{lateral etch width/vertical etch depth})$; and the selectivity $S = (\text{substrate etch rate/mask etch rate})$, Fig. 3.10. Binary optics technology requires anisotropic etching with high selectivity between the mask and substrate to form deep vertical mesas, Fig. 3.10(a). High-fidelity transfer of a preshaped analog mask into the substrate also requires high anisotropy but low selectivity, Fig. 3.10(b). The substrate and mask are etched to completion, i.e. until the mask is completely eroded. Alternatively, the profile can be controlled by varying the lateral etch rates of the mask, Fig. 3.10(c), and the substrate, Fig. 3.10(d). The scanning electron micrographs (SEMs) shown in Fig. 3.11 demonstrate the dependence of the Si etched profile on the etch gas composition; the percentages refer to volume flow rate ratios of gas mixtures of SF$_6$ with O$_2$ (Stern and Medeiros, 1992). To prevent lateral undercut, gas chemistries that enhance formation of in situ sidewall passivation layers can be exploited to achieve directionality (Mogab and Levinstein, 1980; Flamm and Donnelly, 1981). Profiles vary from undercut for 100% SF$_6$, Fig. 3.11(a), to vertical sidewalls for 30% O$_2$/70% SF$_6$, Fig. 3.11(b); at higher O$_2$ concentrations the sidewalls slope outward and micromasking causes 'grass' formation on the substrate surface, Figs 3.11(c) and (d).

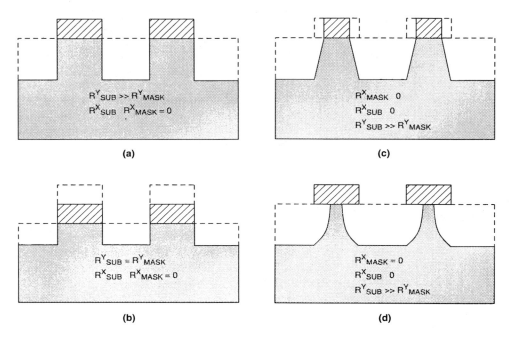

Figure 3.10 Schematic of etched profiles using a rectangular etch mask. (a,b) Substrate: $A = 1$; mask: $A = 1$; and (a) $S \gg 1$, (b) $S \approx 1$. (c) Substrate: $A = 1$; mask: $A < 1$; and $S \approx 1$. (d) Substrate: $A < 1$; mask: $A = 1$; and $S \gg 1$.

3.5.1 Refractive Binary Optics Fabrication

Binary optics technology has been used to fabricate multilevel refractive microlens arrays in both Si and CdTe (Stern *et al.*, 1991b; Stern and Mederios, 1992). The stepwise approximation to an aspheric profile, illustrated in Fig. 3.12, contains no diffractive resets. To minimize scattering from the edges, the phase step heights and number of quantization levels are calculated for an efficient modulo 2π phase structure. For example, to obtain 99% efficiency, the minimum step height is $d_{min} = \lambda_0([16(n_0 - 1)]^{-1})$. As before, iterative steps of UV photolithography and RIE are used to generate the multilevel profile; with the binary coding scheme 64 phase levels require only six masks. Successive mask alignments increase in difficulty due to depth-of-focus limitations incurred by both thick polymer layers and substrate topography. Each iterative process sequence requires $0.25\,\mu m$ or better overlay registration precision and etch depth control of 10–100 nm. For shallow features, a single layer of photoresist suffices. Deeper structures ($>2\,\mu m$) require trilayer resist techniques to planarize the steep topography in order to preserve linewidth fidelity during photolithography and linewidth edge acuity during etching. In the trilayer process, Fig. 3.12, the pattern is defined in a 0.5–$1.0\,\mu m$ thick imaging layer of photoresist and then etched into an approximately 100 nm thick film of PECVD SiO_2, in a CHF_3 plasma. The etched SiO_2 layer serves as the transfer mask to pattern the $4\,\mu m$ thick planarization layer during O_2 RIE; this forms the final mask for the silicon substrate etch. Linewidths of $1\,\mu m$ have been anisotropically etched into the thick planarization layer without

Figure 3.11 SEM micrographs showing effect of adding O_2 to the SF_6 feed gas on the etched Si profile (photoresist mask intact). Etch parameters are 10 mTorr, $-65\,V$ and (a) 100% SF_6, (b) 30% O_2, (c) 32% O_2, (d) 50% O_2.

mask undercutting in both RIE and helicon reactors (Stern *et al.*, 1995). To reduce the O_2 RIE etch time, we use the minimum material that adequately planarizes the surface microstructure. By optimizing the RIE parameters, we have anisotropically etched 1 μm wide lines to a depth of 8 μm in Si with straight sidewalls and smooth surfaces, Fig. 3.13 (Stern and Medeiros, 1992). Figure 3.14(a) shows an eight-level prototype refractive Si microlens for an advanced satellite sensor with etch depths of 2, 4 and 8 μm ($n = 3.5$ at $\lambda = 10\,\mu$m) for a total sag of 14 μm. In an actual device, each of these steps would have been subdivided into smaller quantized steps to produce an efficient microlens. Figure 3.14(b) shows a 16-phase-level CdTe microlens array.

3.5.2 *Continuous Profile Microlenses*

An increasing number of applications for refractive microlens arrays has motivated us to develop processes for high-quality lenslets in infrared and visible materials. (Refractive microlenses are the subject of Chapter 5 in this book.) Although refractive polymer microlenses (Popovic *et al.*, 1988; Hutley, 1991) can be made with good optical quality (Jay *et al.*, 1992; Jay and Stern, 1993), their robustness and spectral range are materials limited. Both RIE and ion beam etching can be used to transfer these preshaped photoresist masters into the substrate, but special attention is needed to maintain the lens profile during etching (Gal, 1993; Gratix, 1993; Stern and Jay, 1993).

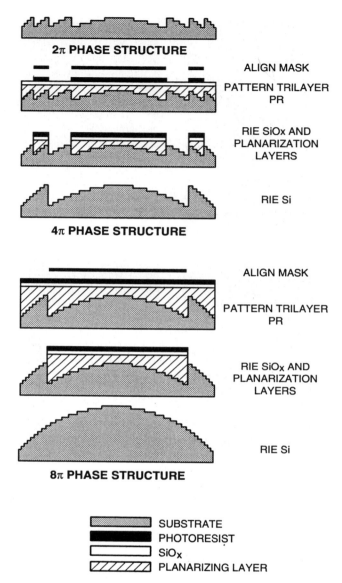

Figure 3.12 Sequence for fabricating refractive binary optics, consisting of iterated steps of photolithography, multilayer resist processing and RIE. Each Si RIE step doubles the previous etch depth, corresponding to sequential 2π, 4π and 8π total phase depth.

Polymer lenslet preforms are fabricated by a photoresist reflow technique that relies on surface tension and plastic flow to reshape cylinders of photoresist, Figs 3.15(a)–(c) (Popovic *et al.*, 1988; Hutley, 1991; Jay *et al.*, 1992); multiple mask exposures to form an aspheric stepped photoresist structure; and direct patterning of an analog lens profile in photoresist using a polar coordinate HeCd laser direct-write system with a radially varying exposure dose (Jay and Stern, 1993).

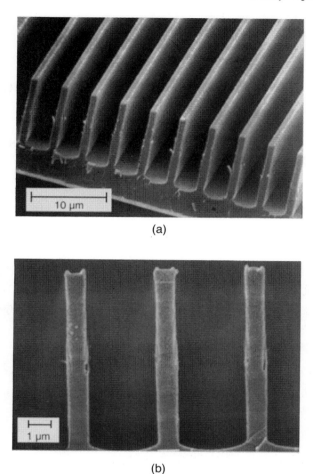

(a)

(b)

Figure 3.13 Anisotropic features etched 8 μm deep in Si. The remaining 0.5 μm photoresist mask is still present.

Thermal reflow is used to smooth out ridges and artifacts in the preshaped resist aspheres. Other direct-write techniques, such as electron beam or focused ion beam lithography, could also be used to form the polymer etch mask (Tanigami *et al.*, 1989). Limitations on the cylinder reflow methods are apparent from Fig. 3.16, which graphs the interferometrically measured wavefront aberrations of 200 μm diameter lenslets against sag height or lens speed. The process is optimized for *f*/2 lenslets with 20 μm sag; these lenslets exhibit less than 0.1 waves of spherical aberration (Jay *et al.*, 1992). Faster lenslets are observed to be slightly prolate with about one wave of spherical aberration, while slower lenslets appear to be oblate ellipsoids and exhibit many waves of spherical aberration. By preshaping the resist prior to thermal reflow, high-quality slower lenslets can be fabricated. For example, we have fabricated 200 μm diameter *f*/4.5 optics with a spherical aberration coefficient of -0.28 using a laser writer to preshape the resist, and arrays of

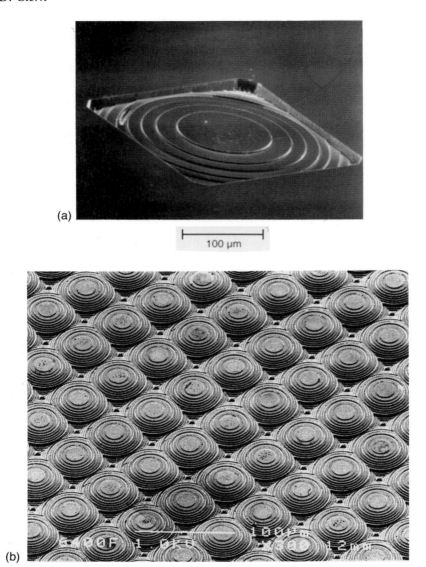

(a)

| 100 μm |

(b)

Figure 3.14   SEMs of (a) eight-phase-level 200 μm diameter refractive silicon microlens with 14 μm sag, and (b) sixteen-phase-level CdTe focal plane microlens array of 55 μm diameter, *f*/0.9 optics with 8 μm sag made by ion beam etching.

260 μm diameter *f*/3.2 lenslets with 1.7 waves of spherical aberration using a multimask/exposure method to form stepped aspherical structures in resist. Without preshaping, 200 μm diameter *f*/3 optics exhibited six waves of aberration!

Once the photoresist master is generated it can be transferred into the substrate by dry etching, as illustrated in Fig. 3.15(d). The desired refractive profile can be directly encoded in the mask for an equal rate etching process or can be compressed in *x* and/or *y* to accommodate the selectivity and anisotropy of the particular etching process. The complexity of the pattern transfer step is reduced if the etching

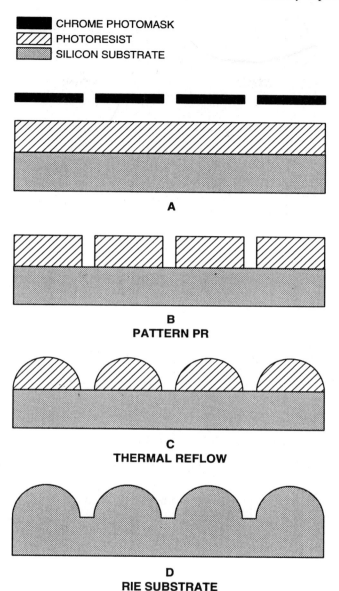

CHROME PHOTOMASK
PHOTORESIST
SILICON SUBSTRATE

A

B
PATTERN PR

C
THERMAL REFLOW

D
RIE SUBSTRATE

Figure 3.15 Schematic illustration of the fabrication of refractive microlenses using a polymer reflow technique and dry etching.

process is designed so that there is no lateral etching of the substrate or the mask. Figure 3.17(a) shows an SEM of a master array composed of $200\ \mu m$ diameter photoresist lenslets with $12.5\ \mu m$ sag. Figure 3.17(b) shows an SEM of a similar array that has been reactive ion etched into Si in a $C_2F_6/SF_6/O_2$ gas mixture. The resultant microlenses have $200\ \mu m$ diameter and $13.5\ \mu m$ sag. For this example, the vertical etch rate is $0.11\ \mu m\ min^{-1}$, the lateral etch rate is approximately 0, and the selectivity is 1.1:1 (Stern and Jay, 1993).

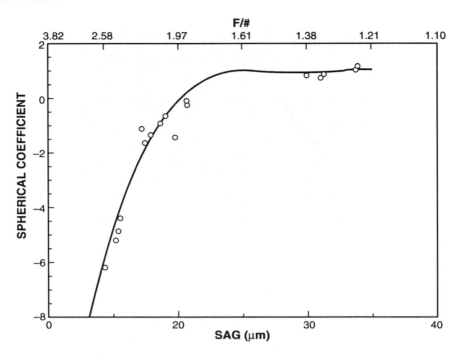

Figure 3.16 Dependence of spherical aberration coefficient on sag and *f*-number for 200 μm diameter lenslets reflowed at 200°C for 10 min. The spherical aberration for the *f*/2 lens is less than 0.1 wave.

3.5.3 *Colour Discrimination Optics*

Colour discrimination can be achieved by taking advantage of the dispersion inherent to a diffraction grating, i.e. dispersion among the first order. Separation of colour in the visible has been investigated with both conventional Bragg gratings and 'echelon'-type gratings (Dammann, 1978; Farn *et al.*, 1993a, 1993b). The first type of grating, when combined with a lens for focusing the individual orders, has been used for entertainment (Chromadepth-3D, Chromatek, Inc., Alpharetta, GA) and for infrared separation and imaging of thermal radiation (Farn *et al.*, 1993b). The colour dispersive Si micro-optic, depicted in Fig. 3.18, is a combination *f*/2 refractive microlens and 17 μm period diffraction grating for colour separation and focusing in the 8–12 μm band. Four mask levels were used to create 16 phase steps with a total phase height or sag of 7.5 μm. There are 64×64 identical pixel elements, each a 100×100 μm square, in this array. The design and performance of this device are detailed elsewhere (Farn *et al.*, 1993b).

Besides the dispersive grating, a unique design based on an echelon-type grating can prove useful when greater control over the placement and intensity of the colour is required (Dammann, 1978; Farn *et al.*, 1993a). The echelon-like grating can be thought of as the superposition of several blazed gratings operating at different wavelengths with each grating blazed efficiently for a different order. This

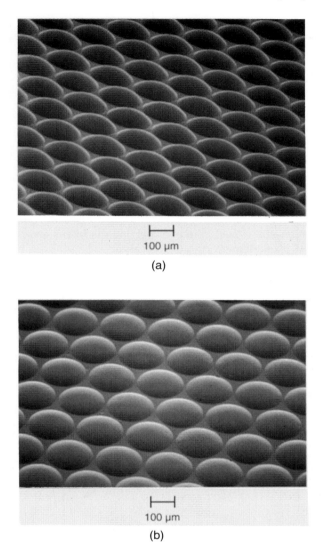

(a)

(b)

Figure 3.17 SEMs of (a) section of an array of 200 μm diameter f/1.3 photoresist microlenses and (b) similar microlenses etched into Si.

allows the zero order of one wavelength to be transmitted, while the shorter and longer wavelengths are diffracted into the -1 and $+1$ orders, respectively, Fig. 3.19. While the phase step height of a conventional N-step blazed grating is $\lambda/N(n-1)$ for the echelon-type grating it is $\lambda/(n-1)$ or N times deeper! Visible echelon-type gratings designed for use in fused silica required etch depths of 1.14 and 2.28 μm for a total etch depth of 3.42 μm. Figure 3.20 shows the measured spectral composition of a four-step echelon with a 43 mm diagonal fabricated in fused silica (Farn *et al.*, 1993a).

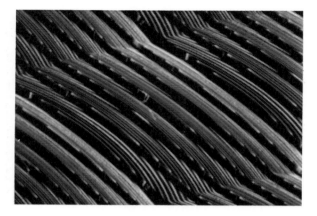

Figure 3.18 Sixteen-phase-level colour dispersing microlens array etched 7.5 μm deep. This is one pixel element of a 64 × 64 array.

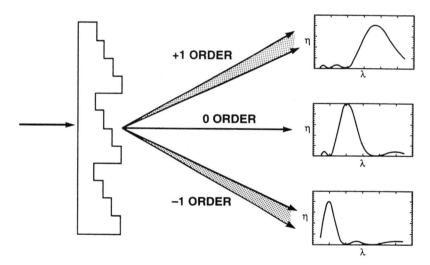

Figure 3.19 Schematic of echelon grating demonstrating spectral separation into different orders.

3.6 Future Directions

3.6.1 *Integrating Micro-optics with Electronic and Mechanical Structures*

Integration of optics, electronics and photonics components will play a key role in exponentially increasing sensor processing capability while reducing the cost of electrooptical systems. Binary optics – which shares a technology base with these components – provides a means for readily integrating micro-optics into these systems. Current research at MIT Lincoln Laboratory centres on extending microlens technology to broad-waveband (more than 20% fractional bandwidth)

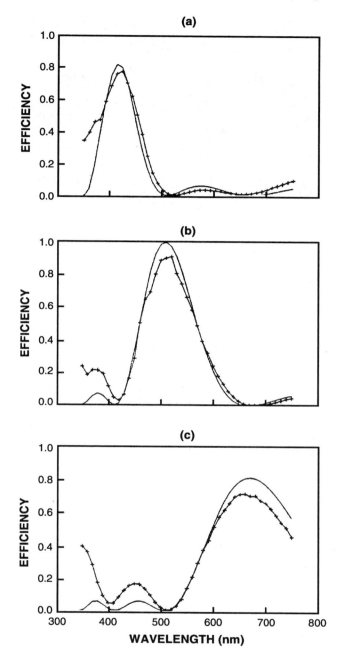

Figure 3.20 Spectral composition of a four-step echelon grating with 16 μm period: (a) −1 order, (b) 0 order, (c) +1 order.

applications; exploring monolithic integration with ICs, OEICs (optoelectronic ICs), and MEMs (microelectromechanical devices); designing and fabricating optics that provide discrimination of incident photons on the focal plane; and developing high-fidelity replication methods. The ultimate potential of these micro-optics arrays awaits the fruition of new 'amacronic' architectures that integrate multiple planes of optics and electronics (Veldkamp and McHugh, 1992; Veldkamp, 1993).

Monolithic integration of active and passive devices on the same substrate can eliminate difficult and time-consuming alignments between discrete planes of devices. Optimal performance of integrated and stacked micro-optics requires accurate registration between the components to prevent vignetting as well as precise control over carrier substrate thickness and flatness to match focus requirements. Monolithically-integrated focal plane array imaging microlenses provide high fill factor, facilitate reduced detector area, decrease noise, and free up space on the focal plane for integrated preprocessing circuitry, Fig. 3.21(a). Dual-sided integration of a microlens array and a focal plane array on a single substrate reduces the system size and weight by combining two elements on one substrate. It also decreases the number of interfaces compared to hybrid schemes and, hence, the associated scattering and reflection losses from high-index materials. Focal plane arrays of CdTe microlenses designed to focus radiation in the 9–12 μm band have been monolithically integrated with epitaxially grown HgCdTe photodetectors on the opposite side of the substrate, Fig. 3.21(b) (Stern *et al.*, 1991a).

3.6.2 *Future Systems – Enabling Technology Integration and Growth*

The complex task of integrating micro-optics, analog circuitry, neural networks, nonlinear materials and photonics technologies requires the forging of collaborations with other disciplines in optical information processing, biology, medicine, data storage and display fields. Technological problems currently impeding the broader use of this technology must be addressed. Limited software is currently available for modelling diffractive optics in the electromagnetic regime. User-friendly ray-trace programs matched to drivers that can write data blocks in polar coordinates need to be developed. The software that drives the pattern generators or e-beam machines at commercial mask foundries, geared to the cartesian coordinate systems used in the microelectronics industry, is cumbersome and expensive for binary optics mask production. For example, vector-scan systems would be better matched to optics applications. The special fabrication concerns of micro-optics fabrication that differentiate it from micromechanics and microelectronics, i.e. making deep structures with submicron resolution and overlay registration, performing dual-sided alignment with submicron precision, and handling of oddly sized and shaped substrates, need to be addressed by both microfabricators and equipment manufacturers. Micro-optics fabrication foundries patterned after ARPA's MOSIS and MEMS facilities could drive down fabrication costs and broaden acceptance of this technology.

Binary optics is the quintessential enabling technology for future electro-optical systems, Fig. 3.22, capable of making components that can increase the overall performance of a larger optical system. Among its more important features are:

- Flexibility in element size, shape, layout and materials
- Reduction in system size, weight and cost

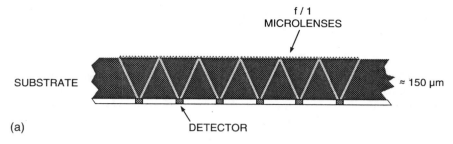

(a)

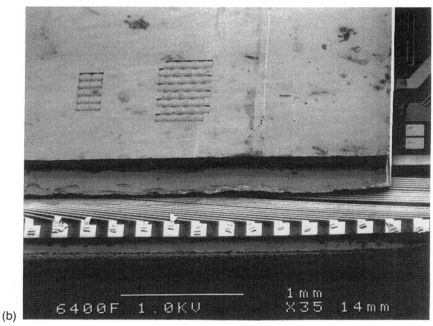

(b)

Figure 3.21 (a) Schematic representation of integrated microlens array/detector system. (b) SEM of a monolithically integrated array of *f*/4 lenslets and HgCdTe photodetectors on opposite sides of a single CdTe substrate.

- Precision manufacture of large, uniform and coherent micro-optic arrays
- Easy incorporation of multiple optical functions into the optical design
- Monolithic integration with microelectronic, photonic and microelectro-mechanical devices
- Batch processing and replication for inexpensive and robust components.

3.7 Acknowledgments

This work was supported by the Defense Advanced Research Projects Agency. This chapter is dedicated to Wilfrid Veldkamp, who founded the Binary Optics program at MIT Lincoln Laboratory. I would also like to acknowledge the contributions of my many colleagues in this effort over the years. These include

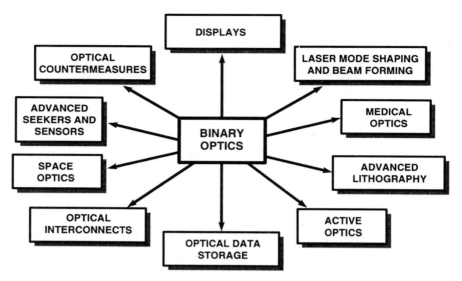

Figure 3.22   Binary optics is an enabling technology.

Gary Swanson, Jim Leger, Michael Holz, Bob Knowlden, Theresa Jay, Mike Farn, Woody Goltsos and Jim Bartley.

References

ANDERSON, E. H. and KERN, D. (1991) Nanofabrication of zone-plate lenses for X-ray microscopy. In Michette, A. G., Morrison, G. R. and Buckley, C. J. (eds) *X-Ray Microscopy III*. Berlin: Springer-Verlag.

ANDERSSON, H., EKBERG, M., HARD, S., JACOBSEN, S., LARSSON, M. and NILSSON, T. (1990) Single photomask, multilevel kinoforms in quartz and photoresist: manufacture and evaluation. *Appl. Opt.* **28**, 4259.

BLOOMSTEIN, T. M. and EHRLICH, D. J. (1991) *1991 International Conference on Solid-State Sensors and Actuators, Digest of Technical Papers*, **507**.

COLE, H. J. and PITTMAN, W. C. (eds) (1993) Conference on Binary Optics, NASA Conference Publication, **3227**; 1989, 1990, 1991, Computer and Optically Generated Holographic Optics Series, *Proc. SPIE*, **1052, 1211, 1555**; 1991, 1992, 1993, Miniature and Microoptics Series, *Proc. SPIE*, **1544, 1751, 1992**.

COX, J. A., WERNER, T., LEE, T. J., NELSON, S., FRITZ, B. and BERGSTROM, J. (1990) Diffraction efficiency of binary optical elements. *Proc. SPIE* **1211**, 116.

D'AURIA, L., HUIGNARD, J. P., ROY, A. M. and SPITZ, E. (1972) Photolithographic fabrication of thin film lenses. *Opt. Commun.* **5**, 232.

DAMMANN, H. (1970) Blazed synthetic phase-only holograms. *Optik* **31**, 95.
  (1978) Color separation gratings. *Appl. Opt.* **17**, 2273.
  (1979) Spectral characteristic of stepped-phase gratings. *Optik* **53**, 409.

FARN, M. W. and GOODMAN, J. W. (1990) Effect of VLSI fabrication errors on kinoform efficiency. *Proc. SPIE* **1211**, 125.

FARN, M. W., KNOWLDEN, R. W., STERN, M. B. and VELDKAMP, W. B. (1993a) Color separation gratings, in Cole, H. J. and Pittman, W. C. (eds) *Conference on Binary Optics, NASA Conference Publication*, **3227**.

FARN, M. W., STERN, M. B., VELDKAMP, W. B. and MEDEIROS, S. S. (1993b) Color separation by use of binary optics. *Opt. Lett.* **18**, 1214.

FLAMM, D. L. and DONNELLY, V. M. (1981) The design of plasma etchants. *Plasma Chem. Plasma Processing* **1**, 317.

FUJITA, T., NISHIHARA, H. and KOYAMA, J. (1982) Blazed gratings and Fresnel lenses fabricated by electron-beam lithography. *Opt. Lett.* **7**, 578.

GAL, G. (1993) *Micro-optics technology for advanced sensors*, presentation at the SPIE Conference on Miniature and Micro-Optics 1992, San Diego, CA.

GALE, M. T. and KNOP, K. (1983) The fabrication of fine lens arrays by laser beam writing. *Proc. SPIE* **398**, 347.

GALE, M. T., ROSSI, M. and SCHUTZ, H. (1992) Fabrication of 2-dimensional continuous-relief diffractive optical elements. *Proc. SPIE* **1732**, 58.

GALE, M. T., ROSSI, M., SCHUTZ, H., EHBETS, P., HERZIG, H. P. and PRONGUE, D. (1993) Continuous-relief diffractive optical elements for two-dimensional array generation. *Appl. Opt.* **32**, 2526.

GOLTSOS, W. and HOLZ, M. (1990) Agile beam steering using binary optics microlens arrays. *Opt. Eng.* **29**, 1392.

GOLTSOS, W. and LIU, S. (1990) Polar coordinate laser writer for binary optics fabrication. *Proc. SPIE* **1211**, 137.

GOODMAN, J. and SILVESTRI, A. (1970) Some effects of Fourier-domain phase quantization. *IBM J. Res. Dev.* **14**, 478.

GRATIX, E. (1993) *Evolution of a microlens surface under etching conditions*, presentation at the SPIE Conference on Miniature and Micro-Optics 1992, San Diego, CA.

HOLZ, M., STERN, M. B., MEDEIROS, S. S. and KNOWLDEN, R. E. (1991) Testing binary optics: accurate high-precision efficiency measurements of microlens arrays in the visible. *Proc. SPIE* **1544**, 75.

HUTLEY, M. (ed.) (1991) *Microlens Arrays*. IOP Short Meeting Series, **30**. UK: IOP Publishing.

JAY, T. R. and STERN, M. B. (1993) Preshaping photoresist for refractive microlens fabrication. *Proc. SPIE* **1992**, 275.

JAY, T. R., STERN, M. B. and KNOWLDEN, R. E. (1992) Refractive microlens array fabrication parameters and their effect on optical performance. *Proc. SPIE* **1751**, 236.

KATHMAN, A. D. (1989) Efficient algorithm for encoding and data fracture of electron beam written holograms. *Proc. SPIE* **1052**, 47.

KIRZ, J. (1974) Phase zone plates for x-rays and the extreme uv. *J. Opt. Soc. Am.* **45**, 301.

KNOWLDEN, R. E. (1987) adapted the original computer code written by S. A. Gaither, *Coupled wave theory of crossed gratings*, unpublished PhD thesis, Massachusetts Institute of Technology.

KORONKEVICH, V. P., KIRIYANOV, V. P., KOKOULIN, F. I., PALCHIKOVA, I. G., POLESHCHUK, A. G., SEDUKHIN, A. G., CHURIN, E. G., SHCHERBACHENKO, A. M. and YURLOV, YU. I. (1984) Fabrication of kinoform optical elements. *Optik* **67**, 257.

LEE, W. H. (1978) Computer-generated holograms: techniques and applications. In Wolf, E. (ed.) *Progress in Optics*, Vol. XVI, pp. 120–232. Amsterdam: North-Holland.

LEGER, J. R. and GOLTSOS, W. C. (1992) Geometrical transformation of linear diode-laser arrays for longitudinal pumping of solid-state lasers. *IEEE J. Quantum Electron.* **28**, 1088.

LEGER, J. R., HOLZ, M., SWANSON, G. J. and VELDKAMP, W. (1988a) Coherent laser beam addition: an application of binary optics technology. *Lincoln Lab. J.* **1**, 225.

LEGER, J. R., SCOTT, M. L., BUNDMAN, P. and GRISWOLD, M. P. (1988b) Astigmatic wavefront correction of a gain-guided laser diode array using anamorphic diffractive microlenses. *Proc. SPIE* **884**, 82.

LEHMAN, H. W. (1991) Plasma-assisted etching. In Vossen, J. L. and Kern, W. (eds) *Thin Film Processes II*, pp. 673–747. San Diego: Academic Press.

LESEM, L., HIRSCH, P. and JORDAN, J. (1969) The kinoform: a new wavefront reconstruction device. *IBM J. Res. Dev.* **13**, 150.

LOGUE, J. and CHISHOLM, M. L. (1989) General approaches to mask design for binary optics. *Proc. SPIE* **1052**, 19.

LOHMANN, A. W. and PARIS, D. P. (1967) Binary Fraunhofer holograms, generated by computer. *Appl. Opt.* **6**, 1739.

MEAD, C. A. (1980) *Introduction to VLSI Systems*. Reading, MA: Addison-Wesley.

MIYAMOTO, K. (1961) The phase Fresnel lens. *J. Opt. Soc. Am.* **51**, 17.

MOGAB, C. J. and LEVINSTEIN, H. J. (1980) *J. Vac. Sci. Technol.* **17**, 721.

MOREAU, W. M. (ed.) (1989) *Semiconductor Lithography: Principles, Practices and Materials*. New York: Plenum Press.

NISHIHARA, H. and SUHARA, T. (1987) Micro Fresnel lens. In Wolf, E. (ed.) *Progress in Optics*, Vol. XXIV. Amsterdam: North Holland.

PALMATEER, S. C., KUNZ, R. R., HORN, M. W., FORTE, A. R. and ROTHSCHILD, M. (1995) Optimization of a 193-nm silylation process for sub-0.25-μm lithography. *Proc. SPIE* **2438**, 445.

POPOVIC, Z. D., SPRAGUE, R. A. and NEVILLE CONNELL, G. A. (1988) Technique for monolithic fabrication of microlens arrays. *Appl. Opt.* **27**, 1281.

RUBIN, S. M. (1987) *Computer Aids for VLSI Design*. Reading, MA: Addison-Wesley.

SHVARTSMANN, F. P. (1993) Replication of diffractive optics. In Lee, S. H. (ed.) *Critical Review on Diffractive and Miniaturized Optics*, **CR49**. Bellingham: SPIE.

SLUISAREV, G. G. (1957) Optical systems with phase layers. *Sov. Phys. Dokl.* **2**, 161.

STERN, M. B. and JAY, T. R. (1993) Dry etching – path to coherent refractive microlens arrays. *Proc. SPIE* **1992**, 283.

STERN, M. B. and MEDEIROS, S. S. (1992) Deep three-dimensional microstructure fabrication for IR binary optics. *J. Vac. Sci. Technol.* B **10**, 2520.

STERN, M. B., HOLZ, M. and JAY, T. R. (1992) Fabricating binary optics in infrared and visible materials. *Proc. SPIE* **1751**, 85.

STERN, M. B., HOLZ, M., MEDEIROS, S. and KNOWLDEN, R. E. (1991a) Fabricating binary optics: process variables critical to optical efficiency. *J. Vac. Sci. Technol.* B **9**, 3117.

STERN, M. B., DELANEY, W. F., HOLZ, M., KUNZ, K. P., MASCHHOFF, K. R. and WLESCH, J. (1991b) Binary optics microlens arrays in CdTe. *Mater. Res. Soc. Symp. Proc.* **216**, 107.

STERN, M. B., PALMATEER, S. C., HORN, M. W., ROTHSCHILD, M., MAXWELL, B. E. and CURTIN, J. E. (1995) Profile control in dry development of high-aspect-ratio resist structures. *J. Vac. Sci. Technol.* B **13**, 3017.

STONE, T. B. and THOMPSON, B. J. (eds) (1991) *Selected Papers on Holographic and Diffractive Lenses and Mirrors, SPIE Milestone Series*, **MS 34**.

SUHARA, T., KOBAYASHI, K., NISHIHARA, H. and KOYAMA, J. (1982) *Appl. Opt.* **21**, 1966.

SULESKI, T. J. and O'SHEA, D. C. (1995) Gray-scale masks for diffractive-optics fabrication: 1. Commercial slide imagers. *Appl. Opt.* **34**, 7507.

SWANSON, G. J. (1989) *Binary Optics Technology: The Theory and Design of Multi-level Diffractive Optical Elements*, Tech. Rep. 854, DTIC#AD-213404, MIT Lincoln Laboratory.

 (1991) *Binary Optics Technology: Theoretical Limits on the Diffraction Efficiency of Multilevel Diffractive Optical Elements*, Tech. Rep. 914, DTIC#AD-A-235404, MIT Lincoln Laboratory.

SWANSON, G. J. and VELDKAMP, W. B. (1989) Diffractive optical elements for use in infrared systems. *Opt. Eng.* **28**, 605.

SZE, S. M. (ed.) (1988) *VLSI Technology*. New York: McGraw-Hill.

TANIGAMI, M., OGATA, S., AOYAMA, S., YAMASHITA, T. and IMANAKA, K. (1989) Low wavefront aberration and high-temperature stability molded micro Fresnel lens. *IEEE Photon. Technol. Lett.* **1**, 384.

TENNANT, D. M., GREGUS, J. E., JACOBSEN, C. and RAAB, E. L. (1991) *Opt. Lett.* **16**, 621.

VELDKAMP, W. B. (1991) Overview of micro-optics: past, present and future. *Proc. SPIE* **1544**, 287.

(1993) Wireless focal planes on the road to amacronic sensors. *IEEE J. Quantum Electron.* **29**, 801.

VELDKAMP, W. B. and McHUGH, T. J. (1992) Binary optics. *Sci. Am.* May, 92.

VELDKAMP, W. B. and SWANSON, G. J. (1983) Developments in fabrication of binary optical elements. *Proc. SPIE* **437**, 54.

VELDKAMP, W. B. and VAN ALLEN, E. J. (1983) Binary holographic LO beam multiplexer for IR imaging detector arrays. *Appl. Opt.* **22**, 1497.

VOSSEN, J. L. and KERN, W. (eds) (1991) *Thin Film Processes II*. San Diego: Academic Press.

WONG, V. V. and SWANSON, G. J. (1991) Binary optic interconnects: design, fabrication and limits on implementation. *Proc. SPIE* **1544**, 123.

(1993) Design and fabrication of a Gaussian fan-out optical interconnect. *Appl. Opt.* **32**, 2502.

4

Direct Writing of Continuous-relief Micro-optics

M. T. GALE

4.1 Introduction

The term continuous-relief micro-optics covers a wide range of refractive and diffractive microstructures offering a variety of optical characteristics ranging from on-axis focusing of a collimated light beam to the formation of complex holographic images. Within the scope of this chapter, we will be concerned with technologies for the fabrication of *planar* continuous-relief micro-optical elements, defined as surface-relief microstructures with a maximum relief modulation of about 5 μm. These microstructures can be fabricated by direct-writing techniques such as the laser (Gale *et al.*, 1994a) or e-beam (Zaleta *et al.*, 1993) exposure of a radiation-sensitive resist film using a computer-controlled scanning system. Accurate control of the exposure beam intensity and processing parameters enables continuous-relief microstructures to be fabricated with lateral feature resolution under 5 μm and height resolution better than 10 nm (see Fig. 4.1). Complex surface-relief microstructures can thus be fabricated directly from design data by a single exposure step followed by appropriate processing. As planar micro-optical elements, the original surface relief can subsequently be mass-produced by replication from an electroformed metal mould as described in more detail in Chapter 6. A major attraction of direct-writing techniques thus lies in the flexible fabrication process, directly producing a planar micro-optical element which is well-suited to the mass-production of low-cost copies by embossing or moulding technology.

Figure 4.1 illustrates the basic structure type and shows some examples of typical continuous-relief planar micro-optical elements. The elements can be dominantly refractive (as microlenslets and lenslet arrays), diffractive (such as kinoforms and grating microstructures) or a combination of both (for example, Fresnel microlenses). Direct writing is of particular interest for the fabrication of combined refractive/diffractive micro-optical elements such as phase-matched Fresnel elements (Kunz and Rossi, 1993), which can exhibit novel and unique optical properties and are difficult to fabricate by other techniques. A further example is the area interlacing of various microstructures to produce micro-optical

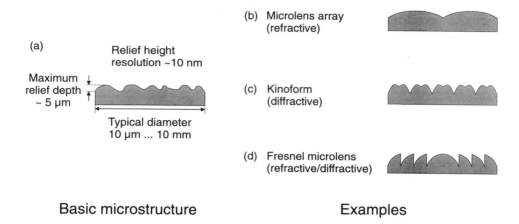

Basic microstructure   **Examples**

Figure 4.1 Continuous-relief micro-optical elements: Basic microstructure (a) of planar phase elements which can be fabricated by direct-writing techniques. Depending upon the microstructure and feature sizes, the elements can have optical properties which are dominantly refractive (b), diffractive (c) or mixed (d).

elements with a combination of optical properties, such as a set of off-axis foci (Rossi and Kunz, 1994); such complex structures can be fabricated with little additional effort using direct-writing techniques.

The distinction between continuous-relief micro-optics and binary optics becomes increasingly blurred as the complexity and number of mask steps in the latter increases. As a rule of thumb, the optical performance (efficiency and image uniformity in particular) of binary optical elements fabricated using three mask steps (eight relief levels) can closely approach that of continuous-relief elements. Current fabrication technology is better developed for binary optical micro-structures, since it is based upon modern high-resolution semiconductor fabrication technology. Direct writing of continuous-relief microstructures is still at the laboratory stage, although progress in recent years has been considerable, and modern commercial laser and e-beam writing machines can be modified for this application. Alternative techniques for the fabrication of continuous-relief micro-optics from laser or e-beam written masks also appear very promising for many types of micro-optical elements; examples include the controlled imaging of half-tone masks using commercial projection aligners (Oppliger *et al.*, 1994) and the reflow of resist structures to produce high numerical aperture microlenslets (Daly *et al.*, 1990; Göttert *et al.*, 1995 and Chapter 5).

An important practical advantage of direct writing of continuous-relief micro-optics becomes apparent in the fabrication of high aperture lenslets. Whereas the alignment and minimum feature size requirements for binary optical structures become very demanding (<0.5 μm for lenslets for the visible), the use of thicker resist films for higher-order phase depths (an 8π phase depth step corresponds to about 5 μm relief height in a typical polymer material at visible wavelengths) enables lenslets with numerical apertures up to 0.5 to be readily fabricated as continuous-relief micro-optical elements with acceptable efficiency (Gale *et al.*, 1994a). The direct-writing approach avoids the need for submicron alignment of multiple exposure steps.

In this chapter we concentrate on the two major current technologies for the direct writing of continuous-relief microstructures – laser beam writing (section 4.2) and electron beam writing (section 4.3). Both techniques have been under development for over 20 years – see for example work by Gale and Knop (1983) and by Fujita *et al.* (1982). In section 4.4 we describe alternative technologies such as diamond turning and laser ablation. All technologies described involve a direct-writing step, by which complex, randomly shaped and profiled microrelief structures can be fabricated under computer control from relief data resulting from optical design procedures. For completeness, fabrication by half-tone mask imaging techniques is also described, although this is not strictly a direct-writing technology. Section 4.5 presents a brief conclusion and comparison of the technologies.

4.2 Laser Beam Writing

The controlled exposure of a photoresist film by direct laser beam writing enables a wide variety of planar micro-optical elements to be produced. The photoresist must be processed such that local thickness of the developed resist film is a continuous (preferably linear) function of the exposure (instead of the high contrast, binary development of resist films for semiconductor mask fabrication). Suitable choice of photoresist and processing conditions allows the use of commercially available materials which have been developed for semiconductor lithography applications, and thus of high quality, reproducible materials. The exposure is generally realized by scanning the substrate under a focused laser beam and synchronously modulating the laser beam intensity to write a continuous, grey-scale exposure pattern in the resist film. Computer control of the writing process then allows a wide range of surface-relief microstructures to be fabricated.

A system for the laser writing of continuous-relief microlens structures was described by Gale and Knop of RCA Laboratories, Zurich (now the Paul Scherrer Institute PSI, Zurich) in 1983. The basic technology and writing process has been continuously improved since then (Gale *et al.*, 1991, 1994a, 1994b) and applied to the fabrication of a variety of different micro-optical elements. Direct laser writing systems have also been described by Haruna and co-workers at Osaka University, Japan (Haruna *et al.*, 1990), by Langlois and co-workers at the National Optics Institute, Quebec (Langlois *et al.*, 1992), by Akkepeddi and co-workers at Hughes Danbury Optical Systems Inc., Danbury, USA (Akkepeddi *et al.*, 1993) and by researchers at the Rochester Photonics Corporation, Rochester, USA (Faklis *et al.*, 1993; Bowen *et al.*, 1994).

This work has been carried out using in-house developed custom laser writing systems. Commercial laser lithographic systems are available for the fabrication of masks for semiconductor lithography[1] and for optical elements[2]; these systems are intended primarily for the fabrication of binary and not continuous-relief photoresist microstructures, but can in principle be modified for the latter. As discussed in section 4.2.3 the scan positioning tolerances for the fabrication of high quality continuous surface-relief profiles are very demanding (typically <100 nm accuracy) and are currently best attained on custom systems.

4.2.1 Laser Writing System Design

The process steps involved in the fabrication of continuous-relief micro-optical elements such as kinoforms, Fresnel lenslets and multi-element arrays are shown schematically in Fig. 4.2. The microstructure to be fabricated is defined by surface-relief data generated by custom or commercial optical design programs. This data is converted to laser beam intensity values using calibration run measurements which define the relationship between the beam intensity (resist exposure) and the corresponding relief height of the resist film after development. The writing process then exposes the resist-coated substrate by scanning the substrate under the focused laser beam, with accurate and synchronized computer control of the stage movement and laser beam intensity. (In principle, the exposure scan could alternatively be performed by deflection of the laser beam; in practice the very demanding positioning tolerances are virtually impossible to achieve over the required areas in excess of $1 \times 1 \, \text{cm}^2$.) The controlled development of the exposed resist film then results in the desired surface-relief microstructure.

An example of a laser writing system which has been described in detail by Gale and co-workers (1991, 1994a) is shown in Fig. 4.3. A (positive) photoresist-coated substrate is exposed by xy-raster scanning under a focused HeCd laser beam ($\lambda = 442 \, \text{nm}$). The translation stage is an air-bearing table driven by linear motors with interferometer position measurement. It has a dynamic positioning accuracy (line straightness) of about $\pm 150 \, \text{nm}$, limited by the control system and by vibrations affecting the optical system (minimized by dynamic anti-vibration elements built onto the granite base). An acousto-optic modulator controls the laser beam intensity (256 intensity levels) and is driven by data stored in a line buffer in the PC. The (8-bit) pixel intensity data are loaded line by line into the buffer and clocked out by the interferometer pulses during each line scan. This ensures accurate synchronization of the laser intensity with the stage position and is relatively insensitive to the constancy of the scan velocity. An autofocus system dynamically holds focus over the scan area during the writing procedure.

Typical writing parameters for the fabrication of micro-optical elements are an interline spacing of $1 \, \mu\text{m}$, a minimum focused spot diameter of $1.5 \, \mu\text{m}$ (at the $1/e$ intensity points) and a writing speed of $10 \, \text{mm sec}^{-1}$. In order to minimize the effects of vibrations and positioning errors, the writing spot size is chosen as large as possible for a given application and minimum feature size (for example of the outer segment of a Fresnel lenslet). For low-aperture microlenses it may vary up to about $10 \, \mu\text{m}$; the larger spot leads to smoother surface profiles but a loss in feature resolution. A discussion of these considerations and the surface roughness achieved in practice is given in section 4.2.3. Writing times using such an xy-scanning system are relatively long (many hours for typical microstructures of $1 \, \text{cm}^2$ area) but, since the photoresist recording is generally converted to a replication mould for producing replicas for use in micro-optical systems, this is in practice not of great significance.

For many applications, anamorphic profiles or arrays of elements are not required and a rotationally symmetric laser writer is more efficient in writing circular symmetric microstructures and in creating larger elements. Previous high precision rotational laser writing systems have emphasized binary processes for making mask elements (Goltsos and Liu, 1990). At the Rochester Photonics Corporation (Rochester, USA), researchers have extended the idea to allow

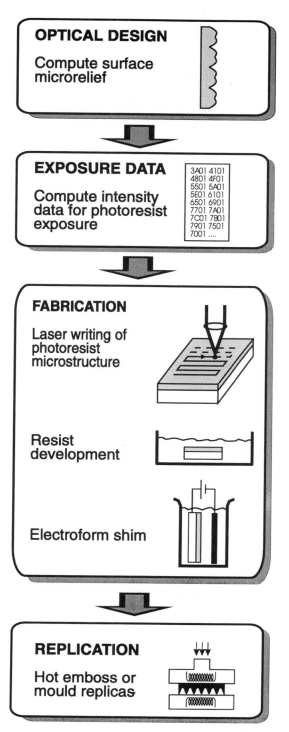

Figure 4.2 Process steps for the fabrication of planar, continuous-relief micro-optical elements.

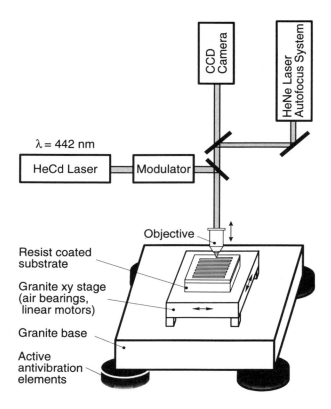

Figure 4.3 Laser writing system using a precision *xy*-scanning stage (Gale *et al.*, 1994a).

generation of high efficiency continuous profiles (Bowen *et al.*, 1994). Although diamond turning can be used to generate masters for moulding processes, rotationally symmetric laser writing allows the surface relief to be generated in a photoresist and then transferred by etching into materials such as silica. Applications include high power beam shaping and achromatization of UV systems. Figure 4.4 is a diagram of the rotational writing system, in which the linear travel allows fabrication of parts up to 250 mm in diameter. The control system computes the necessary exposing power directly from the phase polynomial during writing. The system uses a CD mastering objective with a numerical aperture of 0.91 and a piezoelectric translator to control the objective position and maintain the correct focus position. If the spindle turns at a constant velocity, the exposure power must vary over a wide dynamic range in order to properly expose the photoresist. This is accomplished by using two modulators, one that produces a slow variation in power with radial location and a second that produces the exposure variation for each zone of the 256 levels.

4.2.2 *Photoresist Considerations*

Since the micro-optical element is fabricated in a photoresist film, the quality and processing characteristics of the latter play a key role in producing elements of

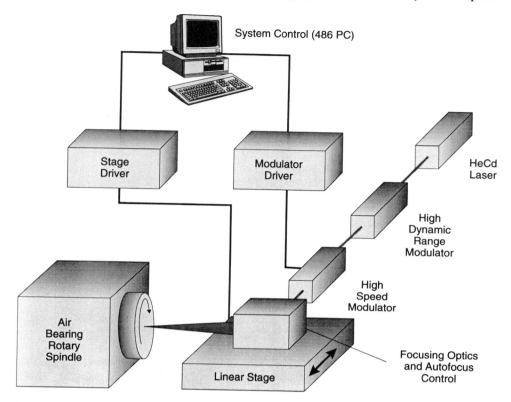

Figure 4.4 Laser writing system using a rotational scanning stage (courtesy of Rochester Photonics Corporation, Rochester, USA).

good cosmetic quality and optimum optical performance. The main requirements of the photoresist material are thus (1) it should be suitable for producing uniform films of high surface quality and reproducible processing parameters, (2) it should be capable of achieving lateral resolution well under 1 μm, and (3) the development characteristic (resist thickness after development as a function of exposure) should be approximately linear to achieve a high dynamic range. Many commercially available photoresists satisfy these criteria. Best results have generally been obtained with positive resists (increasing development etch rate with increasing exposure), often using 'non-standard' processing to achieve a linear development characteristic instead of the 'high-gamma' behaviour required for semiconductor device lithography. Such photoresists are readily available in solutions optimized for the spin coating of high quality films in the 1 to 5 μm thickness range. Typical sensitivities at an exposure wavelength of $\lambda = 442$ nm are in the range 10 to 100 mJ cm^{-2}. In practice, this means that a 15 mW HeCd laser is adequate for an xy laser writing system with writing speeds of cm sec^{-1} and conservative losses in the optical system.

An example of a suitable photoresist for films of up to about 5 μm is Shipley S1828[3] which can be processed to achieve a relatively linear development characteristic using dilute AZ303 developer (Table 4.1). A spinning speed of 2000 rpm (for 30 sec.) gives a film thickness of $t_{film} \simeq 3.5$ μm. Thicker resist films (up

Table 4.1 Example of photoresist preparation and processing parameters for laser beam writing (from Gale, 1994a)

Item/process step	Material/process parameters
Substrate	Schott B270 glass: $50 \times 50 \times 1 \, \text{mm}^3$
Clean	Standard substrate cleaning procedures
Antireflection layer (optional)	Brewer Science Inc. ARC – XL20
Spin coat	30 sec. @ 2000 rpm
Hotplate bake	60 sec. @ 160°C
Photoresist	Shipley S1828: film thickness $\approx 3.5 \, \mu\text{m}$
Spin coat	30 sec. @ 2000 rpm
Hotplate pre-bake	60 sec. @ 115°C
Exposure	Laser writer, HeCd laser ($\lambda = 442$ nm)
Sensitivity	$\sim 50 \, \text{mW cm}^{-2}$
Development	Shipley AZ303 developer: 30 sec. @ 1:10 dilution
Typical development characteristic	$I = a_0 + a_1 t + a_2 t^2 + a_3 t^3 + a_4 t^4$
for these preparation conditions (see	$a_0 = 1.85$
equation (4.3) and Fig. 4.6).	$a_1 = 1.72 \times 10^{-1} \, \text{nm}^{-1}$
t in nm	$a_2 = -1.20 \times 10^{-4} \, \text{nm}^{-2}$
I in arbitrary units 0 ... 255	$a_3 = 4.93 \times 10^{-8} \, \text{nm}^{-3}$
	$a_4 = -6.61 \times 10^{-12} \, \text{nm}^{-4}$

to about 15 μm) can be obtained using lower spinning speeds for shorter times (although this results in some degradation in film quality) or by using other resist formulations. After coating, the films are baked either in an oven at 100°C for 1 hour or on a hotplate at 115°C for 1 minute. The thickness of the as-coated film can be conveniently determined by profilometer measurements at a clean scratch in the film or, if the refractive index is known, by optical measurement of the reflectivity as a function of illumination angle or wavelength; a number of commercial instruments are also available for determining the thickness of photoresist films spin coated onto silicon wafer substrates.

The surface-relief height h of a uniformly exposed area of photoresist after development is a function of the photoresist exposure and development. Since the exposure is proportional to the laser intensity for a given set of scan parameters, this can be expressed as (see Fig. 4.5)

$$h(I) = t_{\text{film}} - t(I)$$

where

    I = laser beam intensity ($\propto$ exposure for a constant scan velocity and interline spacing)

  $h(I)$ = height of (positive) photoresist film remaining after development

  t_{film} = original photoresist film thickness

   $t(I)$ = thickness of resist removed during development.

Since, for many photoresists, even unexposed areas experience a small reduction in the resist film thickness during development the effective modulation $h'(I)$ of the final resist relief (see Fig. 4.5 for a squarewave) is given by

$$h'(I) = h(I) - h(0).$$

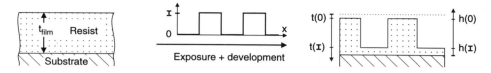

Figure 4.5 Surface relief of developed (positive) photoresist film of thickness t and relief height h. The surface-relief modulation for a squarewave exposure is given by the difference between the heights of the exposed and unexposed areas.

The photoresist development characteristic is a function of the resist film preparation, writing and processing parameters. It can be determined for a given writing system and processing parameters by performing calibration runs and measuring the relief steps of areas of different exposure using a surface profilometer. For the computation of the intensity data, the development characteristic can be conveniently represented as a relationship of the form

$$I = a_0 + a_1 t + a_2 t^2 + a_3 t^3 + a_4 t^4$$

where

$I = $ laser beam intensity

$t = $ relief depth after development (resist thickness removed during development)

$a_0 \ldots a_4 = $ fit parameters (see Table 4.1 for typical values).

The relationship between the final micro-optical element relief profile $h'(x, y)$ and the required laser intensity $I(x, y)$ can be computed from the above equations.

Figure 4.6 shows the measured development characteristic and fit parameters for 30 sec. development of S1828 resist film in 1:10 AZ303 developer. Stronger AZ303 developer concentrations result in a more linear characteristic, but also in significant removal of unexposed resist during the development process which must be taken into consideration when choosing the resist film thickness. For long writing runs, a dependence of the development characteristic upon the time between exposure and development should be taken into account. This effect, a decay in the latent image exposure, becomes significant for writing times of more than about 12 hours; Fig. 4.6 also shows a measured development characteristic 48 hours after exposure. The effect can be compensated for by pre-correcting the exposure data.

The resist preparation and processing conditions are critical steps in the fabrication of micro-optical elements with a given relief profile. Table 4.1 provides a summary of the parameters for preparing and processing an S1828 resist film for laser writing. Good experimental reproducibility of the development characteristic requires careful control over the coating and development procedures. In practice, an accuracy and reproducibility of about ±5% in average relief height is relatively straightforward, whilst a tolerance of ±2% requires considerable effort. It is good practice to write a test calibration pattern beside every micro-optical element and compare the developed depths with the programmed values. For elements requiring a very accurate relief structure, the most practical approach is to write a group of elements with the exposure varied by a small scaling factor (~2%), and to select the best microstructure after development.

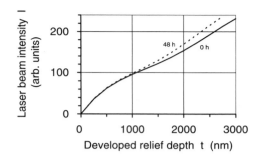

Figure 4.6 Measured development characteristic for S1828 photoresist (preparation as given in Table 4.1): Developed relief depth as a function of the exposure ($\propto$ laser beam intensity). The dotted curve shows the measured latent image decay for a film developed 48 hours after exposure.

4.2.3 *Surface-relief Profile Fidelity*

The ultimate test of the quality of a micro-optical element is its optical performance, which is measured in terms of parameters such as imaging quality, efficiency and scattered light. In the fabrication process, it is useful to consider and monitor physical parameters which are a measure of how well the design surface relief is realized, in particular the absolute relief height, profile form and lateral resolution and the surface roughness. A detailed analysis of these factors has been given by Gale *et al.* (1983, 1994a). The results are summarized here.

The laser writing process can be characterized by the following parameters, illustrated in Fig. 4.7:

$h(x, y)$ = design surface-relief profile height
L = minimum lateral feature of microstructure to be fabricated
D = diameter of focused laser spot (full width at $1/e$ maximum intensity for Gaussian intensity distribution)
b = interline spacing of raster scan along the y-axis
ε = rms variation in x-position of scan line
σ = rms variation of the net exposure from its nominal value, normalized to unity
ρ = rms variation of the relief height h from its nominal value, normalized to unity.

Absolute Surface-relief Height

The precision to which the surface-relief h is reproduced in unstructured, uniform areas is dependent upon the photoresist film sensitivity, laser beam intensity control and the development process. As discussed above, a precision of $\pm 2\%$ can be achieved in practice by good experimental procedure. The implementation of an automatic end-point detection system, for example based on the monitoring of the diffraction efficiency of a calibration grating structure, is in principle capable of effectively compensating for all resist processing parameters and achieving

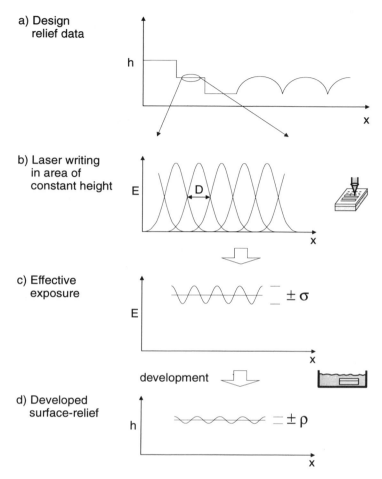

Figure 4.7 Laser writing process characterization: (a) Input : surface-relief data resulting from design; (b) Raster scan exposure in area of constant relief height: individual Gaussian spot exposures for $D/b = \sqrt{2}$; (c) Effective exposure: sum of individual raster line scans with resulting modulation in the local exposure uniformity; (d) Final developed surface relief with residual scan line surface modulation.

reproducible processing with a precision approaching 1% in absolute height control.

Profile Form

In structured areas such as fine Fresnel segments of lateral dimension L, the degree to which the ideal profile form $h(x, y)$ can be realized is limited by the spot diameter D and the interline spacing b. In practice, $D \approx b < L/5$ gives acceptable results in most cases. Typical values for Fresnel lenslets (Gale *et al.*, 1994a) are $L \approx 5 \,\mu\text{m}$, $D \approx b \approx 1 \,\mu\text{m}$. The steepest slope which can be realized, for example at the segment step in a Fresnel lenslet microstructure, is typically about 70° for these values.

97

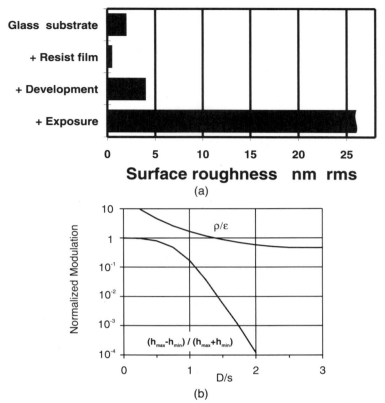

Figure 4.8 Surface roughness for laser written microstructure. (a) Measured rms roughness (from AFM images) for glass substrate, as-coated resist film and for unexposed and uniformly exposed film areas after development. (b) Surface-relief modulation (computed) as a function of D/b (the ratio of laser spot diameter to scan line spacing), assuming a linear development characteristic. An inherent modulation $(h_{max} - h_{min})/(h_{max} + h_{min})$ results from the superposition of the Gaussian line scans. An additional modulation, shown as the ratio of surface-relief modulation to rms line positioning variations (ρ/ε), results from experimental positioning errors (from Gale and Knop, 1983).

Surface Roughness

Figure 4.8(a) shows the measured surface roughness at various stages in the laser writing of a micro-optical element. The glass substrates were standard $50 \times 50 \times 1\ \text{mm}^3$ glass (Schott B270) substrates with a measured roughness of about 2 nm rms. After coating with a 3–5 μm thick layer of Shipley AZ 1400 photoresist (essentially identical to S1828), the surface quality improved to better than 0.5 nm roughness. Development in 1:7 or 1:10 AZ303 developer resulted in a surface roughness in unexposed areas of about 4 nm. This represents the limiting roughness for the microrelief structures.

The roughness in exposed, laser-written areas is in general significantly worse than this limiting value; measured values vary between 25 nm and about 100 nm rms. A detailed analysis of the causes, assuming a Gaussian writing spot profile, has been carried out by Gale and Knop (1983) – Fig. 4.8(b) shows a summary of the results. There are two main sources of the surface roughness:

- Surface modulation introduced by the raster scan
  An inherent surface-relief modulation is introduced by the raster scan exposure corresponding to the superposition of the Gaussian line profiles offset by the interline spacing. The choice of spot size D and interline spacing b are dictated by the experimental set-up and by the minimum feature size L of the micro-optical element to be fabricated. The interline spacing should be at least 5 times smaller than the finest lateral segment size for Fresnel-type microstructures; on the other hand, the writing time is directly proportional to $1/b$. The periodic ('grating') surface-relief modulation decreases as the spot size D increases and is minimized by choosing the largest spot size compatible with the segment resolution, with a limiting size of about $D \simeq L/5$ in practice. Figure 4.8(b) shows the surface modulation $(h_{max} - h_{min})/(h_{max} + h_{min})$, assuming a linear development characteristic, as a function of D/b. A modulation of less than 1% of the relief height is obtained for $D > 1.44s$.

- Surface roughness resulting from positioning error in the individual scan lines
  The above analysis assumes perfect positioning of the scan lines. In practice, significant surface-relief structure is introduced by positioning errors. Figure 4.8(b) also shows the dependence of the relief modulation ρ upon the line positioning error ε. An rms relief roughness of less than 1% also requires line positioning to better than 1%. For an interline spacing of $1\ \mu m$, this corresponds to dynamic ('line straightness') positioning of better than 10 nm along a line scan, which represents a very high-performance positioning system. Table 4.2 shows the experimental data for the laser writing system described above (Gale *et al.*, 1994a). The performance and results are sufficient for the fabrication of a wide range of micro-optical elements for real applications, although a significant improvement in scattered light for high numerical aperture (NA) elements could be achieved with improved positioning accuracy (and should be realized in a next generation writing system under construction).

Lateral Resolution

The focused laser spot size also determines the lateral resolution and the minimum feature size which can be fabricated. Ignoring positioning errors, the resolution is given by the finest spot to which the (Gaussian) laser beam can be focused, which is given by

$$D_0 = \lambda/(2NA)$$

$$D = \sqrt{2}D_0$$

where

   D_0 = spot diameter at the $1/e^2$ maximum intensity points
   NA = numerical aperture of the imaging objective
   D = spot diameter at the $1/e$ maximum intensity points (as used above).

For $\lambda = 442$ nm and an objective with $NA = 0.5$, this gives $D \simeq 625$ nm (full width at $1/e$ intensity points).

Table 4.2 Performance and parameters for experimental laser writing system (from Gale, 1994a)

Step/parameter	Symbol	Parameters	
xy table			
Scan velocity	v	10 mm sec^{-1} (typical)	Air bearing, linear motors, interferometer positioning control
Position measurement resolution		$\lambda/32$ (HeNe) $\simeq$ 20 nm	
Scan interline spacing	b	1 μm/2 μm	(depending upon microstructure)
Scan line positioning accuracy	ε	$\pm$150 nm rms	(measured at 10 mm sec^{-1})
Optics			
Writing wavelength	λ	442 nm	(HeCd laser)
Grey scale		256 levels	(acousto-optic modulator, 8-bit)
Focused spot diameter (I_{max}/e)	D	1.5 to 8 μm	(depending upon microstructure)
Micro-optical elements			
Surface-relief type		Continuous-relief	
Maximum relief height	h (max)	3.2 μm standard	($\sim$6 μm max.)
Maximum area		$\sim$40 $\times$ 40 mm^2	(limited by writing time)
Writing time for 10 mm^2		$\sim$10 hours	($b = 1$ μm, $v = 10$ mm sec^{-1})
Minimum segment size	L	$\sim$5 μm	
Maximum NA for micro-optical elements		0.5	(PMFE with 8π phase step)
Typical surface roughness (rms)	ρ	25 ... 100 nm	(depending upon microstructure and NA)
Optical efficiency		60% to >95%	(depending upon microstructure and NA)

For the fabrication of deep (1–5 μm) micro-optical structures, the depth of focus becomes relevant. For a spot diameter of $D = 1\,\mu$m, the depth of focus z_f for a Gaussian is given by (Yariv, 1976)

$$z_f = 2\pi(D_0^2/4)n/\lambda = \pi D^2 n/4\lambda \simeq 3\,\mu\text{m}$$

where

z_f = distance between points with a spot diameter increased by $\sqrt{2}$
n = refractive index of the resist $\simeq 1.63$ at $\lambda = 442$ nm.

In practice, positioning errors generally limit the diameter of the smallest spot which can sensibly be used in the fabrication process. For the writing system described above, a spot diameter of $D \simeq 1.5\,\mu$m can be used to fabricate continuous-relief segments such as the outer segments of Fresnel lenses with a minimum segment size of about 5 μm (Gale *et al.*, 1994a). This corresponds to maximum $NA \simeq 0.5$ for fabricated lenslets with a phase depth of 8π. The realization of higher resolution microstructures requires an increase in the position accuracy before finer spots with diameters $D \simeq 0.5$–1 μm can result in significant improvements.

4.2.4 *Examples of Fabricated Micro-optical Elements*

In this section we present examples of micro-optical elements fabricated by laser writing using the system constructed at PSI, Zurich (Gale *et al.*, 1994a). The element sizes vary from about 30 μm up to about 1 cm and have maximum relief height of about 5 μm. The surface roughness and resulting scattered light in use varies depending upon the writing parameters. Low NA ($\ll$0.1) lenslets were written with an 8 μm spot size and interline spacing of 2 μm, resulting in relatively smooth surface-relief profiles ($\sim$25 nm rms roughness) with low scatter ($<$5%). Higher NA lenslets were fabricated with a spot size of about 1.5 μm and interline spacing of 1 μm. The resulting surface-relief structures show significant surface roughness of up to about 100 nm rms; with measured efficiencies in the range 60–90%, they are nevertheless of sufficient quality for applications where a certain amount of straylight can be tolerated. The surface roughness is almost totally a result of positioning errors in the writing process, and can be expected to be considerably reduced in future writing systems with improved positioning tolerances.

The capabilities of the laser writing technology are illustrated in Figs 4.9 and 4.10, which show AFM images of micro-optical elements written at PSI. Figure 4.9 shows a section of a phase-matched Fresnel element (PMFE) with segments of varying phase steps of 2π up to 8π at the design wavelength of $\lambda = 633$ nm (the inset shows a further enlargement of a phase step change from 2π to 4π). This lenslet has a size of 250 μm by 300 μm, NA of 0.5 and a measured efficiency of about 60%. Figure 4.10 shows a further selection of micro-optical elements fabricated on the same laser writing system, illustrated by SEM and AFM images of the surface profiles together with data on the optical parameters and performance. Figure 4.11 shows an SEM of a lenslet array fabricated using the laser writing system at the Rochester Photonics Corporation (Bowen *et al.*, 1994).

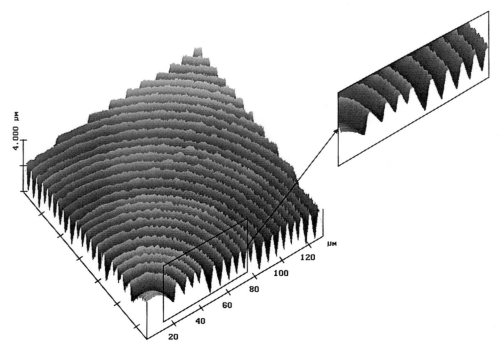

Figure 4.9 Section of laser-written Fresnel lens with NA ≈ 0.5: AFM image showing phase steps varying from 2π up to 8π (from Gale *et al.*, 1994a).

Laser writing technology is highly suited to the fabrication of micro-optical elements with complex, non-spherical profiles such as Fresnel lenslets and kinoform elements. Lenslet arrays with any lenslet shape and essentially zero dead space between lenslets can be fabricated (Gale *et al.*, 1994a). Kinoform microstructures such as fan-out elements with precise continuous-relief profiles have been fabricated from relief-height data computed by optimization routines (Ehbets *et al.*, 1992; Gale *et al.*, 1992, 1993). Fresnel lenslet arrays fabricated at PSI for applications in confocal microscopy have been demonstrated to have a uniformity in focal length along a 200 lenslet line of better than ± 0.2 μm at a nominal focal length of 250 μm (Hessler *et al.*, 1995). Anamorphic Fresnel lenslets (Rossi *et al.*, 1995) and focusing fan-out elements consisting of five spatially interlaced Fresnel lenslets combined in a single planar microstructure (Rossi and Kunz, 1994), both with applications in laser diode-to-fibre coupling, have been fabricated directly by laser writing at PSI. Diffractive elements have been written at the Rochester Photonics Corporation to control the optical transfer function (OTF) of lenses in a desired fashion; specific phase functions can be used in combination with digital post-processing of the image to construct optical systems with an increased tolerance to defocus (van der Gracht, 1995). Elements have also been fabricated for a hybrid diffractive–refractive wide-field eyepiece, offering lower cost and weight with comparable performance to conventional eyepieces (Missig and Morris, 1995). DOEs with non-periodic phase structures have also been fabricated for applications as OTF control filters which can be incorporated into liquid crystal cells to produce electrically switchable optical filters (Stalder and Ehbets, 1994).

4.3 Electron Beam Writing

The direct e-beam writing of continuous-relief microstructures in a film of electron-sensitive resist represents an alternative technique to the laser writing approach described in the previous section. The fabrication of blazed (as opposed to binary) microstructures by e-beam writing was described by Fujita, Nishihara and Koyama of Osaka University, Japan in 1982. Work at the OMRON Corporation (Kyoto, Japan) on the fabrication of continuous-relief Fresnel microlenses has been described by Aoyama and co-workers (1990) and by Imanaka (1992). Other groups working on the direct e-beam writing of continuous-relief micro-optical elements include those at the Chalmers University of Technology, Göteborg, Sweden (Ekberg *et al.*, 1990, 1992, 1994; Larsson *et al.*, 1993, 1994), CSEM, Neuchâtel, Switzerland (Stauffer *et al.*, 1992), the Friedrich Schiller University in Jena, Germany (Kaschlik and Kley, 1992; Kley *et al.*, 1992, 1993) and the University of California, San Diego, USA (Zaleta *et al.*, 1993; Urquhart, 1993, 1994; Däschner *et al.*, 1995).

A comparison of laser and e-beam direct-write lithography has been discussed by Haruna *et al.* (1990). Early work on e-beam writing was carried out using custom writing systems. More recent work has used commercial systems developed for semiconductor binary mask fabrication, with modified software and resist processing to realize the continuous-relief profiles (Kley *et al.*, 1992, 1993). The fabrication using commercial systems benefits considerably from the long-term technology development in this area of semiconductor device production. On the other hand, such commercial e-beam writing systems are relatively expensive (~3–5 $million), typically more than three times the cost of laser writing systems for binary mask fabrication which could, in principle, also be modified for continuous-relief microstructure fabrication.

Electron-beam writing requires exposure under vacuum and an additional conducting film under the resist layer to avoid charging effects. Whereas the minimum spot size for laser writing is in practice about 500 nm, the electron beam can readily be focused to spot sizes under 100 nm, in some systems down to about 2 nm. The basic resolution capability of e-beam writing is thus, in principle, considerably better than that of laser beam writing. However, the broadening of the electron beam spot due to scattering of electrons within the resist film (proximity effect) results in a similar minimum feature size for both approaches in practice, when applied to the fabrication of continuous-relief structures in the range of 1–5 μm relief heights. This minimum feature size is currently about 3–5 μm for typical Fresnel lenslet structures, although further developments in e-beam writing (well-adapted energies, new resist materials, etc.) may lead to significant improvements in the future. For shallow microstructures, with relief depth $\ll 1\,\mu$m, the use of low energy e-beams together with thin resist films (~50 nm) can already produce nanostructures with less than 50 nm feature size.

4.3.1 *System Design and Resist Processing*

Most of the published work on the e-beam writing of continuous-relief micro-optical structures has been carried out on commercial e-beam systems developed

	Element[1]	Application	Size[2]	NA	Focal length	Micrograph[3]	Reference
(a)	PMFE	Optical microsystem	100 μm	0.1	500 μm	AFM	Gale et al., 1994a
(b)	PMFE (anamorphic)	Laser to fibre coupler	250 μm × 300 μm	~0.5[4]		SEM	Rossi et al., 1995
(c)	Lenslet array	Wavefront testing	500 μm	~0.01	22.5 mm	SEM	Gale et al., 1994a
(d)	PMFE array		150 μm	~0.3	250 μm	SEM	Gale et al., 1994a
(e)	Kinoform	9 × 9 fan-out	400 μm			AFM	Gale et al., 1994a
(f)	PMFE	5 × 1 focusing fan-out	100 μm		10 mm	AFM	Rossi and Kunz, 1994

Key
1. PMFE: Phase-matched Fresnel element
2. Diameter or square side, single element for arrays and kinoforms
3. SEM: scanning electron microscope; AFM: atomic force microscope
4. Object side in designed optical microsystem

Figure 4.10 Examples of micro-optical elements fabricated using the PSI laser writing system (Gale *et al.*, 1994a):

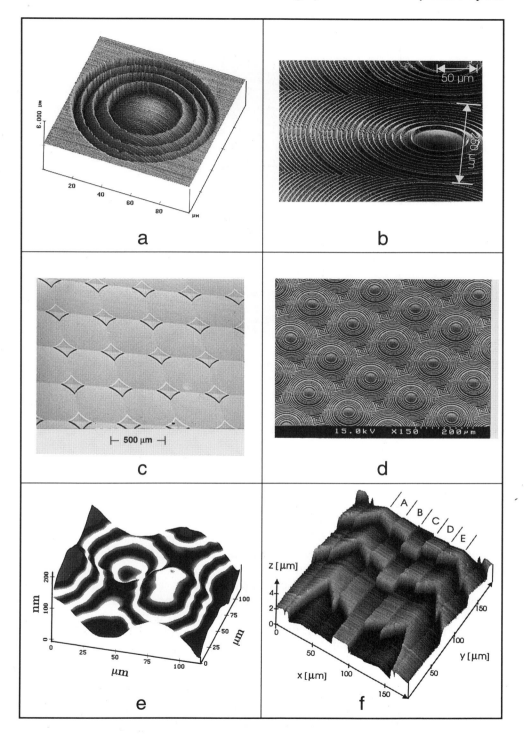

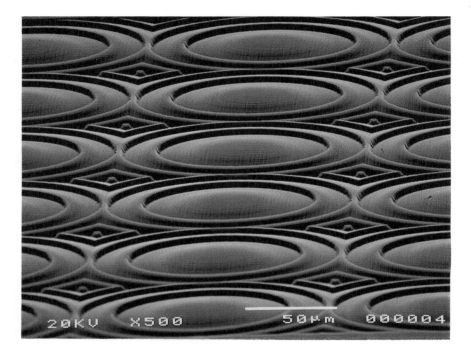

Figure 4.11 SEM of laser-written microlens array (courtesy of Rochester Photonics Corporation, Rochester, USA; for further details, see Bowen *et al.*, 1994).

for semiconductor mask writing, with modifications made in the software and resist processing to enable the fabrication of continuous microreliefs. Whereas standard mask writing typically uses resist thicknesses of about $0.5\,\mu$m, micro-optical elements require significantly deeper relief profiles and thus 'non-standard' resist preparation and processing. Figure 4.12 shows the basic construction of a typical e-beam writing system, consisting of an electron source, electron optics for beam shaping, deflection and focusing, and the resist-coated blank mounted on a high precision xy-stage. A more detailed description of e-beam lithography systems and technology can be found in papers by Kaschlik and Kley (1992), Kley *et al.* (1992, 1993), Verheijen (1993) and Zaleta *et al.* (1993).

There are two principal types of e-beam writers, employing focused and variable-shaped electron beams respectively. In both approaches, the deflection of the electron beam is limited to a few millimetres, so that the writing of most micro-optical elements requires the stitching of multiple scan-fields. Modern interferometer stage control enables stitching with an accuracy of up to about 5 nm.

- Focused beam writers have evolved from scanning electron microscopes. The writing spot is a demagnified image of the electron source and generally has a Gaussian current density profile. The spot diameters can be down to about 1 nm, so that this approach is highly suited to the fabrication of very high-resolution structures as well as curved line features. The disadvantage of the approach is that the exposure is carried out serially in a raster scan and

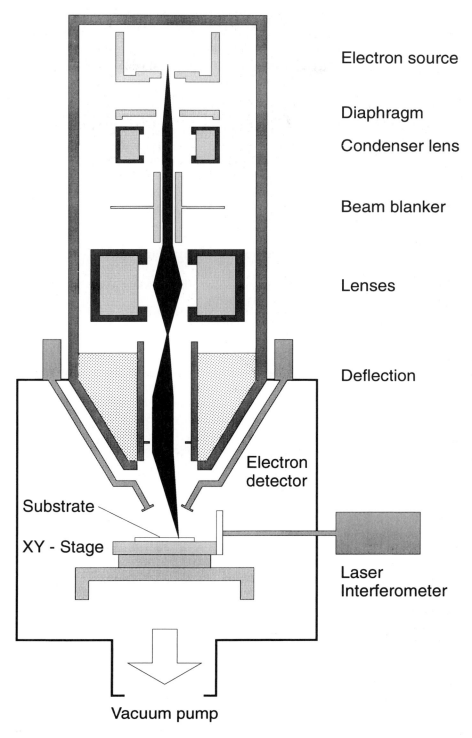

Electron source

Diaphragm

Condenser lens

Beam blanker

Lenses

Deflection

Electron detector

Substrate

XY - Stage

Laser Interferometer

Vacuum pump

Figure 4.12 Schematic of an electron beam writing system, showing the basic components in a typical system.

large uniform areas are still exposed pixel by pixel. The writing speed is thus limited and decreases quadratically with spot size. Vector scan machines reduce the effective writing time by scanning the electron beam only over those areas requiring exposure; this can result in considerable time savings in binary mask fabrication, but is generally not effective for micro-optical elements such as Fresnel lenslets where the whole area must be exposed.

- Variable shaped beam writers have significantly higher writing speeds. The exposure is carried out pulsewise using a demagnified image of a (generally) rectangular aperture with a size that can be changed dynamically. The spot size in the image plane can typically be varied between about 0.1 and 5 μm, with an essentially constant current density cross-section. Using this approach, the writing spot size can be continuously optimized for the local structure element, resulting in significantly faster writing speeds for most structures. The technique is less suitable when curved line structures with very low edge raggedness are required – such structures must be broken down into very small rectangles and the parallel pixel exposure advantage is lost.

The main features and considerations for the e-beam writing of continuous-relief microstructures can be summarized as follows:

Resist

A variety of e-beam resists have been used for micro-optics work; mostly positive resists (increasing development rate with increasing exposure) such as PMMA (Haruna *et al.*, 1990; Stauffer *et al.*, 1992; Larsson *et al.*, 1993), CMS (Haruna *et al.*, 1990), ORWO SK 6000 (Kaschlik and Kley, 1992; Kley and Dorl, 1992), Hoechst AZ PF 514 (Kaschlik and Kley, 1992), Toray EBR-9 (Urquhart *et al.*, 1993) and Shipley SAL-110 (Larsson *et al.*, 1993, 1994; Ekberg *et al.*, 1994). Typical resist sensitivities correspond to a required maximum exposure in the range 10–500 μC cm^{-2} for electron energies of about 20 kV. In order to avoid charging effects, the resist films are typically either spin coated onto conductive substrates such as metal or In$_2$O$_3$ coated glass or themselves coated with a thin conductive film such as Au or CrNi.

Micro-optical elements with typical relief profile depths of up to 5 μm require non-standard resist film thicknesses (semiconductor mask lithography uses resist films of about 0.5 μm thickness). This introduces a number of additional problems. The penetration depth of the e-beam in the resist is dependent upon the electron energy and is about 8 μm for typical 20 kV systems.

Figure 4.13 shows typical development characteristics measured for the Shipley SAL-110 e-beam resist by Larsson and co-workers (1994). In this work continuous-relief resist profiles of up to 2.5 μm were realized with acceptable development characteristics ($\gamma \approx 1$) for reproducible experimental fabrication. The same work showed that successive, stepwise development of the resist does not lead to degradation in the developed microrelief, and thus that post-development of exposed and developed films can be used to improve the absolute depth accuracy of the microrelief structure. Measurements of the development characteristics for a PMMA resist exposed by 20 and 50 kV electrons have been published by Stauffer *et al.* (1992).

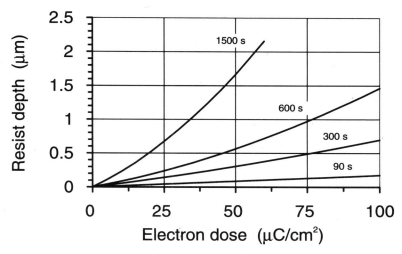

Figure 4.13 Development characteristics (resist depth as a function of electron dose) for the Shipley SAL 110 e-beam resist (from Larsson, 1994). For a given electron dose, the developed resist depth was measured to be a linear function of the development time.

Effective Spot Size

The lateral resolution and minimum feature size for e-beam writing is also dependent upon the e-beam point spread function in the resist. This depends upon a number of factors, including the electron energy, resist film thickness and the resist and substrate materials. It is a superposition of functions resulting from the e-beam focusing, scattering in the resist and back-scattering from the substrate. The widening of the point spread function by scattering contributions is usually referred to as the proximity effect, and results in an effective resolution significantly lower than that corresponding to the focused spot size. The effective spot size D_e can be characterized as the FWHM (full width half maximum) value of the point spread function.

For typical high-resolution e-beam writing systems, D_e is in the range of 50–100 nm. Ekberg and co-workers (1994) have studied proximity effects for the e-beam writing of continuous-relief micro-optical structures such as blazed diffraction gratings. They showed that point spread function can be approximated as a Lorentzian with increasing width as the resist thickness increases. Figure 4.14 shows the estimated dependence for 25 kV electron exposure of the SAL-110 resist. It can be seen that the effective spot size increases from 200 nm for thin films to over 2 μm for films approaching 1 μm in thickness. Similar results were obtained by Haruna *et al.* (1990).

The proximity effect can be reduced by a number of techniques including modelling and compensation in the exposure algorithms (Larsson *et al.*, 1993; Bengtsson, 1994; Ekberg *et al.*, 1994; Nikolajeff, 1995), increasing the writing energy (e.g. from 25 keV to 50 keV) to reduce the spot size (Nikolajeff, 1995) or, conversely, decreasing the energy (e.g. to <5 keV) to reduce the electron penetration (Kley *et al.*, 1994; Brünger *et al.*, 1995). A promising approach also is the use of variable energy exposure, in which structure elements of different

109

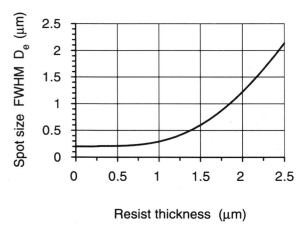

Figure 4.14 Electron beam spot diameter (point spread function, FWHM for 25 keV electrons) as a function of resist thickness for the Shipley SAL-110 positive electron resist (from Ekberg *et al.*, 1994).

relief depth are exposed by an electron beam energy optimized for the corresponding penetration depth (Kley *et al.*, 1994; Brünger *et al.*, 1995).

In summary, these results show that at the present status of e-beam writing of deep (1–5 μm), continuous-relief micro-optical elements, the effective spot size D_e and thus the lateral resolution realized in practice is not significantly better than that for laser writing and is typically about 1 μm. (This does not apply to binary microstructures fabricated in thin resist films and transferred into the underlying substrate by etching.) Techniques such as variable energy exposure have shown promising results and may lead to a significant decrease in the achievable feature size in the future; however, they are not yet fully proven or routine.

Scanning and Spot Shape

The fabrication of continuous-relief microstructures requires the local variation of the net exposure ('variable dose'). This can be achieved in a single pass writing mode by varying the intensity of the writing spot or by varying the dwell time at every pixel. Alternatively, a multiple scan approach, writing over the same area many times to build up a grey-scale exposure, can be used, although this requires precise ion registration of the individual scans (in principle not a major problem in modern systems with overlay accuracy of better than 200 nm) and generally results in a longer writing time.

All commercial e-beam writing systems use a combination of beam deflection and precise substrate translation to cover significant areas. Typical scan field sizes for the electron beam lie in the range 500 μm to 3 mm, depending upon the system and writing spot size. Movement of the substrate xy-table under interferometric control (2.5 or 5 nm resolution) enables adjacent scan fields to be written with better than 100 nm overlay accuracy over areas in excess of 160 mm × 160 mm.

Writing time for scanning a focused spot over the required area can be significant (many hours). Micro-optical structures generally do not benefit from vector scan approaches (instead of an xy raster scan) as commonly used for semiconductor mask fabrication, since typically 100% of the area must be exposed. The writing time can, however, be significantly reduced by employing variable shaped beam techniques, in which the writing spot shape and size is dynamically adjusted to be optimum for the local microrelief pattern. Areas of slowly varying relief or characteristic shape (rectangular features, diagonal lines, etc.) can be written with a shaped spot of suitable size and shape. Kley and Dorl (1992) have described work using a JENOPTIK ZBA 31 electron beam writing system in which a rectangular spot of variable size in the 0.1–6.3 μm range is used to reduce typical writing times by up to an order of magnitude. The fabrication of long, curved structures can also be handled by using continuous path control, in which a constant velocity table scan is combined with beam deflection to compensate for positioning errors and to introduce curvature (Brünger *et al.*, 1995).

4.3.2 Results and Examples

Direct electron-beam writing has been used for the fabrication of a variety of continuous-relief micro-optical structures and elements. Ekberg and co-workers at the Chalmers University of Technology (Göteborg, Sweden) have fabricated 4 μm period transmission blazed grating structures with 67% efficiency using proximity-compensated e-beam writing (Ekberg *et al.*, 1994). Work at OMRON (Kyoto, Japan) has also demonstrated blazed gratings with 2 μm periodicity, as well as Fresnel microlenses with $NA = 2.5$ and 65% focusing efficiency (Aoyoma *et al.*, 1990). The writing of blazed gratings for waveguide input grating couplers has been described by researchers at CSEM (Neuchâtel, Switzerland) – blazed grating couplers with submicron periodicities and depths up to 340 nm were fabricated by direct e-beam writing, as well as reflection gratings with 800 nm periodicity and a measured diffraction efficiency of 75% (Stauffer *et al.*, 1992); Fig. 4.15 shows SEM micrographs of fabricated blazed grating profiles.

Kley and co-workers at the Friedrich Schiller University (Jena, Germany) have used the latest generations of e-beam writing systems from JENOPTIK[4] to demonstrate and fabricate a variety of continuous-relief micro-optical structures, including Fresnel microlenses and microlens arrays. Figures 4.16 and 4.17 show surface profile measurements of structures fabricated on the JENOPTIK ZBA 23A electron beam writer. The lenslets shown in Fig. 4.16 are from a 16×16 array for application in a wavefront sensor. The individual lenslets are 350 μm $\times$ 350 μm in size, 1700 nm deep and have a measured deviation from an ideal spherical surface of less than $\pm$10 nm over the whole surface, with a surface roughness of about 3 nm rms. The structures shown in Fig. 4.17 are from beam-shaping lenslet arrays (kinoforms) designed to produce special image geometries in the focal plane (shown inset). Other micro-optical elements fabricated by this group include arrays of superimposed, crossed cylindrical Fresnel lenslets, a novel fabrication approach enabling a large reduction in the data size and writing time for lenslet arrays.

4.4 Alternative Direct-writing Approaches

In this section we give a short review of alternative direct-writing techniques which have been used to fabricate continuous-relief micro-optical structures.

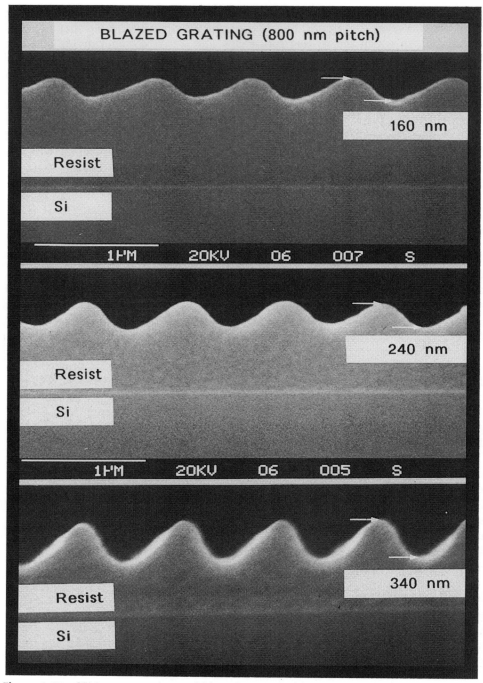

Figure 4.15 SEM cross-sections of blazed gratings fabricated at CSEM by e-beam writing in PMMA (from Stauffer *et al.*, 1992).

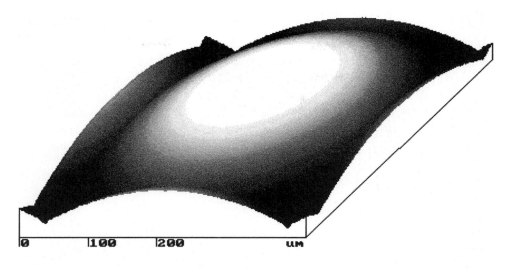

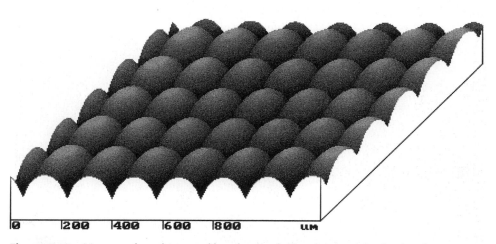

Figure 4.16 Measured surface profile of a single lenslet (top) and a lenslet array (bottom) fabricated by a variable shape electron beam writer (JENOPTIK ZBA 23) for application in a wavefront sensor. The individual lenslets are 350 μm square with a relief amplitude of about 1.7 μm (for further details see Kaschlik and Kley, 1992; Kley *et al.*, 1993).

4.4.1 Focused Ion Beam

Focused ion beam (FIB) writing can be used for directly milling microrelief structures with submicron dimensions into a variety of materials. The basic FIB system is similar to an e-beam writing system (cf. Fig. 4.12) with a vacuum

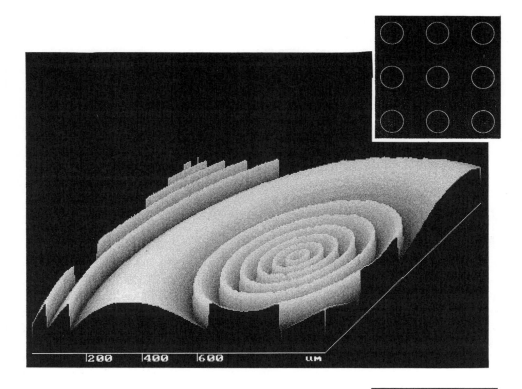

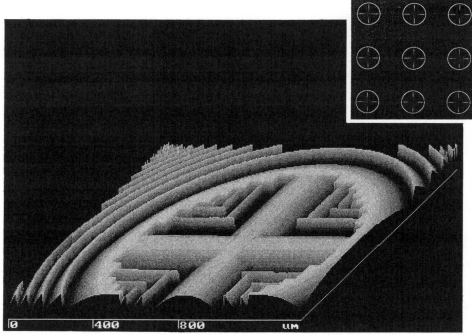

Figure 4.17 Measured surface profile of elements from beam shaping arrays fabricated by Kley and co-workers using a variable shaped electron beam writer (see Fig. 4.16). The reconstructed image in the focal plane is shown inset.

chamber, beam deflection and programmable xy-stage, except that it is equipped with an ion source (typically Ga) with suitable focusing and deflection optics. The focused ion beam has sufficient energy to mill substrates such as glass, quartz, Si or GaAs. The depth to which the substrate material is milled away is related to the ion dose and controlled by the dwell time in a stepwise exposure approach or by the number of passes made over a given area. Continuous-relief microstructures can be fabricated by suitable control of the dose pattern. FIB milling is a direct-writing technique which does not require further processing of the written surface. Most work to date has concentrated on applications in III-V semiconductor devices, for example for cutting laser facets or micromachining integrated optical structures in existing devices.

Harriott and co-workers (1986) at AT&T Bell Laboratories (Murray Hill, USA) have described the FIB milling of microstructures in III-V materials such as InP. The Ga ion beam used had a beam energy of about 20 keV, 160 pA current and was focused to a spot size of 200 nm. Smooth wedges and other continuous-relief microstructures were fabricated by varying dwell time and the number of passes. The removal rate of the InP was about $0.06 \, \mu\text{m}^3 \, \text{sec}^{-1}$, which would correspond to a processing time of about 20 minutes for fabrication of a microlenslet of 10 μm diameter and 1 μm maximum relief depth. The technique is thus relatively slow, but could be speeded up using higher beam currents or by combination with reactive sputtering. It could find applications in the post-processing of surface-relief microstructures to add specific micro-elements at accurately positioned locations. FIB milling of multistep microstructures which approximate to continuous-relief profiles has also been described by Kung and Song (1994) and by Shank and co-workers (1994).

4.4.2 Laser Ablation

UV lasers such as pulsed excimer lasers can be used to directly ablate material from the surface of substances such as polymers and glasses. Energy levels for the ablation of polymer materials are relatively low, typically corresponding to fluences of 0.1 to 1 J cm^{-2}. The energy in the laser spot is less than 100 μJ for typical direct-write applications and can be provided by available small excimer waveguide lasers. A typical writing system has a configuration similar to that required for the laser beam writing in photoresist with a laser, focusing optics and programmable xy-stage (cf. Fig. 4.3), except that the modulator is replaced by a programmable pulse generator for the uv excimer laser. Continuous-relief microstructures can be fabricated, for example, by varying the pulse repetition rate during the scan or by multiple scanning techniques. The effective resolution capability of current laser ablation systems is above 1 μm, limited by the size and roughness of the ablated spots.

The fabrication of continuous-relief microstructures using a KrF waveguide excimer laser ($\lambda = 248$ nm) has been described by Christensen (1994) and by Duignan (1994), both of Potomac Photonics Inc., Lanham, USA, who have demonstrated pyramid structures with a 300 μm base formed in polyimide film. A focused spot size of 10 μm and raster line spacing of down to 2.5 μm was used in this work. The fabrication of shallow, high-resolution gratings (depth ~40 nm, linewidths <100 nm) ablated in polyimide by a holographic reconstruction using a KrF excimer laser has been described by Phillips and Sauerbrey (1993). Work on excimer laser machining at the German Laser-Zentrum Hannover, including a microlens array (300 μm × 300 μm lenslets) machined directly into glass, has been described by Schmidt (1994).

K. Zimmer, of the Insitut für Oberflächenmodifizierung in Leipzig, Germany, has fabricated microlenslets and diffractive microstructures by the ablation of polyimide and AZ 4620 resist using an excimer laser at $\lambda = 248$ nm (Zimmer and Bigl, 1995). The structures were fabricated by the projection of a mask with a contour suitable for the desired topology, for example a semi-circle for a cylindrical topology, and scanning across the substrate surface to generate the required exposure pattern. More complex structures such as microlens arrays were achieved by sequential scanning in orthogonal directions using a special mask, with the same processing parameters for both scans. Figure 4.18 shows the measured surface profiles of a sawtooth microstructure and a microlens array fabricated using this technique.

Laser ablation techniques are still in their infancy and the resolution and fidelity capabilities are marginal for most applications in continuous-relief micro-optics. Their attraction, as in FIB writing, lies in the direct-writing aspects with no further processing required. Their usefulness in the future will depend upon how the technology progresses.

4.4.3 *Laser-assisted Deposition and Etching*

Continuous-relief microstructures can be fabricated by the combination of laser writing and deposition or etching techniques. In this approach, the focused laser beam results in a local enhancement of the deposition or etching rate, and suitable control of the scanning laser beam intensity can produce microrelief structures with submicron resolution. Most techniques of this type require laser scanning in a liquid or gas cell in which the deposition or etching of the substrate takes place.

Jerominek and Pan (1994), at the National Optics Institute, Quebec, Canada, have developed and demonstrated a number of implementations of the basic technique using a vector scan laser writing system integrated with a photochemical reactor cell. Surface-relief microstructures were fabricated on Si substrates by the pyrolithic deposition of polysilicon from a mixture of N_2 and SiH_4. The laser used was an Argon laser operating at $\lambda = 515$ nm and delivering in excess of 1 W power in the focused laser spot of about 2 μm diameter. Fine grating structures and individual lines with a cylindrical profile were written on Si and on amorphous Si films on quartz and glass substrates. Improved control of the surface-relief profile was obtained using laser-enhanced etching. Micro-optical elements such as cylindrical lenslets and blazed grating structures were fabricated by laser writing in a Cl_2 atmosphere.

Laser-assisted etching or deposition produces surface-relief microstructures and requires no further processing. It may find applications in the fabrication of micro-optical elements on processed substrates such as IR CCD imagers or integrated optical circuits.

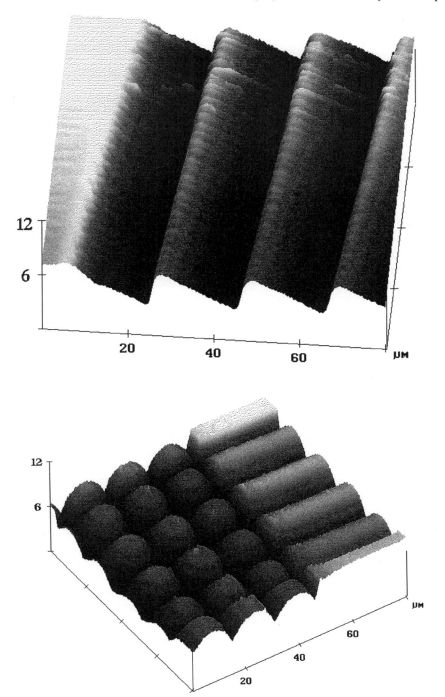

Figure 4.18 Sawtooth and microlens array structures fabricated by direct laser ablation using a scanning mask projection technique (Institut für Oberflächenmodifizierung, Leipzig, Germany; for further details see Zimmer and Bigl, 1995).

4.4.4 Diamond Turning

The fabrication of surface-relief microstructures by diamond turning can be considered to be a direct-write technique in the sense that the microstructure relief is programmable within the limits of machinable profiles. The diamond turning of precision optics is basically a programmable lathe machining technique, using a machine and cutting tool optimized for the fabrication of optical surfaces with minimal roughness and surface damage (McClure, 1991). The lathe turning is limited to the fabrication of rotationally symmetric surfaces, including aspheric and Fresnel microlenses and other types of kinoform. The desired surface is produced directly in the workpiece material and can be electroformed to replication shims for replication into other materials. Best optical surfaces are produced by machining metals such as copper or aluminium (Sweeney and Sommargren, 1995) for which surface roughness of under 1 nm rms can be realized. Glass is too brittle to be machined directly, but good results have been obtained by machining plastics such as acrylics (Hatakoshi *et al.*, 1990; Futhey and Fleming, 1992). The diamond tool can be fabricated in various shapes and sizes, but the tool ultimately imposes limitations on the surface profiles which can be produced. Smoothly varying profiles such as low NA microlenses can be fabricated with very high quality, but micro-optical elements such as high NA Fresnel microlenses with curved blaze profiles represent a major challenge.

Work at the 3M Optics Technology Center (Petaluma, CA, USA) has demonstrated the fabrication of continuous-relief, blazed kinoforms by diamond turning in cast acrylic layers on glass substrates (Futhey and Fleming, 1992). The diamond turning limitations are minimized by the fabrication of a Fresnel lenslet with zones of varying phase steps across the surface radius (referred to as superzone kinoforms). Each zone has a linear blaze profile and the blaze angle in different zones is achieved by changing the diamond tool angle for each zone type. Microlenses of 25 mm diameter and *f*/4 aperture were fabricated with up to 211 zones. The measured performance with a Strehl ratio of 0.8 was somewhat lower than the theoretical prediction (0.997), which was attributed to increased problems in diamond turning the acrylic material compared to copper. Diamond turning in aluminium has been used at the Lawrence Livermore Laboratory (Livermore, USA) to fabricate the master for a 34.75-diopter, 6 mm diameter acrylic lens with a 23 μm relief profile (Sweeney and Sommargren, 1995).

The diamond turning approach can give excellent results in certain cases, and is particularly suited to low NA, rotationally symmetric micro-optical elements. The inherent limitations in the steepness and profile of realizable microstructures, together with requirements for rotational symmetry, restrict its usefulness to a limited group of micro-optical structures.

4.4.5 Half-tone Mask Techniques

Although not strictly a direct-writing technique, the use of half-tone masks for the grey-scale exposure of a photoresist film to obtain continuous-relief microstructures will be briefly described here for completeness. In this approach, the photoresist film is exposed by imaging a grey-tone mask using an optical system such as that in a projection mask aligner. The binary half-tone mask, in which grey-scale is represented by small dots of varying size or density (cf. half-tone printing techniques) can be fabricated by e-beam or laser beam writing. The optical imaging system must effectively incorporate a spatial filter to remove the half-tone dot pattern, so that the image on the photoresist film is a grey-scale exposure which results in a continuous-relief microstructure after development of the photoresist (cf. Section 4.2.2). The removal of the dot pattern by spatial filtering can be achieved by suitable choice of dot frequency and imaging system modulation transfer function (MTF), either by choosing the dot frequency for a given aligner imaging system or by modification of the aligner optics. The half-tone technique allows continuous-relief micro-optics to be fabricated using commercial semiconductor processing equipment for mask lithography and printing.

Early work on the fabrication of continuous-relief micro-optical profiles by the spatial filtering of half-tone images was carried out by Poleshchuk (1989, 1992). The fabrication of Fresnel microlenses and other DOEs using commercial lithographic equipment has been described by Oppliger and co-workers of CSEM, Neuchâtel, Switzerland (Oppliger *et al.*, 1994) and by Gal of the Lockheed Company, Palo Alto, USA (Gal, 1993).

A description of the steps involved in the half-tone fabrication technique is given in the paper by Oppliger *et al.* (1994). Eight-level half-tone masks were generated as 5X reticles using both e-beam and laser beam pattern generators and imaged onto photoresist-coated fused silica substrates using a commercial 5X g-line stepper. After development, the resulting continuous-relief DOE microstructures were transferred into the substrate by proportional etching (plasma etching with a selectivity of about 1:1). Blazed gratings with periods of 8 and 16 μm, and Fresnel microlenses with 0.5 mm diameter and 1.5 mm focal length (NA $\simeq$ 0.17 with a minimum zone width of 4 μm) were fabricated using eight-level half-tone masks. Figure 4.19 shows a section of a Fresnel lenslet fabricated in this way.

Half-tone mask techniques have been shown to produce very good results for micro-optical structures with feature sizes down to about 4 μm. Direct laser or e-beam writing should be capable of ultimately achieving finer features and steeper relief slopes than half-tone imaging techniques, for which the imaging MTF must always be very low at the dot frequency. On the other hand, the use of commercial lithographic equipment represents a significant advantage for commercial production, at least until commercial systems become available for the fabrication of continuous-relief microstructures by laser and e-beam writing.

4.5 Conclusions

This chapter has reviewed techniques for the fabrication of continuous-relief micro-optical structures by direct writing. The technologies are at an early stage of evolution – virtually all of the techniques described must be considered as being at the laboratory stage. No data standards (such as GDSII for binary mask fabrication) exist yet for designing continuous-relief micro-optical elements and manipulating the large data files required for describing continuous-relief microstructures.

Table 4.3 shows a summary and comparison of the different approaches. Laser writing and e-beam writing have been shown to be capable of fabricating micro-optical elements of sufficient performance for real applications. The resist

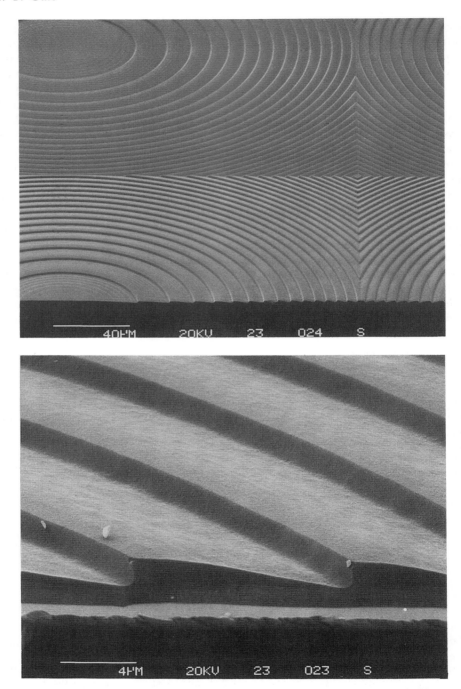

Figure 4.19 Section of a Fresnel lenslet (SEM micrograph) fabricated by half-tone imaging (CSEM, Neuchâtel, Switzerland).

microstructures can usually be evaluated directly in test systems, but are unsuitable for use in commercial systems due to poor mechanical and chemical properties. Replication technology (Chapter 6) enables mass-production of micro-optical elements from a single recording, providing a route from laboratory recordings to industrial usage. Alternatively, the resist surface-relief structure can be transferred into the underlying substrate by proportional etching (Oppliger *et al.*, 1994; Däschner *et al.*, 1995), enabling robust micro-optical elements to be fabricated from individual recordings. Electron-beam writing uses commercially available equipment with modified data handling and control software. Laser writing developments have been achieved mainly using custom-built hardware. With further developments in hardware and process technology, they can be expected to become routine commercial technologies for the fabrication of a wide range of micro-optical elements.

Batch processes for the half-tone mask technique are already available commercially (for example, from CSEM, Neuchâtel, Switzerland), producing micro-optical elements in quartz wafers. Diamond turning is also an established commercial technology for producing a limited range of micro-optical elements. The other techniques described in this chapter are at an early stage of development – their success will depend upon progress and results.

Acknowledgements

Numerous inputs from research groups whose work is described in this chapter are gratefully acknowledged, in particular from J. Bowen and G. M. Morris (Rochester Photonics Corporation, Rochester, USA), E.-B. Kley (Friedrich Schiller University, Jena, Germany), S. H. Lee (University of California, San Diego, USA), J. M. Mayor, Y. Oppliger and P. Sixt (CSEM Centre Suisse d'Electronique et de Microtechnique SA, Neuchâtel, Switzerland), F. Nikolajeff (Chalmers University of Technology, Göteborg, Sweden) and K. Zimmer (Insitut für Oberflächenmodifizierung, Leipzig, Germany). Particular thanks also go to H. P. Herzig, R. E. Kunz and Th. Hessler for valuable comments on the manuscript and to J. Pedersen for help in preparing many of the figures.

The work at the Paul Scherrer Institute PSI, Zurich would not have been possible without the skilled and dedicated contributions of the members of the Optics and other groups in the laboratory; the PSI work was supported by the Swiss Priority Program OPTIQUE.

Notes

1. Commercial laser lithographic systems are available, for example, from Heidelberg Instruments GmbH (Heidelberg, Germany) and Micronic Laser Systems (Taby, Sweden).
2. A rotational laser plotter is available, for example, from the Laser Technology Laboratory, Institute of Automation and Electrometry, Russian Academy of Sciences, Novosibirsk, Russia.
3. Product of Shipley Co. Inc., Newton, MA, USA.
4. JENOPTIK Technologie GmbH, D-07739 Jena, Germany.

Table 4.3 Overview and comparison of direct-writing techniques for the fabrication of continuous-relief micro-optical elements

Direct-write technique	Basic approach	Material(s)	Lateral resolution: Typical minimum feature size (spot diameter)	Typical maximum depth	Comments	References
Laser writing	Scanning, focused laser beam (e.g. HeCd, $\lambda = 442$ nm).	Photoresist	5 μm (~1 μm)	5 μm	Resist microstructure.* Custom hardware.	Gale et al., 1994a Bowen et al., 1994 Akkepeddi et al., 1993 Langlois et al., 1992 Haruna et al., 1990
Electron beam writing	Scanning e-beam: Focused beam or variable shaped beam.	Electron-resist	5 μm (~1 μm effective)	2–5 μm	Resist microstructure.* Commercial hardware.	Kley et al., 1993 Zaleta et al., 1993 Stauffer et al., 1992 Ekberg et al., 1992
Focused ion beam writing	Scanning, focused ion (e.g. Ga) beam.	Glass, quartz, PMMA, Si, InP, GaAs, etc.	Not available [25 . . . 200 nm]	1 μm	Initial work – few published results for continuous-relief microstructures.	Kung and Song, 1994 Shank et al., 1994 Harriot et al., 1986

Technique	Method	Material	Resolution	Depth/height	Comments	References
Laser ablation	Scanning, focused spot or moving mask. Excimer laser (e.g. KrF at λ = 248 nm).	Polyimide, resist, glass	~30 μm (~10 μm)	10 μm	Initial work – few published results for continuous-relief microstructures.	Zimmer and Bigl, 1995; Christensen, 1994; Schmidt, 1994; Phillips and Sauerbrey, 1993
Laser-assisted deposition and etching	Scanning, focused laser beam (e.g. Ar ion laser at λ = 515 nm) in gas cell.	Si	~10 μm (~2 μm)	~10 μm	Initial work – few published results for continuous-relief microstructures.	Jerominek and Pan, 1994
Diamond turning	Precision lathe machining with diamond tool.	Cu, Al, acrylic, plastic	~20 μm (given by tool tip)	~20 μm (deeper possible in some cases)	Established, commercial technology for optical surface fabrication. Applications limited by tool tip size and rotational symmetry.	Sweeney and Sommargren, 1995; Futhey and Fleming, 1992; McClure, 1991; Hatakoshi et al., 1990
Half-tone mask imaging	Controlled imaging of half-tone mask onto photoresist.	Photoresist	4 μm	~2 μm	Commercially available process (not strictly direct write). Commercial hardware. Resist microstructure.*	Oppliger et al., 1994; Gal, 1993; Poleshchuk, 1992

*Resist microstructures are generally suitable only for characterization and testing – they must be transferred into other materials for use by: Electroforming and replication (see Chapter 6) or Proportional transfer by dry etching (see, for example, Oppliger et al., 1994; Däschner et al., 1995).

References

AKKAPEDDI, P., GRATRIX, E. J., LOGUE, J. E. and POWER, M. P. (1993) Review of micro-optics technologies at Hughes Danbury Optical Systems, Inc. In Lee, S. H. (ed.) Diffractive and Miniaturized Optics, Vol. CR49, pp. 98–116. Bellingham: SPIE.

AOYAMA, S., HORIE, N. and YAMASHITA, T. (1990) Micro Fresnel lens fabricated by electron-beam lithography. *Proc. SPIE* **1211**, 175–183.

BENGTSSON, J. (1994) Direct inclusion of the proximity effect in the calculation of kinoforms. *Appl. Opt.* **33**, 4993–4996.

BOWEN, J. P., BLOUGH, C. G. and WONG, V. (1994) Fabrication of optical surfaces by laser pattern generation. In *OSA Technical Digest Series: Optical Fabrication and Testing*, Vol. 13, pp. 153–156. Washington, D.C.: OSA.

BRÜNGER, W., KLEY, E.-B., SCHNABEL, B., STOLLBERG, I., ZIERBOCK, M. and PLONTKE, R. (1995) Low exposure lithography; Energy control and variable energy exposure. *Microelectron. Eng.* **27**, 136–138.

CHRISTENSEN, C. P. (1994) Waveguide excimer laser fabrication of 3D microstructures. *Proc. SPIE* **2045**, 141–145.

DALY, D., STEVENS, R. F., HUTLEY, M. C. and DAVIES, N. (1990) The manufacture of microlenses by melting photoresist. *Meas. Sci. Techn.* **1**, 759–766.

DÄSCHNER, W., LAARSON, M. and LEE, S. H. (1995) Fabrication of monolithic diffractive optical elements by the use of e-beam direct write on an analog resist and a single chemically assisted ion-beam-etching step. *Appl. Opt.* **34**, 2534–2539.

DUIGNAN, M. T. (1994) Micromachining of diffractive optics with excimer lasers. In *OSA Technical Digest Series: Diffractive Optics*, Vol. 11, pp. 129–132. Washington, D.C.: OSA.

EHBETS, P., HERZIG, H. P., PRONGUÉ, D. and GALE, M. T. (1992) High-efficiency continuous surface-relief gratings for two-dimensional array generation. *Opt. Lett.* **17**, 908–910.

EKBERG, M., LARSSON, M., HÅRD, S. and NILSSON, B. (1990) Multilevel phase holograms manufactured by electron-beam lithography. *Opt. Lett.* **15**, 568–569.

EKBERG, M., NIKOLAJEFF, F., LARSSON, M. and HÅRD, S. (1994) Proximity-compensated blazed transmission grating manufacture with direct-writing, e-beam lithography. *Appl. Opt.* **33**, 103–107.

EKBERG, M., LARSSON, M., HÅRD, S., TURUNEN, J., TAGHIZADEH, M. R., WESTER-HOLM, J. and VASRA, A. (1992) Multilevel grating array illuminators manufactured by electron-beam lithography. *Opt. Commun.* **88**, 37–41.

FAKLIS, D., BOWEN, J. P. and MORRIS, G. M. (1993) Continuous phase diffractive optics using laser pattern generation. *SPIE Holography Newsletter* **3**(2), 1.

FUJITA, T., NISHIHARA, H. and KOYAMA, J. (1982) Blazed gratings and Fresnel lenses fabricated by electron-beam lithography. *Opt. Lett.* **7**, 578–580.

FUTHEY, J. and FLEMING, M. (1992) Superzone diffractive lenses. In *OSA Technical Digest Series: Diffractive Optics: Design, Fabrication and Applications*, Vol. 9, pp. 4–6, Washington, D.C.: OSA.

GAL, G. (1993) Micro-optics technology and development for advanced sensors. In Lee, S. H. (ed.) Diffractive and Miniaturized Optics, Vol. CR49, pp. 329–359. Bellingham: SPIE.

GALE, M. T. and KNOP, K. (1983) The fabrication of fine lens arrays by laser beam writing. *Proc. SPIE* **398**, 347–353.

GALE, M. T., LANG, G. K., RAYNOR, J. M. and SCHÜTZ, H. (1991) Fabrication of micro-optical elements by laser beam writing in photoresist. *Proc. SPIE* **1506**, 65–70.

GALE, M. T., LANG, G. K., RAYNOR, J. M., SCHÜTZ, H. and PRONGUÉ, D. (1992) Fabrication of kinoform structures for optical computing. *Appl. Opt.* **31**, 5712–5715.

GALE, M. T., ROSSI, M., PEDERSEN, J. and SCHÜTZ, H. (1994a) Fabrication of continuous-relief micro-optical elements by direct laser writing in photoresist. *Opt. Eng.* **33**, 3556–3566.

GALE, M. T., ROSSI, M., KUNZ, R. E. and BONA, G. L. (1994b) Laser writing and replication of continuous-relief Fresnel microlenses. In *OSA Technical Digest Series: Diffractive Optics*, Vol. 11, pp. 306–309. Washington, D.C.: OSA.

GALE, M. T., ROSSI, M., SCHÜTZ, H., EHBETS, P., HERZIG, H. P. and PRONGUÉ, D. (1993) Continuous-relief diffractive optical elements for two-dimensional array generation. *Appl. Opt.* **32**, 2526–2533.

GOLTSOS, W. and LIU, S. (1990) Polar coordinate writer for binary optics fabrication. *Proc. SPIE* **1211**, 137–147.

GÖTTERT, J., FISCHER, M. and MÜLLER, A. (1995) High aperture surface-relief microlenses fabricated by X-ray lithography and melting. *Proc. EOS Topical Meeting on Microlens Arrays*, Teddington, UK, 11–12 May 1995, pp. 21–25; EOS Topical Meeting Digest Series Volume 5.

HARRIOTT, L. R., SCOTT, R. E., CUMMINGS, K. D. and AMBROSE, A. F. (1986) Micromachining of integrated optical structures, *Applied Physics Letters* **48**, 1704–1706.

HARUNA, M., TAKAHASHI, M., WAKAHAYASHI, K. and NISHIHARA, H. (1990) Laser beam lithographed micro-Fresnel lenses. *Appl. Opt.* **29**, 5120–5126.

HATAKOSHI, G., KAWACHI, M., TERASHIMA, K., UEMATSU, Y., AMANO, A. and UEDO, K. (1990) Grating axicon for collimating Cerenkov radiation waves. *Opt. Lett.* **15**, 1336–1338.

HESSLER, TH., GALE, M. T., ROSSI, M. and KUNZ, R. E. (1995) Fabrication of Fresnel microlens arrays by direct writing in photoresist. In *Proc. EOS Topical Meeting on Microlens Arrays*, Teddington, UK, 11–12 May 1995, pp. 37–43. EOS Topical Meeting Digest Series Volume 5.

IMANAKA, K. (1992) Micro hybrid integrated devices and components: Micro photonic devices. *Proc. SPIE* **1751**, 343–353.

JEROMINEK, H. and PAN, J. (1994) Laser-assisted deposition and etching of silicon for fabrication of refractive and diffractive optical elements. *Proc. SPIE* **2045**, 194–204.

KASCHLIK, K. and KLEY, E. B. (1992) Masken für die Mikrotechnik mit dem Variable-Shaped-Beam Elektronenschreiber. *VDI Berichte* **1012**, 1–28.

KLEY, E.-B. and DORL, W. (1992) Einsatz der Elektronenstrahllithographie zur Herstellung mikrooptischer Bauelement. *VDI Berichte* **1012**, 531–541.

KLEY, E.-B., POSSNER, T. and GÖRING, R. (1993) Realization of micro-optics and integrated optic components by electron-beam-lithographic surface profiling and ion exchange in glass. *Int. J. Optoelectronics* **8**, 513-527.

KLEY, E.-B., THIEME, P., SCHNABEL, B., BÖTTNER, M. and PLONTKE, R. (1994) Elektronenstrahl-lithographische Submikron- und Nanometerstrukturierung mit Energien zwischen 1 und 20 keV. *VDI Berichte* **1172**, 83–94.

KUNG, P. and SONG, L. (1994) Rapid prototyping of multi-level diffractive optical elements. In *OSA Technical Digest Series: Diffractive Optics*, Vol. 11, pp. 133–136. Washington, D.C.: OSA.

KUNZ, R. E. and ROSSI, M. (1993) Phase-matched Fresnel elements. *Opt. Commun.* **97**, 6–9.

LANGLOIS, P., JEROMINEK, H., LECLERC, L. and PAN, J. (1992) Diffractive optical elements fabricated by laser direct writing and other techniques. *Proc. SPIE* **1751**, 2–12.

LARSSON, M., EKBERG, M., NIKOLAJEFF, F. and HÅRD, S. (1994) Successive-development optimization of resist kinoforms manufactured with direct writing, electron-beam lithography. *Appl. Opt.* **33**, 1176–1179.

LARSSON, M., EKBERG, M., NIKOLAJEFF, F., HÅRD, S., MAKER, P. M. and MULLER, R. E. (1993) Proximity-compensated kinoforms directly written by e-beam lithography. In

Lee, S. H. (ed.) Diffractive and Miniaturized Optics, Vol. CR49, pp. 138–161. Bellingham: SPIE.

McClure, E. C. (1991) Manufacturers turn precision optics with diamond. *Laser Focus World*, February, 95–105.

Missig, M. D. and Morris, G. M. (1995) Diffractive optics applied to eyepiece design. *Appl. Opt.* **34**, 2452–2461.

Nikolajeff, F., Bengtsson, J., Larsson, M., Ekberg, M. and Hård, S. (1995) Measuring and modelling the proximity effect in direct-write electron-beam lithography kinoforms. *Appl. Opt.* **34**, 897–901.

Oppliger, Y., Sixt, P., Stauffer, J. M., Mayor, J. M., Regnault, P. and Voirin, G. (1994) One-step 3D shaping using a gray-tone mask for optical and microelectronic applications. *Microelectron. Eng.* **23**, 449–454.

Phillips, H. M. and Sauerbrey, R. A. (1993) Excimer-laser-produced nanostructures in polymers. *Opt. Eng.* **32**, 2424–2436.

Poleshchuk, A. G. (1989) Fabrication of high efficiency elements for diffractive and integrated optics by photorastered technology. In *Proc. 5th National Conference on Optics and Laser Engineering*, Varia, Bulgaria, 18–20 May 1989, pp. 7–8.

——— (1992) Methods for diffractive elements surface profile fabrication. In *OSA Technical Digest Series: Diffractive Optics: Design, Fabrication and Applications*, Vol. 9, pp. 117–119. Washington, D.C.: OSA.

Rossi, M. and Kunz, R. E. (1994) Focusing fan-out elements based on phase-matched Fresnel lenses. *Opt. Commun.* **112**, 258–264.

Rossi, M., Bona, G. L. and Kunz, R. E. (1995) Arrays of anamorphic phase-matched Fresnel elements for diode-to-fiber coupling. *Appl. Opt.* **34**, 2483–2488.

Schmidt, H. (1994) Automated excimer laser carves microstructures. *OLE Opto & Laser Europe*, September, 35–37.

Shank, S. M., Skvaria, M., Chen, F. T., Craighead, H. C., Cook, P., Bussjager, R., Haas, F. and Honey, D. A. (1994) Fabrication of multi-level phase gratings using focused ion beam milling and electron beam lithography. In *OSA Technical Digest Series: Diffractive Optics*, Vol. 11, pp. 302–305. Washington, D.C.: OSA.

Stalder, M. and Ehbets, P. (1994) Electrically switchable diffractive optical element for image processing. *Opt. Lett.* **19**, 1–3.

Stauffer, J. M., Oppliger, Y., Regnault, P., Baraldi, L. and Gale, M. T. (1992) Electron beam writing of continuous-relief profiles for optical applications. *J. Vac. Soc. Technol.* B **10**(6), 2526–2529.

Sweeney, D. W. and Sommargren, G. (1995) Harmonic diffractive lenses. *Appl. Opt.* **34**, 2469–2475.

Urquhart, K. S., Stein, R. and Lee, S. H. (1993) Computer-generated holograms fabricated by direct-write of positive electron-beam resist. *Opt. Lett.* **18**, 308–310.

Urquhart, K. S., Marchand, P., Fainman, Y. and Lee, S. H. (1994) Diffractive optics applied to free-space optical interconnects. *Appl. Opt.* **33**, 3670–3682.

van der Gracht, J., Cathey, W. T., Dowski, E. R. and Bowen, J. (1995) Aspheric optical elements for extended depth-of-field imaging. *Proc. SPIE* **2537**, 279–288.

Verheijen, M. J. (1993) E-beam lithography for digital holograms. *J. Mod. Opt.* **40**, 711–721.

Yariv, A. (1976) *Introduction to Optical Electronics*, 2nd edn, pp. 29–38. New York: Holt, Rinehart and Winston.

Zaleta, D., Däschner, W., Larsson, M., Kress, B. C., Fan, J., Urquhart, K. S. and Lee, S. H. (1993) Diffractive optics fabricated by electron-beam direct write methods. In Lee, S. H. (ed.) Diffractive and Miniaturized Optics, Vol. CR49, pp. 117–137. Bellingham: OSA.

Zimmer, K. and Bigl, F. (1995) Microstructuring of surfaces by excimer laser machining. In *Proc. SENSOR '95*, Nürnberg, 9–12 May 1995, pp. 779–782.

5

Refractive Lenslet Arrays

M. C. HUTLEY

5.1 Introduction

The word 'lens' was probably adopted in the seventeenth century when Robert Hooke, Anton van Leeuvenhoek and others made small hemispheres of glass for use as objective glasses in their microscopes. These had the appearance of lentils and *lens* is the latin word for lentil. In many languages the same word is used for both. It is therefore technically superfluous to refer to 'refractive' lenses or lenslets because it is through refraction at the surface that a lentil-shaped piece of material will form an image. Current practice, however, appears to be to call anything that will form an image a 'lens', so that a Fresnel-zone plate is a 'diffracting lens' and some authors have even referred to concave mirrors as 'reflecting lenses'. Nevertheless, it does no harm to remind ourselves that in studying arrays of 'microlenses' or 'refracting lenslets' we are returning to the origin of the word itself.

In many respects it is easier to make microlenses than it is to make full-size lenses. There are techniques available that will readily give an acceptable result over a small aperture even if the process itself is not really under control. This has enabled researchers to make lenses of modest quality with relatively little effort and, by demonstrating the feasibility of many possible applications, has stimulated and justified a more thorough and controlled approach to manufacturing. Even so, it is probably true to say that some of the most widely used techniques produce good results more by optimizing the parameters that one has at one's disposal than fully controlling the process.

For the sake of simplicity we can distinguish two types of microlens: those which refract light at a suitably shaped boundary of a homogeneous material, and those which refract light throughout the bulk of a material in which there are gradients of refractive index. In practice, those making graded-index lenses have found that it is possible to improve the performance by introducing curvature at the surface (Oikawa *et al.*, 1990), and those making surface-relief lenses have discovered benefits in introducing some gradient of index. It may be that in the future hybrid lenses involving graded index, surface refraction and diffraction will provide the best performance, but for the present we shall consider the two forms of refracting lens independently.

A variety of techniques is now available for the manufacture of microlenses. Some have been developed to the stage where they are used routinely and reliably to make lenses of good optical quality and others have been demonstrated in principle to be capable of producing a device that is capable of forming an image, but require considerable refinement before they could be used seriously in production.

The most important methods available at the time of writing are: ion diffusion to produce graded-index lenses, 'thermal reflow' (surface tension), 'photosculpture' and photolithic techniques using photosensitive glass.

5.2 Graded-index Lenses

There are many forms of lens in which the image-forming properties are brought about by a gradual refraction throughout the bulk of the material due to relatively small variations in refractive index. A small variation of refractive index can only bring about a small angular deviation of a ray, but when this effect is integrated over a significant optical path, it is possible to achieve quite high values of numerical aperture, e.g. 0.25 (Oikawa *et al.*, 1990). Although most of the current interest in microlenses is stimulated by the needs of opto-electronics, and is therefore the domain of physicists and engineers, it is worth bearing in mind that many forms of graded-index lens exist in nature. This is particularly true in the eyes of fish and marine animals, where the shape of the eye has relatively little effect because of the small difference between the refractive index of water and that of the material of the eye. For a review of this aspect of the subject the reader is recommended a paper by Land (1978).

There are various techniques for achieving a suitable refractive index distribution (Moore, 1980) of which the most widely used are: selective exchange of ions in a suitable glass substrate; diffusion polymerization in plastics; and chemical vapour deposition. In ion exchange the substrate is immersed in a bath of an appropriate molten salt for up to a week. Metal ions diffuse into the matrix and replace some of the silicon, sodium or potassium, depending upon the type of glass used. The variation of refractive index depends upon the ions that are exchanged and the type of glass in which it takes place. Table 5.1 shows as an example the change of index induced when sodium in soda-lime glass is replaced by various ions. More detailed studies have been published by Findakly (1985) and Ryan-Howard and Moore (1985).

The rate of exchange and the refractive index depend upon the amount of dopant from the salt present at a given point, so that after an appropriate time there will be a gradation of index from the surface into the bulk of the glass. This process can be accelerated, and to some extent controlled, by applying an electric field in the direction of migration (Houde-Walter and Moore, 1985). Similar diffusion effects can be achieved much more rapidly in plastics by immersing a partially polymerized substrate in a bath of a suitable monomer (Iga and Yamamoto, 1977; Koike *et al.*, 1982).

Two particularly interesting forms of lens made in this way are the rod lens (Fig. 5.1) which is widely used in office copiers and in conjunction with optical fibres

Table 5.1 Change of index in soda-lime
glass for various ion exchanges

Exchange	Δn
Na Ag	0.1
Na Ti	0.15
Na Li	0.01
Na K	0.01

(Iga, 1980; Atkinson *et al.*, 1982) and the planar microlens, with a three-dimensional index distribution (Fig. 5.2). As neither of them depend upon refraction at their external surfaces they may be cemented directly onto other components. This is a significant advantage in that it permits the construction of robust and rigid systems and eliminates Fresnel reflection losses at the surfaces. An axial gradient of refractive index can also be used to correct the aberrations of a spherical surface and could in principle be applied to ion-etched microlenses.

The rod lens, which is typically 2–3 mm in diameter, is similar in many respects to a short length of optical fibre, but is capable of transmitting an image over significant distances by imaging and re-imaging in a periodic manner. They are

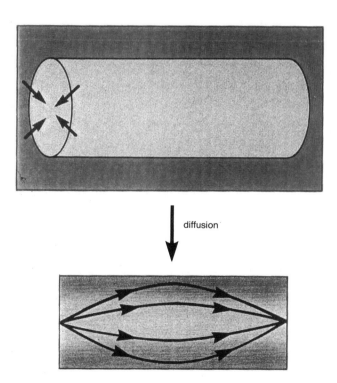

diffusion

Figure 5.1 Ion exchange into a glass rod creates a radial distribution of refractive index producing a graded-index rod lens.

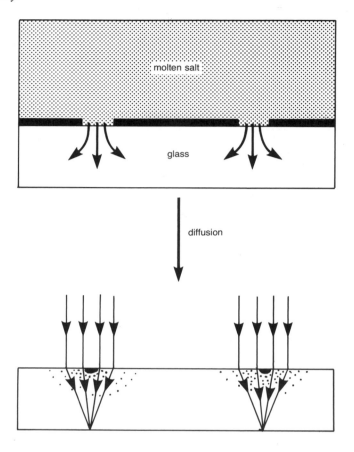

Figure 5.2 Metal ions from a bath of molten salt diffuse through a mask to displace ions of a different metal from the glass. This produces regions with spherical variations of refractive index which behave as small lenses.

most widely available in the form of collimators (quarter of a period long) or for point-to-point imaging (half a period long) which can then be fused or cemented together to form close-packed hexagonal arrays.

Arrays of planar lenses with a three-dimensional distribution of index can be made by selective diffusion of the dopant through a mask such as a layer of titanium which has been evaporated onto the surface and in which there is a series of circular holes (Iga *et al.*, 1974; Oikawa and Iga, 1982). If the size of the holes and the diffusion conditions are carefully chosen, microlenses of good optical quality can be produced with apertures ranging from a few millimetres to 100 microns or less (Hamanaka *et al.*, 1988; Bähr *et al.*, 1994). Figure 5.3 shows an interferogram of a section through a distributed-index microlens and indicates the radial distribution of refractive index centred on the hole through which the diffusion takes place.

A great deal of theoretical and experimental effort has been devoted to modelling the diffusion process and in relating this to the measured properties of the final lens. This work has been very thoroughly reviewed by Iga *et al.* (1984).

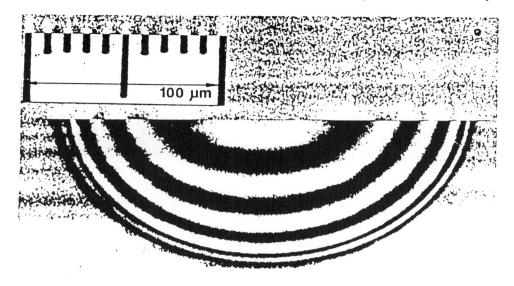

INDEX DISTRIBUTION OF PLANAR MICROLENS

Figure 5.3 Interferogram showing the index variation in a slice of planar microlens (Nippon Sheet Glass Ltd).

For a more detailed summary of the development of gradient-index microlenses the reader is advised to consult the 'GRIN' series of topical issues of *Applied Optics* (1980–1990) and a more recent article by Atkinson *et al.* (1994).

5.3 Photothermal Techniques

Photothermal techniques involve first recording in a photosensitive glass a latent image of the lens array. Exposure to ultraviolet radiation modifies the material in such a way that when it is heated to the softening point, it expands. This then exerts pressure on those areas which were not exposed and forces them into the shape of lenses. This process was developed at the Corning company in the USA (Borrelli *et al.*, 1985) and is shown diagrammatically in Fig. 5.4. Photosensitive glasses were originally developed in order to induce coloration, which was achieved by the absorption of small noble metal colloids such as Ag, Au and Cu. In certain cases it is possible that when the specks of noble metal attain a critical size, they serve as nuclei for the growth of microcrystals and there is a nett increase in density in the exposed material. It is this that, during the thermal cycle, squeezes the soft unexposed glass up out of the surface. Surface tension causes the shape to be spherical.

The height of the glass above the surface is given by

$$\frac{\delta}{T} = \frac{2}{3}\left(1 - \frac{\rho}{\rho_0}\right)$$

(5.1)

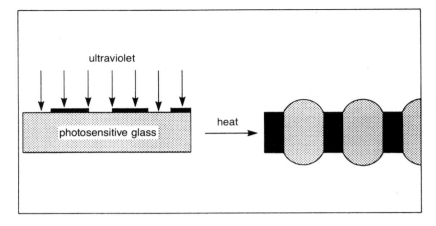

Figure 5.4 Photothermal glass, when exposed to ultraviolet radiation and heated, squeezes the unexposed areas into the shape of small lenses.

where δ is the height of the extruded region, ρ is the density of the unexposed glass, ρ_0 is the density of the crystallized region and T is the thickness of the glass.

From this it can be seen that for a given ratio of densities, the greater the thickness of the glass the shorter will be the focal length of the lens. The process also requires there to be a sufficient interstitial volume to force the unexposed glass into the volume of the lens. In the limit, it would be difficult to make arrays of very closely packed lenses of high numerical aperture on very thin substrates. The manufacturers quote a minimum separation of 15 μm between lenses. For a given pattern of lenses on a substrate of a given thickness, the focal properties are controlled through the ratio of densities which is in turn controlled by the composition of the glass.

One of the great practical advantages of this technique is that in addition to producing the lenses it also renders the interstitial areas opaque. In some applications of lens arrays it is important to attain the maximum possible fill factor, in which case the interstitial regions will be a limitation. In other applications it is the contrast which is more important and in this case this form of lens is particularly appropriate. The fact that the glass is opaque throughout the volume of the interstitial region has the further advantage that crosstalk between light from adjacent lenses is eliminated. In certain display applications this is a very significant advantage.

Kufner and co-workers (1993) have developed a process which bears some resemblance to this, but for making lenses in plastics. They irradiated PMMA through a mask with a beam of protons which caused a reduction of the molecular weight. The material was then immersed in a vapour of the monomer, which resulted in a volume expansion of the irradiated regions and a swelling of the surface to form lenses.

5.4 Thermal Reflow

One of the earliest forms of microlens was developed by Robert Hooke in 1664. He melted in a flame the ends of thin strands of venetian glass, which adopted

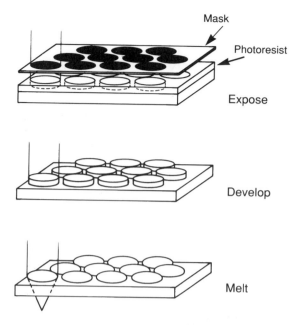

Mask

Photoresist

Expose

Develop

Melt

Figure 5.5 The generation of microlenses by melting small cylinders of photoresist.

a hemispherical form under the action of surface tension. He then mounted these in pitch and polished the backs flat. The same technique can now be applied very conveniently to polymer materials and can be combined with lithography in such a way that it is possible to produce arrays containing a million or more lenses of good optical quality in just a few minutes (Popovic *et al.*, 1988; Daly *et al.*, 1990).

The first stage is to coat a glass substrate with a layer of photoresist. This is then exposed to a pattern of an array of circular masks and developed to form a series of cylindrical islands. These are then heated in an oven or on a hotplate until the resist melts and surface tension draws them into the shape of lenses (Fig. 5.5). These may then be used as they stand or may be processed further. The resist has a deep red colour because it is designed to absorb blue light, and since the layer is quite thick, typically between 5 and 30 microns, the lenses only really transmit at wavelengths greater than 600 nm. In many cases it is therefore preferable to transfer the form of the lens into some other medium either by ion etching into the substrate or by physical replication such as casting, embossing or injection moulding.

In order to calculate the thickness of resist necessary to produce a lens of a given focal length we assume that the lens will have the form of a cap of a sphere and that the volume of the resist remains constant. The paraxial focal length f of a lens consisting of a single spherical surface of radius R in a medium of refractive index n, as shown in Fig. 5.6, is given by

$$f = \frac{R}{n-1} \tag{5.2}$$

133

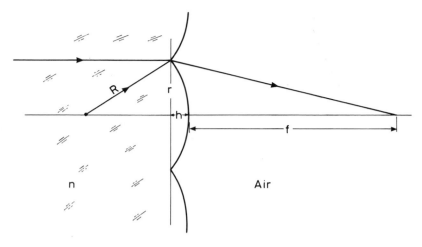

Figure 5.6 Factors determining the focal length of a microlens.

and the height h of the surface undulation of a lens with an aperture radius r

$$h = R - (R^2 - r^2)^{1/2} \qquad (5.3)$$

The volume of a cylinder of resist is $T\pi r^2$ where T is the thickness before melting, and the volume of the lens is $\frac{1}{3}\pi h^2(3R - h)$. It therefore follows that the necessary thickness is given by

$$T = \frac{h}{6}\left(3 + \frac{h^2}{r^2}\right) \qquad (5.4)$$

This leads to thicknesses of resist coatings that are significantly greater than those generally encountered in microlithography and to put this into context we list in Table 5.2 the relevant parameters for typical microlenses based on a refractive index of 1.6. For lenses of a given numerical aperture all dimensions expand linearly with the diameter of the lens and because it is difficult to achieve a good uniform coating of thick resist (e.g. 100 μm) this effectively places an upper limit on the diameter of lens that can be produced in this way.

The melting process works well for lenses of a relatively high numerical aperture where the lens constitutes a significant fraction of a hemisphere. For lenses of long focal length the cylinder of resist is in the form of a thin plate before melting, and rather than forming a spherical surface, tends to sag in the middle. In practice it is found that acceptable lenses can be made with aperture ratios between $f/1$ and $f/3$. This is borne out in Fig. 5.7 which shows the profiles before and after melting of lenses with a variety of thicknesses. This figure also shows that the process is very tolerant of the state of resist before melting. Good-quality lenses can be made with relatively crude lithography although greater consistency and reliability is achieved only if the process is carried out efficiently.

The precise form of the lenses, and hence their focal properties, are determined by the effects of surface tension. In particular the contact angle of the softened resist with the surface of the substrate will strongly influence the shape of the lens. If this angle matches the tangent of the ideal shape one would expect no distortion

Table 5.2 Various parameters for 100 μm diameter lenses made in photoresist of refractive index 1.6. Measurements in micrometres

Focal length	Radius of curvature R	Pole height H	Coating thickness
100 (*f*/1)	60	27	14.8
200	120	10.9	5.3
400	240	5.5	2.6

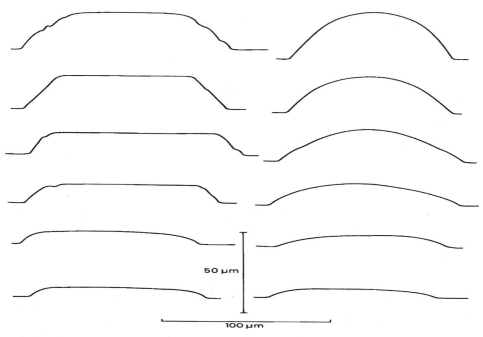

Figure 5.7 The profile of the resist before and after melting.

but if this is not so, spherical aberration will be introduced. The shape of the surface can be described quite well by a third-order polynomial. The coefficients can be determined by assuming constant volume of the material, setting the contact angle as a boundary condition and minimizing the surface area. Calculations based on these assumptions agree qualitatively very well with the observation that thick lenses can be formed relatively easily but thin lenses tend to sag at the centre. However, a full theoretical description will have to take into account the dynamics of the process. It is observed experimentally that not only does the temperature affect the result, but so do the rate and manner of the temperature change, e.g. melting in an oven and on a hotplate can produce markedly different results.

In effect we can say that for a given diameter of lens there will be one value of focal length for which the tangent of the ideal surface matches the contact angle, and the aberrations will be minimized. This is borne out in Fig. 5.8 which shows the measured phase profile of a series of lenses made from resist of different thicknesses. Under optimum conditions the wavefront may be of excellent quality and exceed the Rayleigh quarter wavelength criterion. However, it would be misleading to suggest that this is typical of the performance of all lenses made by this technique. A real application may demand a lens of a focal length and diameter which does not match the ideal conditions.

In order to produce a good-quality lens of a given specification, it is necessary either to exercise some control over the process or modify the lenses after they have been formed. Popovic *et al.* (1988) formed their lenses on preformed pedestals in order to restrict the spreading of the resist, but also used a restricted aperture in the centre of the lens. More recently Göttert *et al.* (1995) have used X-ray lithography in the so-called LIGA technique to generate hemispherical lenses on very thick pedestals (Fig. 5.9). It has also been suggested that the contact angle can be controlled simply by coating the substrate with different materials (Haselbeck *et al.*, 1993) but a systematic study of this approach has yet to be published.

5.5 Photoresist 'Sculpture'

The normal way in which photoresist is used is to expose it to a binary pattern, so that after development, resist is either left intact or it is removed completely. However it is also possible to use it in an 'analogue' mode in which the amount of resist that is removed depends upon the exposure at that point. The response is not usually linear but it can be measured so it is possible in principle to create any desired surface structure by controlling the spatial distribution of exposure. There are various ways in which this technique can be used to generate arrays of microlenses.

The first involves scanning a focused laser beam in raster fashion across the surface of the resist as one would when writing a photomask (Gale and Knop, 1983). In this case, however, the intensity is fully modulated rather than being switched on and off. This has the advantage of extreme flexibility and control over the form of the lenses. It does require an operation equivalent to the writing of a photomask for every exposure but this is not a particular disadvantage if the result is to be used as a master from which large numbers of replicas are to be generated. Since this technique has already been fully described in the previous chapter we shall not consider it further.

An alternative but more restricted approach is to generate square arrays of lenses by crossing two arrays of cylindrical lenses. Cylindrical lenses (which have applications in their own right) can be generated by a blanket exposure to a suitable binary mask which translates across the surface in a controlled manner. The mask may be in the form of a grating of narrow slits, as shown in Fig. 5.10(a), or in the shape of a cross-section of the desired profile, modified to take account of the non-linearity of the resist, and is scanned along the axis of the cylinder, as in Fig. 5.10(b) (Artzner, 1992).

In the first case the mask must be translated in a direction perpendicular to the slits and the required exposure profile is achieved by suitable variation of the speed. If necessary it may then be smoothed by thermal reflow. In the second case the mask is translated at constant speed in the direction parallel to the axis of the cylinder and the correct exposure profile is achieved through the shape of the

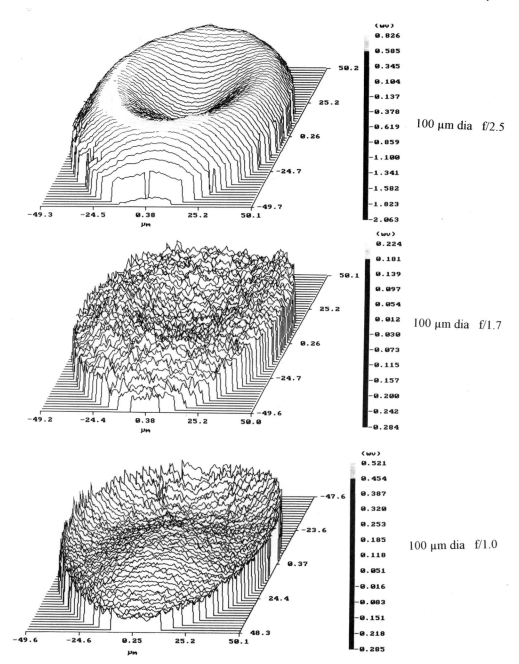

Figure 5.8 Wavefront error for lenses of the same diameter but different focal ratios.

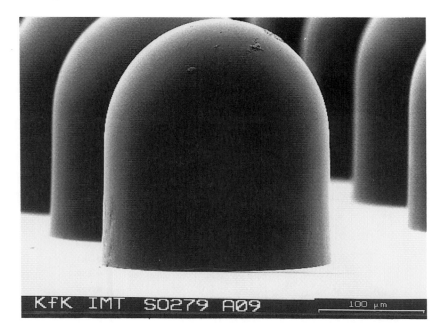

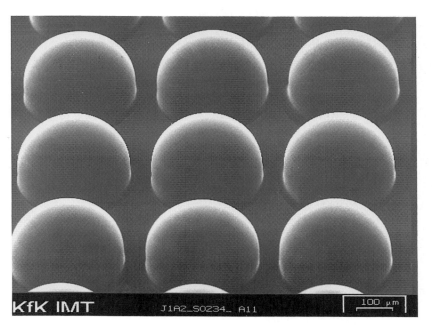

Figure 5.9 Microlenses of very high numerical aperture made using the LIGA technique (Göttert *et al.*, 1995).

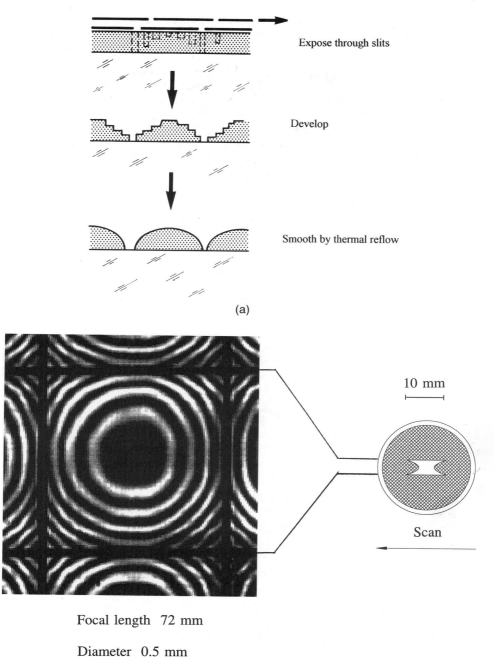

Expose through slits

Develop

Smooth by thermal reflow

(a)

10 mm

Scan

Focal length 72 mm

Diameter 0.5 mm

(b)

Figure 5.10 (a) Generation of cylindrical microlenses by exposing photoresist through a moving array of slits. (b) Exposure mask designed to generate cylindrical lenses when scanned across photoresist and fringes of equal thickness of crossed cylindrical lenses produced by this method (Gex).

mask. In both cases care must be taken to avoid discontinuities where lenses join. Both techniques offer full control over the profile of the lens, are not restricted in numerical aperture, and allow for a high fill factor since the crossed cylinders generate lenses with a square aperture. The cylindrical exposure pattern is recorded in orthogonal directions before developing the resist. The two cylinders are therefore coincident and the resulting lens does not suffer from the astigmatism that is introduced when combining separate cylinders to emulate a spherical lens.

These techniques are particularly appropriate where it is necessary to combine a long focal length with a high fill factor. In particular the method shown in Fig. 5.10 has been applied very effectively to produce lens arrays for the Shack–Hartman wavefront sensors used in astronomy to compensate for atmospheric turbulence (Artzner, 1992). The form of these very shallow lenses can conveniently be monitored by observing the 'Newton's rings' fringe patterns generated by interference between light reflected from the surface of the lens and from the photoresist/glass interface. An example of such a pattern is shown in Fig. 5.10(b).

A further method of generating an arbitrary exposure profile is to expose through a static so-called 'grey-scale' mask in which the optical density varies as a function of position. This can be achieved in two ways: (1) the mask is written in the form of a complete array and copied into the resist using a standard projection system with typically a ×10 reduction in size (Purdy, 1994); (2) a single grey-scale mask may be photoreduced using an array of microlenses (Davies and McCormick, 1993). In either case the shape of the resist surface that is produced is more or less fixed by the density distribution on the mask, but the vertical scale can be varied by changing the overall exposure.

The writing of a grey-scale mask on the same scale as the array is a significantly more demanding task than writing the lens directly into the resist and controlling the shape through the intensity or the dwell time.

The mask is not a true grey scale but a binary mask in which, on a sufficiently small scale, the mean density is determined by a dot pattern with varying proportions of the area rendered opaque. This can be achieved either by varying the size of the dots or by varying their separation. The largest pitch that can be used is determined by the need to preserve a smooth surface on the resist image. The smallest pitch that can be used is determined by the diffraction limit of the final copying system.

Gal and colleagues (1994) have successfully used this approach not only to generate conventional microlenses but also miniature Fresnel lenses and lenses combined with gratings and Fresnel-zone plates. In this work they used an electron beam writing system in order to achieve adequate resolution and varied the size of the spots on a regular grid in order to reduce the amount of data required in the computer program. Examples of their results are shown in Fig. 5.11.

When making complex objects it is important to work at the highest possible resolution. This is particularly true of diffracting components because the sharpness of the edge of a grating groove is critical in achieving high efficiency. However, this is not the case when making smooth objects such as lenses and it is possible to achieve satisfactory results using far simpler technology. An array of microlenses can be produced by recording the array of images of a single grey-scale mask generated at the focal plane of a microlens array (Davies and

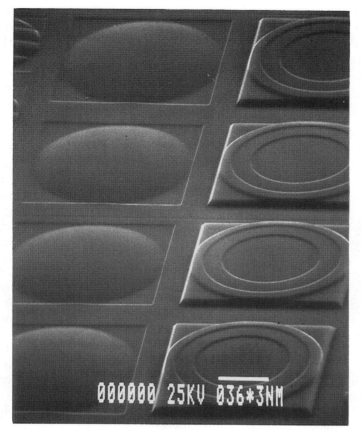

Figure 5.11 Micrographs of lenses generated using a single exposure through a digital grey-scale mask (Purdy, 1994, GEC Marconi Infrared).

McCormick, 1993; Hutley, 1995). Since the resolution in the images is not of prime importance, this lens array may be of a lesser optical quality than that which is ultimately required. In particular, it is possible to use an array of lenses made by melting photoresist in order to produce a lens array with a specification which would not have been possible using that technique alone. In practice it is necessary to use a transparent replica of the photoresist lens array because the resist has a very low transmission at short wavelengths.

The chief advantages of this approach are its simplicity and versatility. The initial grey-scale mask may typically be a few centimetres in diameter, which is a convenient size for output from a computer plotter. All that is then required is a light source, diffusing screen, collimating lens and lens array as shown in Fig. 5.12(a). The size of the lenses produced can be controlled by choosing the appropriate magnification, but the pitch of the array will of course be controlled by the pitch of the 'camera lens' array. It is therefore possible to vary the fill factor at will. If the mask is at the focus of the collimator the pitch will be copied exactly but if the mask is at a finite distance from the lens array the pitch of the new array will be different.

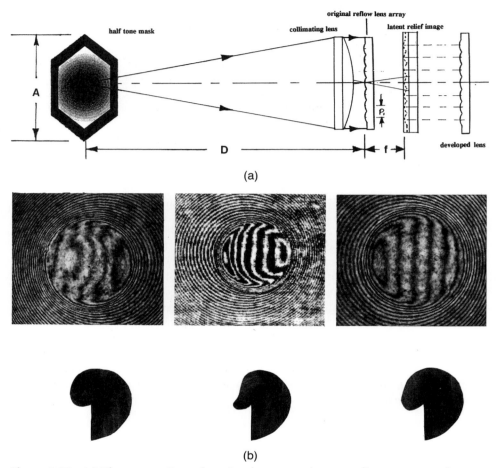

(a)

(b)

Figure 5.12 (a) The generation of a microlens array by recording an array of images of a single grey-scale mask (Davies and McCormick, 1993). (b) The use of a rotating solid mask instead of a stationary grey-scale mask. Control of aberrations by modifying the shape of the mask.

For centrosymmetric shapes such as a spherical lens, the grey-scale mask may be replaced by a solid mask of the appropriate shape and rotated during exposure. By trimming the mask to correct errors in the measured wavefront, it is possible to generate lenses with reduced or controlled aberration, as demonstrated in Fig. 5.12(b).

5.6 Grinding and Polishing

The techniques we have described so far have all been appropriate to the manufacture of very small lenses and in many cases would not be appropriate for larger lenses. However, it should be borne in mind that conventional glassworking techniques can also be applied to the manufacture of microlenses with radii of curvature down to about 1 mm. Indeed, once Hooke had formed his spherical surface by 'thermal reflow' he still had to polish the rear surface flat using pitch and Tripoli compound.

The limitation of conventional glassworking techniques to the manufacture of microlenses lies not in the limitations of the process but in the cost. It is feasible to grind and polish lenses a fraction of a millimetre in diameter. The main problems are in mounting, centring and edging and these require the extreme skill of an experienced glassworker who can only work on one lens at a time. For this reason conventional glassworking techniques are not usually appropriate except for the manufacture of individual lenses.

5.7 Secondary Techniques

Some of the techniques we have described so far are somewhat restricted in the range of lenses that can be made or produce lenses that are only suitable for a restricted range of applications. This is particularly true of those made in photoresist, because of the absorption and softness of the resist. Photoresist, in the thicknesses used for making typical microlenses, has a deep red colour. It transmits well wavelengths longer than about 600 nm and is therefore quite suitable as it stands for many optoelectronic applications, but it is of little value for use with white light. Although it is a fairly soft material, it does become significantly harder when baked at a high temperature, and lenses made by the melting technique are sufficiently robust for laboratory use. Nevertheless there are various advantages in using a photoresist lens as a starting-point from which to generate lenses in a different form.

5.7.1 *Extension of Focal Length*

We have seen in Fig. 5.7 that thermal reflow techniques can provide good-quality lenses over a limited range of focal ratios around $f/1.5$ to $f/3$. A simple way of extending this range is to encapsulate the lens in a dense medium of a chosen refractive index. Since the power of the lens is proportional to the difference of refractive index across the curved boundary, in principle any focal length longer than that of the original lens can be obtained. In the limit, if the surrounding medium has the same index as the material of the lens there will be no refraction and the 'lens' will have an infinite focal length. In practice it is convenient to achieve this by cementing a cover glass onto the lens array with a suitable optical cement. Available cements range in index from about 1.4 to 1.6. So by combining the range of radii of curvature available from the melting technique with the range of difference of index available from these materials, it is possible to cover the range of focal ratios from $f/1$ to $f/$infinity. (Though for low f/numbers there is no guarantee of optical quality!)

An additional advantage of cementing a cover glass to the lens array is that it may then be cemented directly to other components without affecting the optical power. This permits the construction of robust optical systems and reduces the losses through Fresnel reflection at the surfaces. These advantages are, of course, shared by the planar graded-index lenses described earlier.

143

5.7.2 *Replication*

There are several advantages in using a replica derived from a lens array rather than using the original. In the case of photoresist lenses, the ability to choose a material that is transparent over the required wavelength range is the most obvious. Furthermore, by choosing a material of the appropriate refractive index, the focal length available from a given radius of curvature can be controlled. However, from an economic point of view the most important factor is that large numbers of lenses of a consistent optical quality and focal properties can be produced relatively cheaply. This is of particular value in the case of a master that has been tailored at some expense to meet a particular specification or when very large quantities are required.

There are three main methods of replication: resin casting, embossing and injection moulding, and all have been successfully applied to the production of microlens arrays. With resin casting a double replication is necessary to preserve the vergence of the lens but with embossing and injection moulding this occurs as part of the process. These techniques are considered in more detail in Chapter 6.

5.7.3 *Ion Milling (Dry Etching)*

One of the attractive features of making a lens in photoresist is that it can then be transferred to the underlying substrate by ion milling. This enables lenses to be made in virtually any substrate, thereby extending the range of wavelengths for which lens arrays can be made. The lenses thus made are as robust as the substrate and the process offers the opportunity to modify and control the form of the lenses themselves.

Two techniques are available: ion beam etching and RF sputtering (Gordon *et al.*, 1991; Mersereau *et al.*, 1993; Stern and Jay, 1993). In each technique the specimen is placed in a vacuum and bombarded by energetic ions which eject atoms from the surface of the specimen. In ion beam etching, argon ions from either a saddle field source or from a Kaufman gun are accelerated towards and collimated by a charged grid. In an RF system a discharge is established between two parallel plates. This causes free electrons to oscillate and to collide with molecules of argon. Some of these will gain sufficient energy to be ionized and to sustain a plasma.

In both of these techniques the removal of material arises as a result of a physical process in which the energy of the incoming ion is transferred to atoms on the surface and is sufficient to overcome the energy binding the atom to the surface. At ion energies between 100 and 500 eV the sputter yield increases linearly between 10^{-1} and 10 atoms per ion. At greater energies the yield increases more slowly because ions penetrate too far into the surface. Etch rates are therefore slow, typically 10 or 20 nm min^{-1}, and vary with angle of incidence (Wilson, 1991). In ion beam etching the etch rate can be increased by inclining the specimen at the optimum angle. However, even with a parallel plate etcher the variation of etch rate with angle is significant because different regions of the microlens surface will be inclined at different angles to the incoming ions, will etch at different rates and the shape of the lens will be distorted as it is transferred into the substrate.

The shape will also be distorted by the redeposition of material and by trenching which occurs when ions are reflected off the side of a steeply inclined region of the specimen and increase the ion flux at a neighbouring region.

Two parameters which are of great importance when transferring a surface profile from one material into another are the etch rate and the relative etch rate. The etch rate in the substrate effectively controls the time taken to etch to a given depth. The relative etch rate determines the vertical magnification of the profile. For example if the substrate etched at half the rate of the photoresist, by the time all of the resist had been removed, the substrate would have been etched to a depth equal to half the height of the original photoresist lenses and the focal length would have doubled. By this means it is possible to produce lenses with specifications that would not have been possible to fulfil with original resist lenses. Furthermore, if by changing the etching conditions during the etch, it is possible to vary the relative etch rate, the basic shape of the lens can be changed.

Etch rates and relative etch rates can be changed by introducing chemically reactive gases into the chamber during the etch process. Oxygen, for example, will have the effect of increasing the rate of removal of resist but will have little effect on a silica substrate. Fluorine-based compounds will on the other hand increase the rate of removal of silica but will have relatively little effect on photoresist. By a suitable choice of gas it is therefore possible both to increase and decrease the depth of modulation and hence the optical power of microlens arrays. Under these circumstances a variety of mechanisms may be involved in the removal of material. Reactive gas molecules may chemically etch the surface while the inert ions bombard the surface and remove material, such as oxides, which would otherwise impede the chemical etching. The argon gas may combine with gas molecules which would not themselves react with the specimen but form ion species which do. In the case of ion beam etching one has the choice of introducing the reactive gas either into the gun, in which case it is known as 'reactive ion beam etching' or into the vacuum chamber near to the substrate, in which case it is known as 'chemically assisted ion beam etching'. In a parallel plate system the introduction of a reactive gas is simply referred to as 'reactive ion etching'.

In order to achieve an array of ion etched lenses of a desired shape it is necessary to have a theoretical model of the whole process to include both physical and chemical effects. It is also necessary to characterize fully the etching parameters and to have facilities for the accurate measurement of the surface (Gratrix, 1993).

5.8 Testing and Specification

Many of the processes we have considered for the manufacture of microlenses are not fully understood and controlled. It is rather the case that one generates a lens array and then measures the performance, hoping that by accurately repeating the conditions one may achieve a consistent result. To achieve the optimum conditions it is therefore crucial to measure the physical and optical parameters of the lenses with some accuracy (Hutley *et al.*, 1991; Mersereau *et al.*, 1993; Schwider *et al.*, 1993).

There are essentially two requirements: to measure the form of the lens (or the refractive index distribution) and to measure its optical performance. The

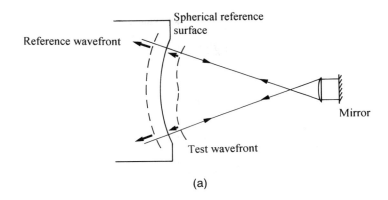

(a)

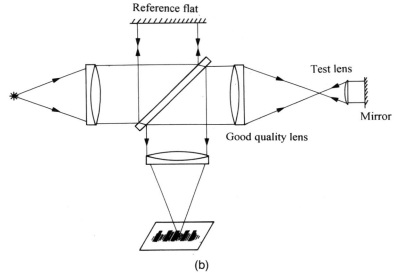

(b)

Figure 5.13 Interferometers for testing microlenses: (a) Fizeau, (b) Twyman–Green, (c) wavefront shearing (after Schwider *et al.*, 1993), (d) Mach–Zehnder.

optical performance can be measured and expressed in two ways i.e., the shape of the transmitted wavefront or the image-forming properties. Measurements of the form of the lens and of the transmitted wavefront are of particular importance to the manufacturer because these provide information in the form that allows errors to be identified and corrected. The user on the other hand is usually more concerned with focal properties such as modulation transfer function, point-spread function and encircled energy function.

The most convenient method for measuring the form of the lens is interferometry. It is possible to employ a stylus measuring instrument but this is tedious and difficult to align and only provides information about one section of the lens. The direct measurement of the index distribution in a graded-index lens is particularly inconvenient because it involves cutting a slice through the lens and measuring the variation of optical thickness with an interferometer.

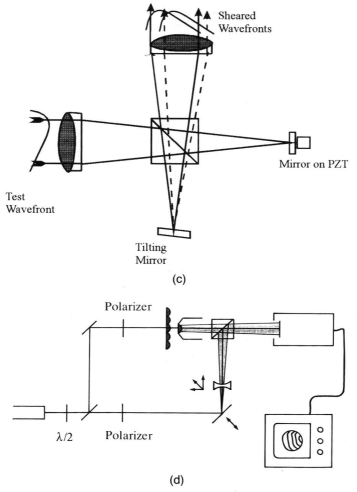

Figure 5.13 (*continued*)

In order to test both lenses and moulds it is necessary to measure both concave and convex surfaces in both transparent and opaque materials.

Of the various forms of interferometer available, the Fizeau, Twyman–Green and lateral shearing types are most readily adapted to the purpose and these are shown diagrammatically in Fig. 5.13. In order to avoid the effects of spurious diffraction at the edges of the aperture it is important that the instrument images the aperture of the lens onto the detector. Unfortunately the geometry of microlenses is often such that with many commercially available instruments this cannot be achieved. In particular, interference microscopes which have a plane reference surface are unable to cope with the steeply sloping sides of lenses of high numerical aperture. It is therefore necessary to design instruments specially for the purpose. The Fizeau and Twyman–Green interferometers are in many senses equivalent. The Fizeau has the advantages of a common path interferometer, but with the Twyman–Green it is possible to equalize the optical paths and use a source of low temporal coherence. Lateral shearing interferometers have the advantage that no reference wavefront is required but they require the analysis of two interferograms per surface and the accuracy depends upon an accurate knowledge of the shear (Sickinger *et al.*, 1994).

All three interferometers can be used to analyze the wavefront transmitted by the lens. The Fizeau and Twyman–Green require the light to pass twice through the lens, which doubles the sensitivity of the instrument. However, the need for an extra mirror and the mechanical means of adjusting it can be a significant inconvenience, particularly for example when measuring off-axis a single lens of a large array. The lateral shearing interferometer can work in both single and double pass configurations and can be adapted to measure both in transmission and reflection. It has the disadvantage that it cannot measure to the edge of the field (where aberrations are often greatest) and its sensitivity depends upon the degree of shear.

A particularly convenient alternative for this application is the Mach–Zehnder interferometer shown in Fig. 5.13(d). The main practical advantage is that it allows greater access to the lens array under test. The microscope objective forms an image of the lens at the detector array. The transmitted light is therefore diverging as it leaves the objective and it is therefore necessary to introduce a similar amount of divergence into the reference wavefront. An additional advantage of this design is that it may be incorporated into the carcass of a microscope. Indeed, if the reference beam is blocked off it can operate as a simple microscope. By re-focusing on the focused spot it is possible to measure directly the point spread function and by measuring the displacement between this position and that when focused on the lens the focal length can be determined. It is therefore a very convenient all-round instrument for measuring microlenses.

A variety of computer software packages exist which will interpret the fringe pattern from an interferometer and display the shape of the wavefront. These will then also calculate the performance of the lens in the image plane and present the data in the form of point spread function, modulation transfer function and encircled energy function as shown in Fig. 5.14. With an interferometer it is then possible with a single instrument to provide data in the form required by the manufacturer and by the user. The great advantage of interferometry from the point of view of the manufacturer is that defects in the lens are immediately identifiable whereas focal plane data only describe the performance integrated over the whole area of the lens.

However, it is sometimes an advantage to measure the performance directly at the focal plane if only because it provides the reassurance that the lens is being tested in the way it will be used in practice. One approach is to measure the intensity distribution in the plane of the image of a point source. This can be achieved very conveniently by displaying a line intensity scan of the image recorded on a ccd or vidicon camera attached to a microscope. The position of best focus may easily be determined in this way and one can see at a glance the effects of spherical aberration in enhancing the secondary rings in the Airy pattern. An alternative approach is to use an instrument dedicated to the measurement of modulation transfer function. Most commercial instruments project an image of a line source which is then imaged by the optical system under test onto a detector array. From the line-spread function thus measured it is possible to calculate the

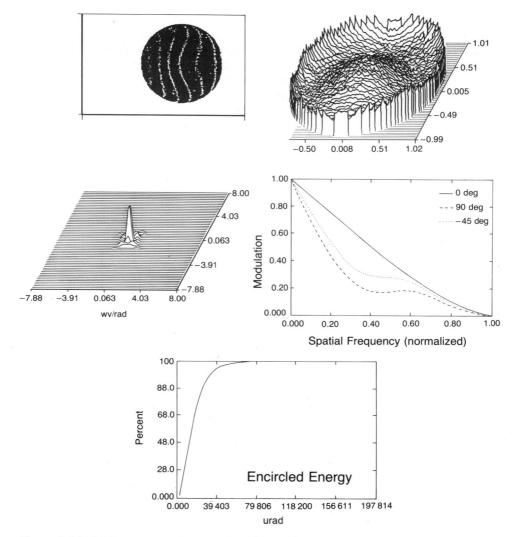

Figure 5.14 Various ways of presenting the performance of microlenses:
(a) wavefront interferogram, (b) isometric projection of the wavefront errors,
(c) point spread function, (d) modulation transfer function, (e) encircled energy
function.

MTF. The great advantage of this approach is that the performance of the lens
may conveniently be measured at a variety of conjugates (Moisel, 1992). Most
interferometers measure at something close to infinite conjugates and would
require significant rearrangement to measure under different conditions. In
practice microlenses will frequently be used at finite conjugates and it is important
to be able to measure their performance as used.

It is important that measurements in the focal plane should be photometric.
Many published papers simply show the ability of a lens to form an image without
any quantitative evaluation and this is of very limited value. Although a

photograph may show fine detail in an image this does not necessarily indicate that the lens is of particularly good quality because contrast can easily be enhanced by the photographic and subsequent reprographic process. Furthermore, it should be borne in mind that the width of the point spread function is not a sensitive indicator of quality. It is the amount of energy in the wings of this function that tends to degrade images and cause crosstalk in electrooptical systems. The modulation transfer function and the Strehl ratio take this into account and are therefore more relevant.

5.9 Summary

It is not possible to provide a precise and simple intercomparison of the capabilities of the various methods of making microlenses because the various relevant parameters interact in different ways. Furthermore, as there are no agreed norms for the presentation of the data it is often difficult to find. For example, one author may claim to be able to make $f/1$ lenses but omit to mention that they have severe aberrations. Another may be more modest in his claims but in fact make better lenses. The comments that follow are a combination of data derived from the literature and the personal experience of the author's group in measuring lenses from a variety of sources. This information must be used with caution because there is no guarantee that the best or even typical samples will have been measured. Before acquiring microlenses one should always obtain from the manufacturer the maximum amount of data relevant to the application.

- Graded-index rod
  Not generally available with diameters less than 1 mm.
  Numerical apertures between 0.1 and 0.6 ($f/5$–$f/0.7$)
  Wavelength transmission 380–2000 nm

- Graded index planar arrays
  Diameters between 10 μm and 1 mm
  Numerical apertures up to 0.25
  Chromatic aberration tends to be worse than for surface refracting lenses
  Particularly suitable for stacked optics

- Photothermal
  Diameters 70 μm to 1 mm
  Minimum separation between lenses 15 μm
  Maximum numerical aperture 0.35
  Interstitial areas are opaque
  Wavelength transmission 200 nm–4.5 μm

- Thermal reflow
  Diameters 15 μm–500 μm can be extended to 1 mm with variable results
  Numerical apertures 0.15–0.45 ($f/3$–$f/1$); can be extended to 0 (f/∞) by immersion
  Wavelength transmission >600 nm for photoresist
  Usually regarded as a master for replication in transparent medium or for transferring into substrates of various materials by ion etching
  Aberrations <$\lambda/4$ for lenses of optimum parameters

- Photosculpture
  Aperture size not limited
  Suitable for very long focal lengths
  Aberrations may be controlled
  Suitable for high fill factors
  Same restraints on wavelength transmission as for Thermal Reflow
  Same possibilities for replication as for Thermal Reflow

References

Applied Optics 'GRIN' topical issues: April 1980, **19**(7); March 1982, **21**(6); February 1983, **22**(3); June 1984, **23**(11); December 1985, **24**(21); February 1988, **27**(3); October 1990, **29**(28).

ARTZNER, G. (1992) Microlens arrays for Shack–Hartmann wavefront sensors. *Opt. Eng.* **31**, 1311–1322.

ATKINSON, L. G., KINDRED, D. S. and ZINTER, J. R. (1994) Practical GRINS: materials, design and applications. *Opt. and Photonics News*, June, 28–31.

ATKINSON, L. G., MOORE, D. T. and SULLO, N. J. (1982) Imaging capabilities of a long gradient index rod. *Appl. Opt.* **21**, 1004–1008.

BÄHR, J., BRENNER, K.-H., SINZINGER, S., SPICK, T. and TESTORF, M. (1994) Study of index distributed planar microlenses for 3-d micro-optics fabricated by silver-sodium ion-exchange in BGG35. *Appl. Opt.* **33**, 5919–5924.

BORELLI, N. F., MORSE, D. L., BELLMAN, R. H. and MORGAN, W. L. (1985) Photolytic technique for producing microlenses in photosensitive glass. *Appl. Opt.* **24**, 2520–2525.

DALY, D., STEVENS, R. F., HUTLEY, M. C. and DAVIES, N. (1990) The manufacture of microlenses by melting photoresist. *Meas. Sci. Technol.* **1**, 759–766.

DAVIES, N. and McCORMICK, M. (1993) Three dimensional optical transmission and micro optical elements. *Proc. SPIE.* **1992**, 247–257.

FINDAKLY, T. (1985) Glass waveguides by ion exchange: a review. *Opt. Eng.* **24**, 244–250.

GAL, G., HERMAN, B., ANDERSON, W., SHOUGH, D., PURDY, D. and BERWICK, A. (1994) Development of the dispersive microlens. *Pure Appl. Opt.* **3**, 97–101.

GALE, M. T. and KNOP, K. (1983) The fabrication of fine lens arrays by laser beam writing. In *Industrial Applications of Laser Technology*, SPIE Vol. 398, pp. 347–353.

GORDON, T., JONES, C. L. and PURDY, D. J. (1991) Applications of microlenses to infrared detector arrays. *Infrared Physics* **31**, 599–604.

GÖTTERT, J., FISCHER, M. and MÜLLER, A. (1995) High aperture surface relief microlenses fabricated by X-ray lithography and melting. In *Microlens Arrays*, EOS Topical Meetings Digest, Vol. 5, pp. 21–25.

GREATRIX, E. J. (1993) Evolution of a microlens surface under etching conditions. *Proc. SPIE* **1992**, 266–274.

HAMANAKA, K., NEMOTO, H., OIKAWA, M. and OKUDA, E. (1988) Aberration properties of the planar microlens array and its application to imaging optics. In *Micro optics*, SPIE Vol. 1014, pp. 58–65.

HASELBECK, S., SCHREIBER, H., SCHWIDER, J. and STREIBL, N. (1993) Microlens fabrication by melting photoresist on a base layer. *Opt. Eng.* **32**, 1322–1324.

HOOKE, R. (1664) *Micrographia*. London: Royal Society.

HOUDE-WALTER, S. N. and MOORE, D. T. (1985) Gradient-index profile control by field-assisted ion exchange in glass. *Appl. Opt.* **24**, 4326–4333.

HUTLEY, M. C. (1995) Simple 'photosculpture' technique for making arrays of lenses with controlled aberration. In *Microlens Arrays*, EOS Topical Meetings Digest, Vol. 5, pp. 44–49.

HUTLEY, M. C., DALY, D. and STEVENS, R. F. (1991) The testing of microlens arrays. In *Microlens Arrays*, IOP Short Meetings Series No. 30. London: Institute of Physics.

IGA, K. (1980) Theory for gradient-index imaging. *Appl. Opt.* **19**, 1039–1043.

IGA, K. and YAMAMOTO, N. (1977) Plastic focussing fiber for imaging applications. *Appl. Opt.* **16**, 1305–1310.

IGA, K, KOKUBUN, Y. and OIKAWA, M. (1984) *Fundamentals of Microoptics*. Tokyo: Academic Press.

IGA, K., HATA, S., KATO, Y. and FUKUYO, H. (1974) Image transmission by an optical system with a lens-like medium. *Jpn. J. Appl. Phys.* **13**, 79–86.

KOIKE, Y., KIMOTO, Y. and OHTSUKA, Y. (1982) Studies on the light-focussing plastic rod. *Appl. Opt.* **21**, 1057–1062.

KUFNER, M., KUFNER, S., FRANK, S., MOISEL, J. and TESTORF, M. (1993) Microlenses in PMMA with high relative numerical aperture: a parameter study. *Pure Appl. Opt.* **2**, 9–19.

LAND, M. F. (1978) Animal eyes with mirror optics. *Scient. Am.* **239**(6), 126–133.

MERSEREAU, K., NIJANDER, C. R., FELDBLUM, A. Y. and TOWNSEND, W. P. (1993) Fabrication and measurement of fused silica microlens arrays. *Proc. SPIE* **1751**, 229–233.

MERSEREAU, K. O., CRISCI, R. J., NIJANDER, C. R., TOWNSEND, W. P., DALY, D. and HUTLEY, M. C. (1993) Testing and measurement of microlenses. *Proc. SPIE* **1992**, 210–215.

MOISEL, J. (1992) Charakerisierung der Abbildungseigenschaften von Mikrolinsen. Diplomarbeit Erlangen.

MOORE, D. T. (1980) Gradient-index optics, a review. *Appl. Opt.* **19**, 1035–1038.

OIKAWA, M. and IGA, K. (1982) Distributed-index planar microlenses. *Appl. Opt.* **21**, 73–77.

OIKAWA, M., NEMOTO, H., HAMANAKA, K. and OKUDA, E. (1990) High numerical aperture planar microlens with swelled structure. *Appl. Opt.* **29**, 4077–4080.

POPOVIC, Z. D., SPRAGUE, R. A. and NEVILLE-CONNELL, G. A. (1988) Technique for the monolithic fabrication of microlens arrays. *Appl. Opt.* **27**, 1281–1284.

PURDY, D. (1994) Fabrication of complex micro-optic components using photo-sculpting through halftone transmission masks. *Appl. Opt.* **3**, 167–175.

RYAN-HOWARD, D. P. and MOORE, D. T. (1985) Model for the chromatic properties of gradient index glass. *Appl. Opt.* **24**, 4356–4366.

SCHWIDER, J. *et al.* (1993) Production and control of refractive and diffractive microlenses. *Proc. SPIE* **1992**, 102–113.

SICKINGER, H., FALKENSTÖRFER, O., LINDLEIN, N. and SCHWIDER, J. (1994) Characterization of microlenses using a phase-shifting shearing interferometer. *Opt. Eng.* **33**, 2680–2686.

STERN, M. B. and JAY, T. R. (1993) Dry etching – Path to coherent refractive microlens arrays. *Proc. SPIE* **1992**, 283–292.

WILSON, S. J. (1991) *The use of ion beam etching in the manufacture of diffracting optical components*. Report MOM 99, National Physical Laboratory, Teddington, UK.

6

Replication

M. T. GALE

6.1 Introduction

The physical copying of surface-relief microstructures by replication techniques such as hot embossing, moulding or casting is expected to become a key technology for the low-cost mass-production of diffractive optical elements (DOEs) and other micro-optical elements with micrometre- or nanometre-sized features. These replication processes are capable of achieving nanometre resolution over large areas and the cost of replicated microstructures is relatively independent of the complexity of the detailed structure. Replication technology is already used for the commercial production of submicron grating structures (hot embossed diffractive foil and security holograms) and data storage microstructures (injection moulded compact discs). The commercial replication of micro-optical elements requires further development to routinely handle the deeper and finer microstructures typical for elements such as refractive and diffractive microlenses, binary optics and other phase DOEs.

The basic replication process (Fig. 6.1) involves the copying of a surface-relief microstructure from a metal form, the replication shim or stamper, into a formable material such as a thermoplastic or curable polymer. A summary of the main polymer materials used in replication technology, their acronyms and relevant optical properties is given in section 6.3 and listed in Table 6.1. The original microrelief can be fabricated in virtually any material using the techniques described elsewhere in this book, or using one or a combination of suitable high-resolution lithographic processes such as those which have been developed for the fabrication of semiconductor devices (Moreau, 1988). This original surface microstructure is then electroformed to produce a first-generation metal copy, usually in nickel (Ni), from which further generations can be electroformed. The Ni copy can then be used for the replication into polymer or other material. An overview of the hot embossing replication technology for diffractive foil has been published by Kluepfel and Ross (1991). The hot embossing of micrometre-sized structures for integrated optical components has been described by Baraldi (1994) and Gale *et al.* (1994a). The fidelity of replicated nanometre- and micrometre-sized

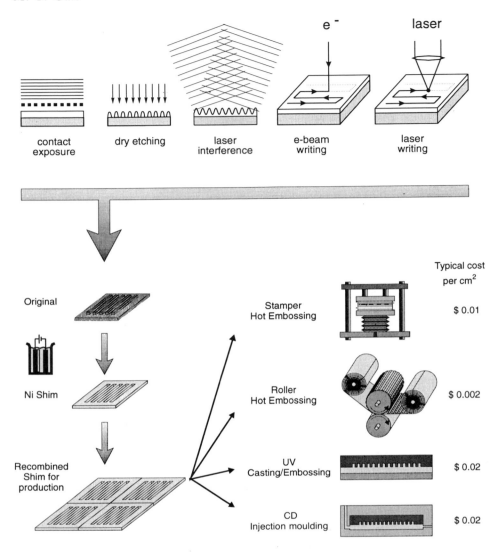

Figure 6.1 Basic concepts for the fabrication of surface-relief micro-optical elements by replication. The original element is fabricated by one or a combination of a variety of microstructuring technologies. It is then electroformed to a nickel shim and recombined, if necessary, to a large-area production shim which can be used in one of a number of high-resolution replication technologies.

microstructures is very dependent upon the aspect ratio (feature width to height ratio) and as a general 'rule of thumb' it can be stated that, independent of the absolute feature size, an aspect ratio of $1:1$ can be replicated easily, $1:5$ with care and $1:10$ only with great difficulty.

The major replication techniques are also shown in Fig. 6.1. They can be summarized as:

Table 6.1 Optical properties of polymer materials used in replication technology

Acronym	Material	Replication technology	Refractive index $n(\lambda = 633$ nm)	Typical wavelength transmission range (nm)		Glass transition temperature T_g (°C)	Manufacturer(s) (not exhaustive)	References to use in micro-optics
				λ_{min}	λ_{max}			
PC	Polycarbonate	Injection moulding; hot embossing; roll coating	1.58	380	1600	145	Röhm GmbH; Dow Plastics; GE Plastics; Bayer	Gale *et al.*, 1994a,b; Baraldi, 1994
PMMA	Polymethyl-methacrylate	Injection moulding; hot embossing	1.49	400	1100	94–108	Röhm GmbH; Degussa	Chou *et al.*, 1995; Bremer *et al.*, 1993; Göttert *et al.*, 1993; Müller *et al.*, 1995b
PVC	Polyvinyl-chloride	Injection moulding; hot embossing; roll coating	1.54	400	2200	75–105	Hoechst; BF Goodrich	Knop and Gale, 1980
NOA 61	Photopolymer	UV-casting	1.56	350	3000	—	Norland Prod. Inc.	

(1) Hot embossing. Stamping, roller or reciprocating systems are used for the hot embossing of surface-relief microstructures into thermoplastic foil. Hot embossing technology is routinely used for reproducing diffractive optical microstructures such as holograms and submicron grating patterns for diffractive foil (Kluepfel and Ross, 1991). Commercial roll embossing systems achieve embossing speeds of up to 1 m s^{-1} in polymer foil of up to 2 m width; typical polymers include PVC (polyvinyl chloride) and PC (polycarbonate), usually, but not necessarily, precoated with a thin metal film. Typical costs for hot embossed foil are below 0.001 cm^{-2}. Although commercial roller systems are fully capable of submicron resolution, microstructures with relief amplitudes in excess of about $1 \mu\text{m}$ present considerable problems. Better results are obtained for deeper microstructures using a reciprocating or stamping hot press, which is also capable of relatively high throughput. Improved results can also be obtained by embossing into a thin, customized thermoplastic layer coated onto a stable substrate film such as polyester.

(2) Injection moulding. Injection moulding is the standard replication technology for fabricating compact discs (CDs) with micron pit relief structures and is being investigated by various groups for applications in micro-optics. Customized injection moulding equipment has been developed for fabricating PMMA (polymethyl methacrylate) and PC microstructures, for example for integrated optical devices (Neyer *et al.*, 1993). Injection moulded replicas are of higher cost than those fabricated by hot embossing (typical CD fabrication costs are $\sim$$0.01 \text{ cm}^{-2}$), but the technology has the potential of producing higher quality, deeper microrelief structures in stable, thick (>1 mm) PMMA or PC substrates.

(3) Casting. Casting techniques have long been used for producing very high fidelity replicated spectroscopic gratings in epoxy materials. Of particular interest is the replication into a thin film of UV-curable material coated onto a rigid substrate such as glass. A related technology is replication using a roller press into a UV-curable film on a plastic substrate film, with rapid curing at the contact stage or shortly afterwards (so-called 'UV-embossing'). Such techniques have the advantage of producing a replicated film which can be highly resistant to a subsequent solvent coating step. They are particularly suited to the replication of deep or high aspect ratio microstructures.

A number of other replication techniques are also used commercially in the fabrication of certain plastic foil and sheet products. Solvent evaporation, in which a solution of the polymer is poured onto the replication shim and allowed to dry before separation, can produce very high quality replicas and is often used in the preparation of samples with sub-nanometre resolution for electron microscopy. The calendering process, in which molten polymer is squeezed between two rollers, one of which contains the surface-relief structure, is used for the mass-production of surface-relief patterned plastic foil.

In the following sections, we give an overview of the main replication technologies applicable to micro-optics and DOE production. References to published work are given wherever possible, although replication technology has been researched and developed mainly by industrial companies and much of the expertise is still confidential and unpublished.

6.2 Electroforming

The basis of modern technology for the electroforming of a metal replication tool from an original surface-relief recording in lacquer, photoresist or some other suitable material was developed by the audio disc industry in the first half of the 20th century, and subsequently refined for the higher-resolution microstructures required for diffractive elements. In the early 1970s RCA developed technology for the production of embossed video hologram tape (Hannan *et al.*, 1973) and surface-relief diffractive microimages (Knop and Gale, 1980). ABN and other embossed holography pioneers built upon this work to commercialize embossed holography for security features and diffractive foil (Kluepfel and Ross, 1991). Video disc technology, involving the electroforming of a Ni disc from an original recording in photoresist, was also developed in the 1970s (Palermo *et al.*, 1977). CD mastering technology is now a routine, widespread industrial process.

The vast majority of microstructure electroforming for the fabrication of replication tools is carried out by Ni electroplating, for which suitable low-stress plating bath formulations are well-known. Figure 6.2 shows the basic steps. The surface of the original recording, which is generally in an electrically non-conducting material, must first be coated with a thin film of metal to form a conductive coating for subsequent electroplating. This metal film has a typical thickness in the range of 1 to 50 nm and must adhere sufficiently well to the recording to remain intact during the electroplating process, but must then separate perfectly with the Ni from the original. A variety of different processes can be used, including:

- Electroless silver (Ag) deposition: widely used in the embossed holography industry (Kluepfel and Ross, 1991). The substrate is first spray coated with a sensitizer and then sprayed with an Ag solution, resulting in the formation of a thin film of Ag on the substrate surface. An alternative is the use of electroless Ni deposition (Feldstein, 1970).

- Evaporation of gold (Au) or silver (Ag). This is widely used for video disc master fabrication.

- Sputter deposition of Au or Ag, producing somewhat better adhesion and sidewall coverage than evaporation techniques.

The conductive substrate is then electroformed in a Ni-plating bath to produce a durable Ni copy which can be separated from the original without damage. Ni shims for the embossed hologram industry have typical thicknesses in the range 50–100 μm and can be commercially fabricated in areas in excess of 50 cm × 50 cm (see, for example, product list of 3D Ltd[1]). CD replication stampers are significantly thicker ($\sim$0.5–1 mm) and require additional machining and tool housing support for the injection moulding machines. Suitable Ni-plating solutions are supplied by most electroplating chemical houses. The basic components of a Ni-plating system suitable for microstructure and DOE electroforming are shown in Fig. 6.3.

Copies of Ni shims and stampers can be produced with no significant loss of quality and feature resolution (see below) by 'passivating' the surface (for Ni, typically by immersion in dichromate solution) and repeating the electroforming

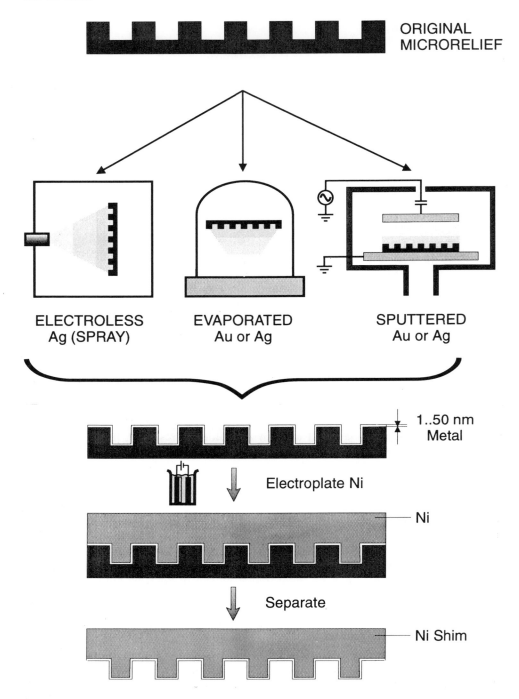

Figure 6.2 Essential steps in the fabrication of the Ni embossing shim or mould.

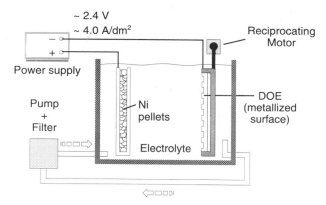

Figure 6.3   Basic arrangement of a Ni-plating system for DOE microstructure electroforming. A reciprocating motion continuously moves the metallized element during plating. Typical thicknesses for hot embossing shims are between 50 and 150 μm.

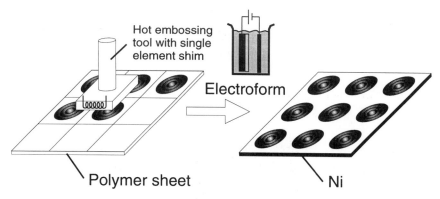

Figure 6.4   Recombination process for the fabrication of large shims for production press. One or more smaller shims are replicated into a lacquer-coated substrate to form a large-area recombination which is then electroformed to give a single large shim (courtesy of 3D Ltd CH-6314 Unterägeri).

step (Kluepfel and Ross, 1991). The passivation serves to produce a surface which is sufficiently conductive for electroplating, but which enables subsequent easy and damage-free separation of the two pieces. In this way, second and higher-generation copies can be fabricated, a fan-out procedure enabling multiple copies of replication shims and stampers to be produced from a single original recording.

An important technique applicable to DOE and micro-optical element replication is the recombination of small-area shim structures to produce large-area shims. This can be achieved by replicating individual elements into a large-area polymer sheet foil and electroforming a new Ni shim from the composite (Fig. 6.4). Such services are offered by commercial shim fabrication houses[1] for embossed holography, and further development of the technology for deeper microstructures, with relief amplitudes in excess of 1 μm, is under way.

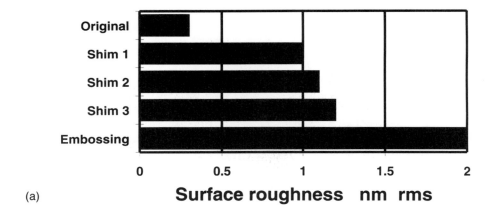

(a)

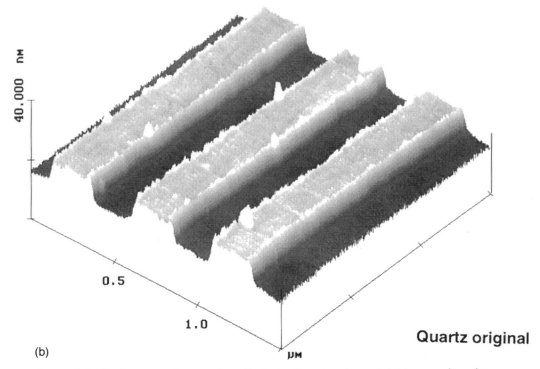

(b)

Figure 6.5 Surface roughness of replicated microstructure: (a) Measured surface roughness of original, electroformed shims and a hot embossed polycarbonate replica. (b) AFM micrographs of a shallow grating etched into quartz, (c) the corresponding shim and (d) the replica in PC. The measured surface roughnesses along the top of the grating lines are 0.5, 0.7 and 0.8 nm rms respectively.

The resolution and fidelity of the electroforming process is extremely high – typical surface-relief microstructures for DOEs and other planar micro-optical elements are copied with nanometre-scale resolution. It is generally safe to assume that the major dimensional changes in the replication chain occur in the embossing

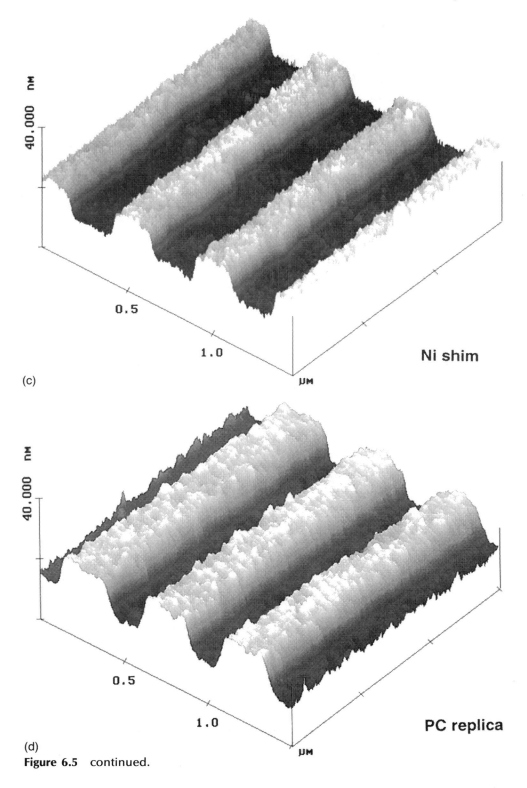

(c)

Ni shim

(d)

Figure 6.5 continued.

or moulding step (in particular in demoulding) and that any loss in feature resolution in the electroforming step can be neglected. Lateral dimensions are typically held to better than 1% in the combined shim fabrication and replication process.

Measurements of the surface roughness of electroformed shims fabricated from a quartz substrate with an rms surface roughness of <0.3 nm (Fig. 6.5(a)) show a measured roughness of about 1 nm (Gale *et al.*, 1994b). Subsequent generations of electroformed shims have marginally higher roughness, about 0.2 nm per generation for the electroforming technology investigated here. Replicas fabricated by hot embossing into polycarbonate (PC) film using a first-generation shim also showed a roughness of about 2 nm. Figure 6.5(b) shows other measurements on a shallow grating of about 400 nm period and 10 nm depth. The AFM-measured surface profile is shown for grating microstructures etched into quartz and for subsequent shim and PC replica copies. A deterioration of the profile and an increase in the surface roughness can be seen; the measured surface roughness on the grating lines is about 0.5 nm for the quartz original and 0.7 and 0.8 nm for the shim and replica respectively. The fabrication of electroformed mirrors with a measured surface roughness below 0.3 nm rms (evaporated Au film on Ni) for X-ray astronomy has been described by Altkorn and co-workers (1992).

6.3 Materials for Replication

The most common materials suitable for the replication of micro-optical elements in general are thermoplastic polymers such as polycarbonate and PMMA for hot embossing and moulding, or epoxy materials which can be cured by heat or radiation. Table 6.1 shows the optical properties of a selection of materials which have been used for replication of micro-optical elements for use at visible and near infra-red (IR) wavelengths. All of these materials are, in principle, suitable for transmissive elements in the wavelength range in which they are transparent. Reflective elements are produced by subsequently coating the replicated surface with a thin metal film, for example aluminium or gold. Further information on plastic materials and processing can be found in literature such as the *Modern Plastics Handbook* (Toensmeier, 1995).

Metallized polyester film is widely used in the large-scale commercial replication of holograms and diffractive foil by hot embossing; measurements show that the surface relief is partly replicated into the (Al) metal film and partly into the polyester, which is not strictly a thermoplastic material. Reflective elements for use with high power lasers (such as CO_2 lasers for machining) have been fabricated directly as electroformed structures in a metal such as copper, giving a more robust element which can be optimally bonded to a heat sink or cooling mount (Budzinski and Tiziani, 1995).

6.4 Hot Embossing

The surface-relief microstructure of a Ni shim can be copied with nanometre resolution and fidelity by a variety of hot embossing techniques. In the laboratory, the simplest approach is to use a hot plate press such as that shown in Fig. 6.6.

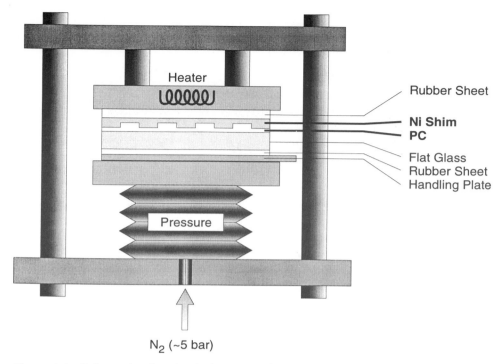

Figure 6.6 Schematic of a simple hot press for the replication of DOE microstructures. The Ni shim and polycarbonate (PC) foil lie on a flat glass block, sandwiched between the pressure plates and sheets of heat-resistant rubber.

A sheet of plastic foil is sandwiched between the Ni shim and an optically smooth backing plate (for example a polished flat glass block, as shown, or a shim of an optically polished glass substrate) and heated under pressure to significantly (typically >50°C) above the softening temperature T_g of the plastic (about 90°C and 150°C for PVC and PC respectively). After removal of the pressure and cooling to under T_g, the plastic can be separated from the Ni shim to give a high quality copy of the planar microstructure. This technique works extremely well for shallow microstructures (relief depths of less than about 1 μm) and for deeper structures with no features and high aspect ratio (<1 depth to width ratio). Figure 6.7 shows sections of a phase-matched Fresnel lenslet replicated into PC foil of about 250 μm thickness – the fine structure visible on the surface is, as far as can be determined by optical and profilometer measurements with nanometre resolution, an exact replica of that present in the original recording by laser writing in photoresist (Gale *et al.*, 1994b).

The hot embossing of sub-25 nm features with 100 nm depth from an SiO_2 mould into a PMMA film has been described by Chou and co-workers (1995). Such embossing can be carried out in the laboratory using a relatively unsophisticated hot press. Figure 6.8 shows a very simple hot press constructed with everyday materials – the potential of the hot embossing process is underlined by the fact that this crude press readily achieves nanometre feature replication over areas in excess of 100 cm^2!

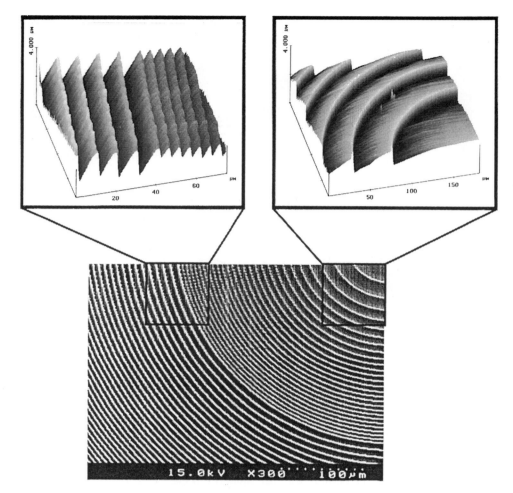

Figure 6.7 Atomic force microscope (AFM) surface-profile measurements of a micro-optical element replicated into polycarbonate foil (from Gale, 1994b).

Higher aspect ratio features present increasing problems in separation ('demoulding'). Knop and Gale (1980) have described work on diffractive colour microimages in which a grating microstructure with periodicity $1.4\,\mu m$, linewidth $0.7\,\mu m$ and line depths in excess of $3\,\mu m$ was successfully hot embossed into PVC foil, although separation of the shim and the foil was only possible by pulling parallel to the grating lines. Such microstructures can be successfully hot embossed in the laboratory, but not by the majority of today's industrial mass-production embossing technologies. Modifications of the hot plate press, including the direct heating of the Ni shim by passing a high current, short duration heating pulse directly through the Ni shim, are also described in the same reference.

Separation of shim and hot embossed foil can be considerably eased by coating the surface of the Ni shim with a very thin (a few nanometres) release film, such as of PTFE (Teflon-like) material, deposited by sputtering (Lehmann *et al.*, 1978)

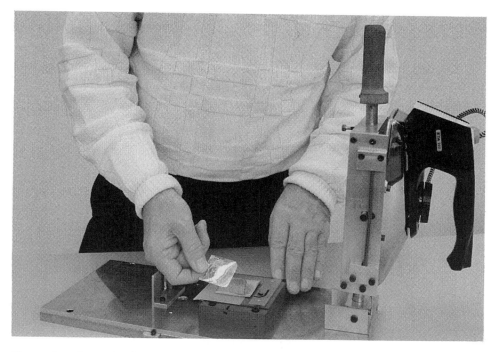

Figure 6.8 Low-cost hot press capable of nanometre feature reproduction.

or plasma reaction processes (Knop and Gale, 1980). The deposition of ultrathin Teflon-like polymer film for DOE replication has been described by Gröning and co-workers (1995).

Commercial hot embossing is widely used for the production of diffractive foil and security features (Kluepfel and Ross, 1991). Typical Ni shims required for commercial production machinery are about $25\,cm \times 35\,cm$ in size. Flat bed embossing (cf. Fig. 6.6), usually releasing the pressure after cooling to below the T_g temperature, is relatively slow (cycle times of seconds to minutes), but can yield very high quality holograms. Higher speeds can be obtained using a reciprocating press. Roll embossing requires relatively expensive machinery, but can achieve very high production rates, in excess of $1\,m\,s^{-1}$ for plastic foil of up to $2\,m$ width. Figure 6.9 shows a commercial high pressure roll embosser[2]. These commercial systems, in particular roll embossers, are optimized for the hot embossing of diffractive microstructures of relief depth up to about $1\,\mu m$. High aspect ratio microstructures and deeper microstructures such as Fresnel microlenses are best replicated using the slower approaches such as flat bed or reciprocating presses. Figure 6.10 shows a Ni shim containing various diffractive and micro-optical surface-relief structures together with a roll of (metallized) hot embossed replicas. The fidelity of the replicated optical elements is, with due care, such that only minimal deterioration in the optical performance and efficiency (after correction for any change in refractive index) is observed in comparison with the original recording. The surface roughness of replicated PC foil has been measured to be about 2 nm rms, an increase of about 1 nm over the Ni shim.

Figure 6.9 Commercial high-pressure roll embosser for the replication of holograms and other security features (photograph courtesy of Hologram Industries SA, Paris).

Transfer holograms for applications such as security features are produced as hot stamping foil, typically a polyester carrier with a wax release layer coated with a metallized, embossable lacquer layer (Kluepfel and Ross, 1991). Figure 6.11 shows the basic layer structure. The hologram is embossed either into the lacquer layer which is then metallized, or alternatively directly into the metallized lacquer layer. The structure is then coated with a hot metal adhesive. For application to plastic cards or paper, a heated plate presses onto the carrier, the hot metal adhesive layer bonds the embossed metal/lacquer layers to the underlying material card and the carrier foil separates at the wax release layer. The result, a very thin hologram bonded to the card or paper surface, is widely used for holograms on credit cards and paper documents. This approach has potential for the application of reflective DOEs to mounting or component structures in microsystems, although its use for transmissive devices is not as promising due to the optical requirements imposed upon the adhesive layer.

The hot embossing of deep microstructures using a custom press has also been developed as part of the LIGA (German acronym for Lithographie, Galvanik und Abformung) technology for replicating high aspect ratio microstructures (Becker *et al.*, 1986; Göttert *et al.*, 1993). An example, shown in Fig. 6.12, is the replication of a miniature LIGA spectrometer in the form of a multi-mode waveguide with a vertical grating structure at one end (Hagena *et al.*, 1995; Müller *et al.*, 1995a,b). Typical parameters for the hot-embossed grating structure are periodicity $\sim 2\,\mu m$, groove depth $\sim 300\,nm$ and height $\sim 100\,\mu m$. This example shows that the hot

Figure 6.10 Ni shims and hot embossed replicas (replicas courtesy of 3D Ltd, CH-6314 Unterägeri).

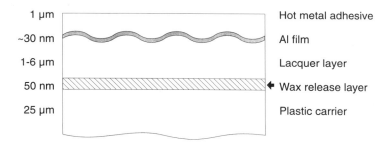

1 μm	Hot metal adhesive
~30 nm	Al film
1-6 μm	Lacquer layer
50 nm	← Wax release layer
25 μm	Plastic carrier

Figure 6.11 Layer structure of hot stamping foil. After application (at the upper adhesive layer) onto plastic cards or paper, the structure separates at the wax release layer (see arrow) to leave the metallized holographic microstructure with a lacquer overcoat.

embossing of very high aspect ratio features is possible, although cycle times of tens of minutes are typical for current technology.

Replication of surface-relief microstructures into materials other than plastics is also possible. In particular, the (room-temperature) embossing into partially cured films of sol–gel materials such as alkoxides of SiO_2, TiO_2 and mixed oxides has been used for the fabrication of diffractive microstructures. Grating waveguide couplers have been fabricated with submicron grating periods for applications in waveguide elements (Dannberg *et al.*, 1994) and evanescent wave sensors

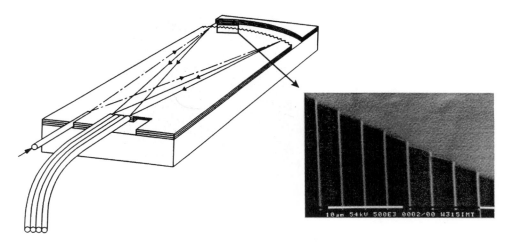

Figure 6.12 Miniature spectrometer produced by hot embossing and enlargement of the replicated vertical grating microstructure, which has a periodicity of about 2 μm and a total height of about 100 μm (courtesy of Forschungszentrum Karlsruhe, Germany; see Müller *et al.*, 1995a,b).

(Tiefenthaler *et al.*, 1983). After embossing, the films are fully cured to produce hard, stable replicas of the microstructure. Popall and co-workers (1994) have described the moulding of similar materials for fabricating microlenses and other micro-optical components.

6.5 Moulding

Plastic moulding techniques are used today for the mass-production of a vast range of consumer and industrial plastic components. Examples in optics include moulded Fresnel lenses for large-area light collecting systems such as condensers and field lenses in cameras and overhead projector systems, and compact discs (CDs) for the digital reproduction of audio and software data. The moulding of high aperture Fresnel lenses was described in the early 1950s by Miller of Eastman Kodak (Miller *et al.*, 1951). Reviews of moulded plastic micro-optics have been given by Pasco and Everest (1978) and by Teyssier and Tribastone (1990).

Moulding involves the stepwise forming of molten plastic under high pressure in a temperature-controlled mould containing the relief microstructure to be replicated (Fig. 6.13). The plastic is then cooled and solidified in the mould before removal, resulting in a very high fidelity copy of the microstructure. Cycle times vary considerably, from minutes down to seconds for modern CD moulding equipment. In injection moulding, the plastic material is injected as a pre-heated molten resin into the mould; this results in short cycle times such as required for CD production. Compression moulding involves introducing the plastic in either powder or sheet form and pressing between two heated dies, resulting in longer cycle times but better reproduction of deeper structures (e.g. for Fresnel lenses). Many combinations of the basic moulding process also exist, such as injection–compression moulding in which the mould is fully clamped after the injection

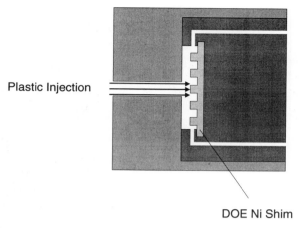

Plastic Injection

DOE Ni Shim

Figure 6.13 Schematic of injection moulding process.

of molten plastic. The most common materials used for the moulding of high-resolution optical microstructures (such as CDs) are polycarbonate and PMMA.

Moulding technology is, in principle, capable of very high-resolution replication and the separation of replicated plastic from the mould only after cooling is an advantage for the highly accurate reproduction of submicron and deep features. However, current commercial moulding technology has in general not yet reached the combination of very high resolution ($\ll 1\,\mu$m) and structure depth ($>1\,\mu$m) requirements of many DOE microstructures. Individually, both types of structure are routinely produced commercially.

Very high resolution 'moth-eye' surface-relief grating microstructures with a grating periodicity of about 250 nm are injection moulded by PLASMON Data Ltd[3] for WORM (Write Once, Read Multiple) data storage disks (Gale, 1989). Figure 6.14 shows equipment and micrographs of submicron moth-eye microstructures produced at PLASMON by a modified CD injection moulding process. The stamper and moulded CDs of test DOE microstructure replicated by a standard CD moulding process is shown in Fig. 6.15.

Deep optical microstructures ($\gg 1\,\mu$m relief depth) are injection moulded as LIGA process components (Rogner and Ehrfeld, 1991; Brenner *et al.*, 1993; Göttert *et al.*, 1993). The injection moulding of integrated optical structures such as channel waveguide circuits with micrometre lateral and depth dimensions is being developed and commercialized by a number of research groups (Rogner, 1992; Neyer *et al.*, 1993; Kalveram *et al.*, 1995). The injection moulding of shallow ($<1\,\mu$m modulation depth) DOE microstructures has been described by Coops (1990), who showed that high quality optical elements can be mass-produced by this technology.

6.6 Casting and UV-replication

High quality spectroscopic grating replicas have long been fabricated by casting techniques, in which a resin is applied to the surface of a master form and allowed

Figure 6.14 Injection moulding equipment and details of submicron, moth-eye grating microstructures fabricated for optical disk data storage applications (courtesy of Plasmon Data Ltd, Melbourn, England). The planarized areas are data written by locally heating the grating microstructure with a focused laser beam, resulting in flow and levelling of the surface relief.

to set (Hutley, 1982). The application of release and metal layers to the form can assist separation and the curing process can be speeded by heating, although the long curing time of typically many hours is a disadvantage for many applications. The technique is, however, well established and capable of producing very high fidelity copies of high-resolution, deep and high aspect ratio gratings and other microstructures.

UV-replication combines this approach with rapid curing by exposure to ultra-violet (UV) radiation. The basic approach is illustrated in Fig. 6.16. The Ni shim is coated with a thin layer of UV-curable material (photopolymer) and pressed against a glass substrate. The sandwich is then cured under pressure and separated after hardening. A thin release coating such as a sputtered Teflon-like layer can be applied to the shim to assist separation. Optical UV-curable materials include 'adhesives' such as Norland NOA 61.[4] Since this is applied as a liquid and hardened in contact, the technique is highly suited to the accurate replication of

Figure 6.15 Master nickel stamper (lower centre) and injection moulded CDs produced by standard CD injection moulding in a test fabrication run (carried out within EUREKA project EU 922, FOTA in collaboration with Baumer Electric AG, Frauenfeld, Switzerland and OMD, Diessenhofen, Switzerland).

UV - radiation

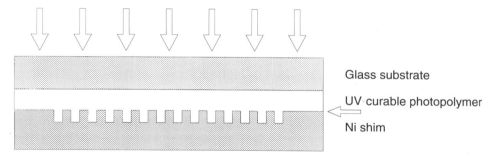

Figure 6.16 UV-embossing process. The photopolymer is cured in contact with the Ni shim and subsequently separated (see arrow) leaving the microstructure replicated into a thin polymer film on the substrate.

high resolution ($\ll 1\,\mu$m) and deep ($\gg 1\,\mu$m) microstructures. Lateral shrinkage of materials such as NOA 61 on curing is about 1%. Suitable UV lamps are available commercially. The technique can be applied in mass-production, for example using a shim mounted onto a roller to replicate into UV-curable material on a plastic tape. Figure 6.17 shows a system used at EPIGEM[5] for the rapid and large-volume replication of diffractive and integrated optical microstructures.

Figure 6.17   Commercial roller UV-embossing system (courtesy of EPIGEM Ltd, Wilton, England).

Coops (1990) has described the use of UV-curable lacquers for the replication of DOE grating microstructures of about 500 nm modulation depth, for applications in CD optical pick-ups. Diffraction efficiencies equal to those of injection moulded replicas were demonstrated for elements produced from both quartz and Ni moulds.

Dry photopolymers have also been developed especially for the replication of DOEs. A dry photopolymer material SURPHEX is marketed by Du Pont (Shvartsman, 1993); it can be laminated onto a substrate, embossed and UV-cured under pressure and separated after hardening. Accelerated life-testing has shown excellent stability and lifetime for the optical replication material. Shvartsman (1993) has also given an overview of the various types of photopolymer replication materials and properties.

UV-replication offers many attractive features for the replication of DOEs. The process can be high speed in continuous, roller embossing systems. The material cost is higher than for hot embossing, but improved fidelity for very fine, deep microstructures with high depth to width ratios can justify this for many DOE types.

6.7 Comparison of Replication Technologies

A meaningful comparison of the various replication technologies for surface-relief micro-optics is difficult due to the wide variation of microstructure dimensions involved. A number of general observations can, however, be made:

- The major industrial replication technologies (in terms of volume production) for diffractive surface-relief microstructures are hot roller embossing and injection moulding. They are well established and capable of reproducing high-resolution, shallow microstructures with lateral feature sizes down to about 300 nm and relief depths up to about 1 μm, provided that the feature width to depth ratio is greater than about 1.

- The replication of microstructures with deep relief (>1 μm) is not yet an established mass-production process, but has been demonstrated for diffractive elements using a variety of replication technologies. Examples include UV-embossing and customized injection moulding. As the market for such elements increases, it can be expected that these technologies will also become established for this application.

- Injection moulding approaches can also replicate mounting, alignment and housing structures in the same fabrication step. They thus have great potential for the production of partial or complete micro-optical systems and can significantly lower the net production cost of the assembled system.

- All replication technologies are, in principle, capable of very high fidelity reproduction ($\ll$1 μm feature size). Diffraction efficiency and uniformity closely approaching that of the original microstructure (after correcting for any change in refractive index) can be achieved. The main limitation is the effective cycle time required for the replication and the compromises involved in reducing this to a commercially interesting value.

- It can be expected that the best high-resolution replication results will be obtained using techniques in which separation takes place after hardening of the polymer (e.g. UV-embossing or customized injection moulding).

- The release of the polymer from the shim or mould is a critical process for high fidelity optical structures which can affect the surface roughness and feature resolution on the nanometre scale. The use of release coatings and additives will play a significant role in minimizing this source of error.

- Long-term relaxation and other ageing effects have not yet been investigated in detail, but initial tests indicate that they should not be a problem for diffractive microstructures.

Table 6.2 summarizes the main replication technologies and gives a qualitative comparison of their strengths and weaknesses. It can be expected that replication technology in general, with its potential for low-cost, high-volume production, will play an increasingly important role in the commercialization of micro-optical elements and their widespread introduction into products.

Acknowledgements

Numerous inputs from industries and university research groups are gratefully acknowledged and referenced at the appropriate places in the text. Particular thanks go to Baumer Electric AG (Frauenfeld, CH), CSEM (Neuchâtel, CH), EPIGEM (Wilton, GB), Hologram. Industries (Paris, F), IMM (Mainz, D), Dr

Table 6.2 Replication technologies for micro-optical elements

Technology	Typical micro-optical elements	Strengths	Weaknesses	References
Hot embossing: Flat bed	DOEs in general, Fresnel elements, Integrated optical microstructures	Good replication for high resolution and deep microstructures	Slow (typical cycle times of minutes)	Chou et al., 1995; Baraldi, 1994; Knop and Gale, 1980
Hot embossing: Reciprocating press	DOEs, Fresnel elements	Faster than flat bed (typical cycle times of seconds)	Resolution and aspect ratio generally inferior to flat bed embossing	Kluepfel and Ross, 1991
Hot embossing: Rotary press	Holograms	Established technology, high speed, continuous, roll-to-roll	Limited depth (max. ~1 µm at present) and aspect ratio	Kluepfel and Ross, 1991
Hot embossing: LIGA press	High aspect ratio microstructures	Demonstrated technology for LIGA microstructures such as micro-spectrometer	Slow (typical cycle times ~tens of minutes)	Müller et al., 1995b; Göttert et al., 1993
Injection moulding: CD type	CDs, shallow DOEs and grating microstructures	Established technology with automated production lines, high speed (cycle time ~5 s)	Limited aspect ratio and depth (~1 µm)	Gale, 1989
Injection moulding	Micro-optical lenses, LIGA microstructures, DOEs, integrated optical microstructures	High aspect ratios, simultaneous moulding of alignments and housing features	High cost of mould (typically 50–100 k$)	Teyssier and Tribastone, 1990; Brenner et al., 1993; Neyer et al., 1993; Coops, 1990; Rogner and Ehrfeld, 1991
Casting	Spectroscopic gratings	High resolution and aspect ratios possible, good dimensional stability	Slow (typical curing times of several hours), expensive	Hutley, 1982
UV-embossing	DOEs, Fresnel microlenses, integrated optical microstructures	High resolution and aspect ratios, high speed potential	Costs (more expensive than hot embossing)	Shvartsman, 1993; Coops, 1990; EPIGEM[5]

A. Neyer (University of Dortmund, D), OMD (Diessenhofen, CH) and PLAS-MON Data Ltd (Melbourn, GB). Work at PSI Zurich would not have been possible without the skilled and dedicated contributions of the members of the Optics and other groups in the laboratory, as well as the support of the Swiss Priority Program OPTIQUE and other programs of the Board of the Swiss Federal Institutes of Technology.

Notes

1. 3D Ltd, Gewerbestrasse 17, CH-6314 Unterägeri, Switzerland.
2. Hologram. Industries SA, 42-44, rue de Trucy, F-94120 Fontaine-sous-bois, France.
3. Plasmon Data Limited, Whiting Way, Melbourn, Royston, Hertfordshire SG8 6EN, England.
4. Norland Products Inc., PO Box 7145, N. Brunswick, NJ 08902, USA.
5. Epigem Ltd, Wilton Materials Research Centre, Wilton, Middlesbrough, Cleveland TS90 8JE, England.

References

ALTKORN, R., CHANG, J., HAIDLE, R., TAKACS, P. Z. and ULMER, M. P. (1992) Electroform replication of smooth mirrors from sapphire masters. *Appl. Opt.* **31**, 5153–5154.

BARALDI, L. G. (1994) *Heissprägen in Polymeren für die Herstellung integriert-optischer Systemkomponenten*, PhD Thesis ETH Nr. 10762 (in German), Zürich: ETH, Eidgenössische Technische Hochschule Zürich.

BECKER, E. W., EHRFELD, W., HAGMANN, P., MANER, A. and MÜNCHMEYER, D. (1986) Fabrication of microstructures with high aspect ratios and great structural heights by synchrotron radiation lithography, galvanoforming and plastic moulding (LIGA Process). *Microelectron. Eng.* **4**, 35–36.

BRENNER, K.-H., KUFNER, M., KUFNER, S., MOISEL, J., SINZINGER, S., TESTORF, M., GÖTTERT, J. and MOHR, J. (1993) Application of three-dimensional micro-optical components formed by lithography, electroforming and plastic molding. *Appl. Opt.* **32**, 6464–6469.

BUDZINSKI, CH. and TIZIANI, H. J. (1995) Radiation resistant diffractive optics generated by micro-electroforming (in German). *Laser und Optoelektronik* **27**, 54–61.

CHOU, S. Y., KRAUSS, P. R. and RENSTROM, J. (1995) Imprint of sub-25 nm vias and trenches in polymers. *Appl. Phys. Lett.* **67**, 3114–3116.

COOPS, P. (1990) Mass production methods for computer-generated holograms for CD optical pickups. *Philips J. Res.* **44**, 481–500.

DANNBERG, P., BRÄUER, A., KARTHE, W., WALDHÄUSL, R. and WOLTER, H. (1994) Replication of waveguide elements using ORMOCER material. In Reichl, H. and Heuberger, A. (eds) *Microsystem Technologies '94*, pp. 281–287. Berlin: VDE-Verlag.

FELDSTEIN, N. (1970) Two room-temperature electroless nickel plating baths – properties and characteristics. *RCA Rev.* **31**, 317–329.

GALE, M. T. (1989) Diffraction, beauty and commerce. *Phys. World* **2**, 24–28.

GALE, M. T., BARALDI, L. G. and KUNZ, R. E. (1994a) Replicated microstructures for integrated optics. *Proc. SPIE* **2213**, 2–10.

GALE, M. T., ROSSI, M., PEDERSEN, J. and SCHÜTZ, H. (1994b) Fabrication of continuous-relief micro-optical elements by direct laser writing in photoresist. *Appl. Opt.* **33**, 3556–3566.

GÖTTERT, J., MOHR, J. and MÜLLER, C. (1993) Examples and potential applications of LIGA components in micro-optics. In Ehrfeld, W., Wegner, G., Karthe, E., Bauer, H.-D. and Moser, H. O. (eds) *Integrated Optics and Micro-Optics with Polymers*, pp. 219–247. Stuttgart, Teubner Verlag.

GRÖNING, P., SCHNEUWLY, A. and SCHLAPBACH, L. (1995) Deposition of an ultrathin Teflon-like polymer film with an extremely low surface energy. In *Proc. 6th European Conf. on Applications of Surface and Interface Analysis ECASIA 95*, Montreux, Switzerland, 9–13 October 1995.

HAGENA, O. F., BACHER, W., HECKELE, M., MOHR, J., MORITZ, H. and MÜLLER, C. (1995) Erfahrungen beim Aufbau und Betrieb einer Kleinseriefertigung für LIGA-Spektrometer. In *Proc. 2. Statuskolloquim des Projektes Mikrosystemtechnik*, Forchungszentrum Karlsruhe (Germany), Wissenschaftliche Berichte Vol. FZKA 5670, pp. 41–44.

HANNAN, W. J., FLORY, R. E., LURIE, M. and RYAN, R. J. (1973) Holotape: A low-cost pre-recorded television system using holographic storage. *J. SMPTE* **82**, 905–915.

HUTLEY, M. C. (1982) *Diffraction Gratings*, pp. 125–127, London: Academic Press.

KALVERAM, S., GROSS, M. and NEYER, A. (1995) Injection moulded 2×8 couplers for optical communications. In *Proc. 7th European Conf. on Integrated Optics ECIO '95*, Delft, NL, 3–6 April 1995, pp. 77–80.

KLUEPFEL, B. and ROSS, F. (eds) (1991) *Holography Market Place*. Berkeley, CA: Ross Books.

KNOP, K. and GALE, M. T. (1980) *Surface-relief images for colour reproduction*. London: Focal Press.

LEHMANN, H. W., FRICK, K., WIDMER, R., VOSSEN, J. L. and JAMES, E. (1978) Reactive sputtering of PTFE films in argon–CF_4 mixtures. *Thin Solid Films* **52**, 231–235.

MILLER, O. E., McLEOD, J. H. and SHERWOOD, W. T. (1951) Thin sheet plastic Fresnel lenses of high aperture. *J. Opt. Soc. Am.* **41**, 807–815.

MOREAU, W. M. (1988) *Semiconductor Lithography: Principles, Practices and Materials*. New York: Plenum Press.

MÜLLER, C., KRIPPNER, P., KÜHNER, T. and MOHR, J. (1995b) Leistungsfähigkeit und Anwendungsgebiete von UV-VIS und IR-LIGA-Spektrometern. In *Proc. 2. Statuskolloquim des Projektes Mikrosystemtechnik*, Forchungszentrum Karlsruhe (Germany), Wissenschaftliche Berichte Vol. FZKA 5670, pp. 175–179.

MÜLLER, C., FROMHEIN, J., GÖTTERT, J., KÜHNER, T. and MOHR, J. (1995a) Microspectrometer based on integrated optic components in polymers as spectral detection system for the VIS- and NIR-range. In *Proc. 7th European Conf. on Integrated Optics ECIO '95*, Delft, NL, 3–6 April 1995, pp. 491–494.

NEYER, A., KNOCHE, T., MÜLLER, L., KLEIN, R., FISCHER, D. and YUNKIN, V. A. (1993) Low cost fabrication technology for low loss passive polymer waveguides at 1300 nm and 1550 nm. In *Proc. ECOC '93*, Montreux, Switzerland, 12–16 September 1993, pp. 337–340. Zurich: Swiss Electrochemical Association.

PALERMO, P., KORPEL, A., DICKINSON, G. and WATSON, W. (1977) Video disc mastering and replication. *Optics and Laser Tech.* **9**, 169–174.

PASCO, I. K and EVEREST, J. H. (1978) Plastics optics for opto-electronics. *Optics and Laser Tech.* **10**, 71–76.

POPALL, M., KAPPEL, J., SCHULZ, J. and WOLTER, H. (1994) ORMOCERs, inorganic–organic polymer materials for applications in micro systems technology. In Reichl, H. and Heuberger, A. (eds) *Microsystem Technologies '94*, pp. 271–279, Berlin: VDE-Verlag.

ROGNER, A. (1992) Micromoulding of passive network components. *Proc. Conf. on Plastic Optical Fibres and Applications*, Paris. IGI Europe.

ROGNER, A. and EHRFELD, W. (1991) Fabrication of light guiding devices and fiber coupling structures by the LIGA process. *Proc. SPIE* **1506**, 80–91.

SHVARTSMAN, F. P. (1993) Replication of diffractive optics. In Lee, S. H. (ed.) *Diffractive and Miniaturized Optics*, Vol. CR49, pp. 117–137. Bellingham: SPIE.

TEYSSIER, C. and TRIBASTONE, C. (1990) Plastic optics: challenging the high-volume myth. *Lasers & Optronics*, December, 50–53.

TIEFENTHALER, K., BRIGUET, V., BUSER, E., HORISBERGER, M. and LUKOSZ, L. (1983) Preparation of planar optical SiO_2–TiO_2 and $LiNbO_3$ waveguides with a dip coating method and an embossing technique for fabricating grating couplers and channel waveguides. *Proc. SPIE* **401**, 165–173.

TOENSMEIER, P. (ed.) (1995) *Modern Plastics Encyclopaedia '95* and *Modern Plastics Encyclopaedia '95 Handbook*. New York: McGraw Hill.

7

Planar Integrated Free-space Optics

J. JAHNS

7.1 Why Integration?

Recent research on the use of free-space optics for computing and switching applications (Midwinter, 1993; Jahns and Lee, 1994) has led to, amongst other things, the insight that new packaging techniques are required that allow the building of compact, stable and relatively inexpensive systems with high alignment precision. A typical requirement often encountered in photonic applications is an alignment precision of a few micrometres or even less. Furthermore, mechanical and thermal stability has to be guaranteed. Experience has shown that conventional optomechanics is not able to satisfy these requirements simultaneously. Although laboratory systems have been demonstrated for the use of free-space optics for interconnection applications, such systems are usually bulky and expensive. The cost is due to a large extent to the amount of human manpower required in the assembly of a complex system.

An alternative to optomechanical technology is the use of micro-optics. Micro-optical elements fabricated by lithography and etching have been known for about 20 years (Dammann and Görtler, 1971); different types of refractive and diffractive micro-optical elements have been developed during the past couple of years. Many of them are presented in the other chapters of this book, therefore we do not discuss them here. Instead we will look at the use of micro-optics at a systems level. Specifically, we will discuss an approach that allows the building of integrated free-space optical systems which results in a considerable reduction in the volume and cost of an optical system. This is achieved by monolithic integration of the optics on a single substrate. This, in turn, also yields systems with much improved stability, high alignment precision, and considerably less cost than conventional optomechanical systems.

Two approaches to micro-optical integration have been suggested: the 'stacked approach' and the 'planar approach' (Fig. 7.1). The use of stacked micro-optics was originally proposed by Iga *et al.* (1982) and later pursued by Brenner (1991; see also Chapter 8 of this book) and Hamanaka (1991) for interconnection purposes, and by Veldkamp (1993) for sensor applications. Conceptually, the

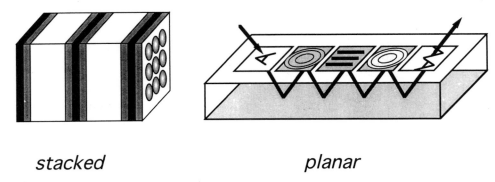

stacked   planar

Figure 7.1 Approaches to micro-optical integration of free-space optical systems.

stacked approach is similar to conventional free-space optics: passive and active micro-optical components are placed behind each other in a three-dimensional fashion as indicated in the figure.

In the planar approach (Jahns and Huang, 1989), micro-optical elements are put on one or two sides of an optical flat using a two-dimensional layout. Light propagates inside the substrate along a zigzag path. The main purpose of the two-dimensional geometry is the fabrication compatibility with standard technology that is used in the manufacturing of electronic integrated circuits. Interconnection experiments using light propagation inside glass substrates have also been performed by Hase (1987), Sauer (1989), Kostuk *et al.* (1989) and Prongué and Herzig (1989).

In the remainder of this chapter, we describe in detail the planar optics approach to building integrated free-space optical systems. First, we present a general description of the approach in section 7.2. Then, we describe various demonstration experiments (section 7.3), consider system properties (section 7.4), and discuss several potential applications (section 7.5).

7.2 Planar Optics

In the planar optics concept, micro-optical elements (diffractive, refractive or a combination thereof, mostly used in reflection) are placed on an optical substrate using lithographic techniques (Fig. 7.2). The positioning of the elements is achieved with a very high precision, typically in the submicron range even for large substrates with an area of several square inches. The light signals travel inside the substrate following a zigzag path. The areas where the light hits the substrate surfaces are coated with reflective layers. The substrate has a thickness of several millimetres to allow sufficient lateral propagation of the signal. The area of one micro-optical element is typically 1 mm^2. Integrated optical 4f-systems have been demonstrated in which the optics covered a total area of approximately 1×10 mm^2 (Jahns, 1990).

As already mentioned above, one of the virtues of the planar optics approach is the two-dimensional layout that makes it compatible with standard lithographic and surface-mount techniques. This is desirable because it makes available a whole range of equipment and technology. Unlike in conventional optomechanics with

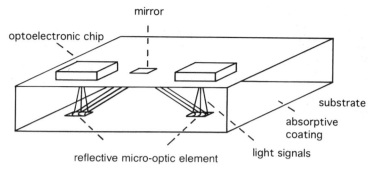

Figure 7.2 Planar optical system consisting of a glass plate with integrated optics and optoelectronic chips.

its time-consuming sequential assembly, the optical elements are fabricated simultaneously on the same substrate. Micro-optical elements with a surface topology may be preferable for the implementation of the optics, since they can be replicated using inexpensive embossing techniques (Gale *et al.*, 1994). The replication of whole systems is an interesting step towards making inexpensive planar optical interconnects.

Another interesting feature is the fact that the optical substrate not only serves as the interconnection medium, but also as a motherboard for optoelectronic chips. These are surface mounted onto the substrate using hybrid integration techniques such as flip-chip bonding (Miller, 1969). The chips are mounted independently of each other so that tolerances inherent in the mounting process do not add up in complex systems. The lateral alignment precision of the flip-chip bonding process is in the range of 1–2 micrometres (Goodwin *et al.*, 1991). The fact that the chips are mounted on the surface also makes them accessible for testing, repair and cooling purposes. The substrate can be made of various materials. Fused silica (SiO_2) lends itself to this application because high quality substrates are readily available in many sizes. However, other glasses or semiconductor materials (Si or GaAs) are also of interest provided they are transparent at the wavelength used. In general, the requirements for substrate quality are easily obtained with standard optical flats. However, special requirements with respect to the precision of the thickness and coplanarity of the surfaces have to be considered and will be discussed later.

The zigzag propagation of the light inside the substrate is a challenge to the design of planar optical systems. There are two reasons for this: first, optical designs have to be found that allow the imaging of a reasonably large number of pixels, despite aberrations such as astigmatism and coma that occur when going off-axis. Second, it is necessary to make efficient coupling elements for sending light into the substrate under reasonably large angles. Diffractive elements can be used for this purpose. However, technology and physics impose certain limits for making these elements and limit the coupling angle α to values of 5–25°. The resolution of conventional optical lithography often limits one to relatively small angles, even in the case of elements with only two phase levels. Larger coupling angles of 40–50° can be achieved, for example, using advanced direct-write electron beam lithography, at least for gratings.

The surface-relief structure of etched diffractive elements can also cause problems because of the so-called 'shadowing effect' which may result in a decrease of the light efficiency (Hutley, 1982). Larger coupling angles can be achieved if advanced lithography techniques are used to make the coupling elements such as direct electron beam writing (Nishihara and Suhara, 1987; Shiono *et al.*, 1987) or lithography with a reduction stepper (Walker *et al.*, 1993). Diffractive elements made of dichromated gelatine allow one to realize large coupling angles (Haumann *et al.*, 1990); however, they are difficult to make and are not compatible with subsequent fabrication steps that are necessary to build an integrated opto-electronic system.

7.3 Planar Optical Interconnections

In this section, we describe three demonstration experiments which show the usefulness of planar optics as an interconnection technology. These are: optical signal distribution (section 7.3.1); optical imaging in planar optical systems (7.3.2); and the integration of the passive optics with optoelectronic chips such as surface-emitting microlasers and modulator arrays (section 7.3.3). Other potential applications of planar optics will be mentioned later in section 7.5.

7.3.1 *Signal Distribution*

The distribution of a signal is one of the tasks that may be of interest for interconnection applications. Specific examples are clock distribution and the broadcasting of signals. Here, we will consider the task of clock distribution. In order to guarantee synchronicity between the various parts of a complex system such as a computer or a switching system, a clock signal is distributed from the central clock to the processors or stations (Fig. 7.3). Physically, these may be racks, boards, chips, or even various parts of one chip. A planar optical implementation may be useful at the board-to-board level down to the intra-chip level.

The concept of a planar optical clock distribution is shown in Fig. 7.4. An incoming collimated laser beam which carries the clock signal is split several times by a cascade of beam splitters. The beam splitters can be implemented with high efficiency as binary phase gratings (Walker and Jahns, 1992; Walker *et al.*, 1993) (Figs 7.5 and 7.6). Beam deflector gratings are used to control the light propagation of the substrate. If the angle of propagation is large enough inside the substrate, then again binary phase gratings can be used for the implementation of the beam deflector gratings. In this case, it is assumed that higher orders are evanescent. If this is not the case, blazed gratings can be used which are optimized for sending a maximum amount of light into a specific diffraction order (Fig. 7.5). Such blazed gratings can be fabricated, for example, by direct electron beam lithography. An example of such a grating is shown in Fig. 7.7 (Walker *et al.*, 1993).

A large fanout of the input beam can be obtained by cascading several stages of the basic 1×4 system (Walker and Jahns, 1992). The fanout, i.e. the number of generated beams, grows exponentially with the number of stages. The size of the cascade is ultimately limited by power budget considerations, i.e. by the power of the input beam, the sensitivity of the detectors, and the light efficiency of the

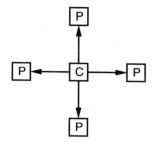

Figure 7.3 Scheme of optoelectronic processor with central clock and processor modules.

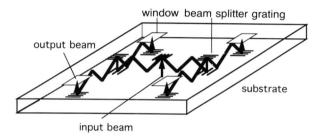

Figure 7.4 1 × 4 signal distribution using planar optics.

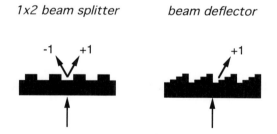

Figure 7.5 1 × 2 beam splitting and beam deflection using diffraction gratings.

optical system. The efficiency of the optical system, in turn, is largely determined by the efficiency of the mirrors. Very high reflectivities of approximately 99% are required. Such values can be achieved with silver mirrors or with dielectric mirrors. It is estimated that clock distribution systems with a fanout of 64 or more may be feasible (Walker and Jahns, 1992).

The propagation of a light beam over large distances imposes specific demands on the quality of the substrate. Parameters such as substrate thickness and coplanarity have to be manufactured with high precision, otherwise the light propagation may become instable; i.e. the output beam may not appear at the designed output position. However, a suitable tolerant design of the optics allows the tolerances to be increased significantly (Jahns, 1994). A more detailed treatment of the stability of light propagation and the parameter tolerances are discussed later (section 7.4).

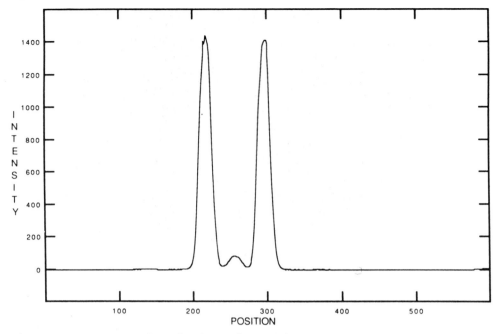

Figure 7.6 Experimental result of 1 × 2 beam splitter grating.

7.3.2 *Imaging*

The routing of a single light beam through a system is not a challenging task optically, since by using a suitable design a single beam can be imaged without aberrations. Things become more complicated when we want to image a large number of beams through the same optical system. In this case, aberrations put a limit on the number of channels. The specific problem of planar optics is the fact that the light signals impinge on the optical elements under non-normal incidence, leading to aberrations such as astigmatism and coma.

In order to keep the influence of aberrations small, we can benefit from the fact that interconnection applications require only the handling of light beams that originate from discrete light sources, as indicated in Fig. 7.8. This is true, for example, in the case of 'smart pixels'. These are optoelectronic devices with optical input and output positions separated by a pitch d. It has been pointed out by Lohmann (1991) and McCormick *et al.* (1991) that in this case a conventional 4f-imaging system makes inefficient use of the optical space-bandwidth product, because large areas between the light sources are imaged even though they are not of interest. The use of a lenslet-based imaging system improves the situation since every light source gets its own miniature imaging setup and light from between the sources is purposely lost because of vignetting. The advantage is gained, however, at the expense of propagation distance. In order to avoid crosstalk between the channels caused by diffraction, the distance Δz between the lenslet arrays must not be larger than $(\pi/\lambda)d^2$ (Leggatt and Hutley, 1991; Sauer *et al.*, 1994). Here, we have assumed that the aperture of the lenslets is identical to the pitch d.

Figure 7.7 High frequency blazed grating fabricated by direct electron beam writing.

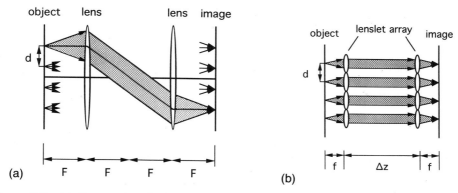

Figure 7.8 (a) Conventional $4f$ system. F = focal length of the lenses. (b) Microchannel system using lenslet arrays. f = focal length of the microlenses, d = pitch, Δz = separation of the microlens arrays.

If we are only interested in simple imaging, a solution to this problem is possible by a technique called hybrid imaging, which combines the $4f$ and the lenslet approach (Lohmann, 1991). Figure 7.9 shows an integrated version of such a system (Jahns and Acklin, 1993). In the hybrid imaging system, lenslets are used to collimate the light from the various sources. Then a $4f$-setup images the lenslet array on the input side to a second array on the output side. The lenslets in the second array focus the light to detectors or modulators. In the hybrid system, the imaging task is split in a favourable way between the lenslets that provide the high resolution and the lenses in the $4f$-part. The $4f$-system allows the propagation of light over large distances. However, it does not have to provide high resolution, and so, relatively slow lenses can be used (Jahns *et al.*, 1994). This leads to two beneficial effects: first, the influence of aberrations is reduced, and second, the depth of focus is increased. Both effects are beneficial to a planar optical

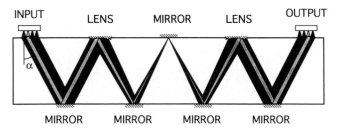

Figure 7.9 Integrated hybrid imaging system. The 4*f* part forms an image of the input lenslet array onto the output array.

implementation: the first effect solves our aberration problem, the second increases substrate tolerances. The reduction of the aberrations is demonstrated in Fig. 7.10. Here the spot diagrams for a 4*f*-system and for a hybrid system are shown. An experimental result for an integrated hybrid imaging system is shown in Fig. 7.11 (Jahns and Acklin, 1993).

At this point, we would like to mention that, based on imaging operations, one can also 'synthesize' space-variant interconnections if the space-variance is introduced in the image plane. An example is the crossover interconnection network (Jahns and Murdocca, 1988) where special retroreflective arrays are used to shuffle the light beams around. A planar optical implementation which is based on the use of micro-machined retroreflectors was recently suggested by Song *et al.* (1992).

7.3.3 *Hybrid Integration of Passive Optics with Optoelectronic Chips*

In order to be able to build systems with high functionality, we must consider ways of integrating passive optics with optoelectronic components. Flip-chip solder bump bonding is an available technique that allows the surface-mounting of chips on a substrate and is therefore ideally suited for the planar optics concept. The process itself is visualized in Fig. 7.12(a). The chip and the substrate have matching arrays of solder bumps. The chip is aligned relative to the substrate until the solder bumps are in good alignment. The subsequent bonding is done either thermally, ultrasonically, electrostatically or with conductive adhesives. Several parameters are of interest: the pitch p of the solder bumps; the height t of the solder bumps; and their size or diameter d. The larger the number of bumps, the larger the pressure has to be during the bonding procedure. The lateral alignment precision after the bonding process δx is another important parameter. For optoelectronic applications, typical values are $p = 100 \, \mu m$, $t = 10 \, \mu m$ and $\delta x \leq 2 \, \mu m$. The alignment precision can be further increased by a subsequent reflow operation. During this step, the solder is heated up beyond its melting point. Under the influence of the surface tension in the liquid solder, the bumps tend to self-align. This effect can be used to achieve submicron alignment precision (Goodwin *et al.*, 1991). The pitch can also be reduced to smaller values if necessary. This might be the case, for example, for optoelectronic interconnections if one wants to achieve a high interconnection density. The height of the solder bumps depends

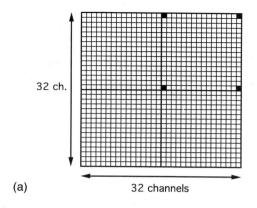

(a)

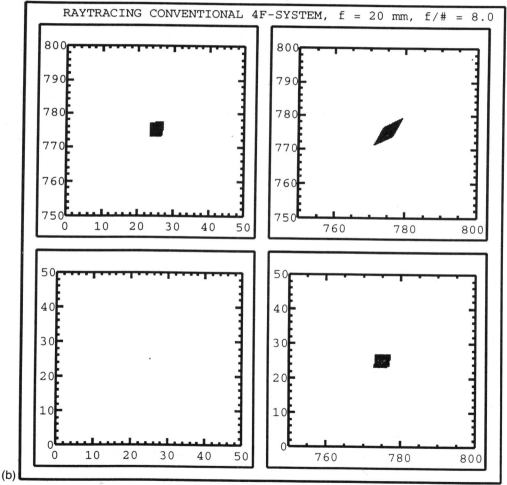

(b)

Figure 7.10 (a) Array with 32 × 32 pixels. For the positions that are hatched, spot diagrams are shown in (b) and (c). (b) Diagrams calculated for a conventional 4*f*-setup. Part (c) continued over.

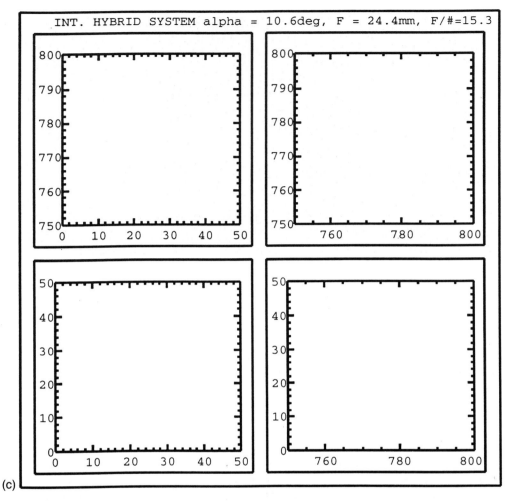

(c)

Figure 7.10 (c) Diagrams calculated for an integrated hybrid imaging system. Parameters for both systems are given in the figures.

on the choice of the solder material, which in turn depends on the choice of the bonding process. If thermocompression is used, relatively soft materials such as gold or indium or a legation of lead and tin are used as solder materials.

Solder bump bonding offers two functions that are interesting for planar optics: first, it provides electrical contacts between the substrate and the chip, and second, it provides a precise alignment between the chip and the substrate.

The use of flip-chip bonding for optoelectronic applications has been demonstrated by Goodwin *et al.* (1991) and von Lehmen *et al.* (1991). To demonstrate its applicability to the planar optics concept, we bonded a chip with an array of vertical-cavity surface emitting lasers (VCSELs) on a glass substrate (Jahns *et al.*, 1992a,b). The integrated optics consisted of two one-dimensional arrays of diffractive lenslets on a 3 mm thick glass substrate. The optics routed the light from each laser to a well-defined output position. This concept is shown in Fig. 7.12(b). In our experiment the pitch of the lasers was $p = 507 \mu$m, and the

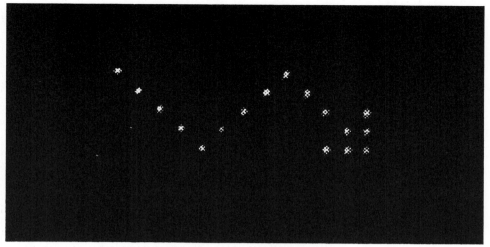

Figure 7.11 Experimental result for the ingegrated hybrid imaging system with 32 × 32 channels. The channel spacing is 50 μm in each direction. Seventeen of the 1024 channels were illuminated using sequential recording by moving a single mode fibre from one position to the next.

solder bumps were made of In with an area of 90 μm × 90 μm and a height of approximately $t = 5$ μm. The separation between the input and output positions in the x-direction was 1.2 mm.

7.4 Tolerance Considerations

As in a conventional optical system, it is important to consider the influence of imperfections such as misalignment, temperature changes, wavelength variations, etc. on the performance of the optics. Here, we consider the influence of substrate imperfections and temperature changes. The effect of wavelength variations depends on the type of optical elements that are used. For diffractive and refractive–diffractive systems, some considerations and calculations can be found in Jahns *et al.* (1992a) and in Sauer *et al.* (1994).

7.4.1 *Substrate Imperfections*

We consider two effects: variations in the substrate thickness and the case in which the two surfaces of a substrate are not perfectly coplanar. We consider the effect of these imperfections on the position of a collimated light beam. This case allows a relatively simple analysis that gives useful insights. A practical case where it is important to consider the influence of substrate imperfections on the position of a beam is the clock distribution system of section 7.3.1.

First, we consider the case where the thickness of the substrate may vary (Fig. 7.13(a)). We assume that a light beam enters the substrate under normal incidence. It is then redirected by means of a grating and bounces several times before being

189

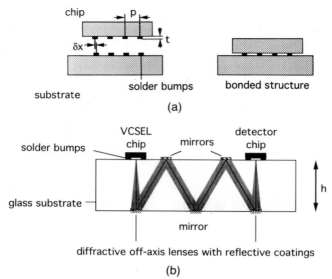

(a)

(b)

Figure 7.12 (a) Principle of flip-chip solder bump bonding. (b) Hybrid integration of optoelectronic chips on a planar optical substrate.

redirected again until it hits the upper surface normally. The position s of the beam and the tolerance Δs of the lateral output position can be easily calculated as a function of the propagation angle α, the ideal substrate thickness h, and the tolerance of the substrate thickness Δh. Obviously there is a linear relationship between Δs and Δh. The lateral position s of the beam is given by

$$s = 2(h + \Delta h)\tan(\alpha). \tag{7.1}$$

When the light beam bounces many times, the offset between the ideal and the actual position grows, as shown in Fig. 7.15(a). The curves in Fig. 7.15 show the offset of the centre position of a collimated light beam from an ideal position in a plane where the beam hits a focusing lens. This lens focuses the light onto a photodiode. An offset in the position of the beam can be tolerated if it is not too large. It is assumed that the beam has a typical diameter of 1 mm (so that diffraction effects can be neglected) and that the lens diameter is 2 mm. Hence an offset of several hundred micrometres could be tolerated.

A similar relationship to equation (7.1) can be derived for the case of a uniform wedge between the two surfaces (Fig. 7.13(b)). The wedge angle is denoted by $\Delta\alpha$. In this case, a beam that hits the bottom surface under an angle α is reflected under an angle $\alpha + 2\Delta\alpha$. This effect accumulates, so that the angle becomes increasingly larger with the number of bounces. In addition, the thickness of the substrate varies with the lateral position. Therefore, the equations for the position of the beam are more complicated than in the case of thickness variation. For the situation shown in Fig. 7.13(b), the beam position is given as

$$s = h\frac{\tan(\alpha) + \tan(\alpha + 2\Delta\alpha)}{1 - \tan(\alpha)\tan(\Delta\alpha)}. \tag{7.2}$$

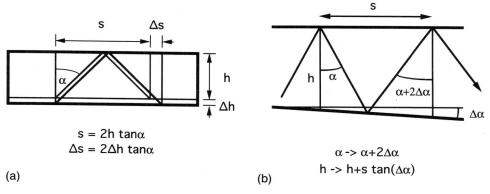

$$s = 2h \tan\alpha$$
$$\Delta s = 2\Delta h \tan\alpha$$

(a)

$$\alpha \rightarrow \alpha + 2\Delta\alpha$$
$$h \rightarrow h + s \tan(\Delta\alpha)$$

(b)

Figure 7.13 (a) Thickness variation of a substrate and resulting offset of a light beam. (b) Wedge between the two surfaces.

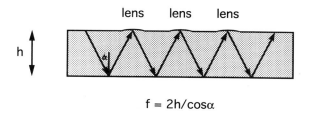

$$f = 2h/\cos\alpha$$

Figure 7.14 Lenses at equidistant positions can improve the stability of light propagation inside the substrate. The focal length f of the lenses is chosen such that the action of a confocal resonator is simulated.

Again, if we allow many bounces the error in the beam position Δs grows, as shown in Fig. 7.15(b).

A possibility for correcting this effect is indicated in Fig. 7.14. Here lenses are placed in well-defined positions on the top surface. The focal length of the lenses is chosen to be $2h/\cos\alpha$. In this case, the lenses have the same effect as the mirrors in a confocal resonator (Siegman, 1986). The confocal resonator is known to be the most stable laser configuration and is very immune to misalignment of the mirrors. Despite the fact that in a planar optical substrate the number of bounces is considerably smaller than in a laser resonator, we have an analogous situation with the stability of the light propagation. Consequently, we apply the same solution as in laser physics and use the configuration shown in Fig. 7.14 to improve the stability. For both cases that were considered above, Figs 7.15(a) and (b) show the improvement in the lateral positioning of the beam with the corrective influence of the lenses. The deviation from the ideal position Δs in the corrected system is now typically of the order of 100 μm or less. This offset is small enough to collect most of the light onto the detector by using a suitable optics.

It is important to note for practical reasons that the required lenses have low numerical apertures. As an example, we assume a beam diameter $D = 1$ mm which

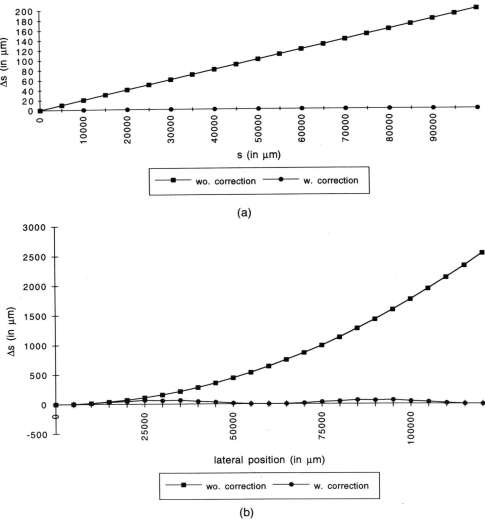

Figure 7.15 Improvement of the stability using the scheme of Fig. 7.14(a) shows the situation where the substrate thickness is off by $\Delta h = 10\,\mu$m. The curve with the squares shows the offset Δs in the case where no correction lenses are used. The round dots show the case with correction lenses. (b) Analogous calculation for a wedge with an angle of $\Delta\alpha = 1$ arcmin.

corresponds to the lens diameter, a substrate thickness $h = 5$ mm, and an angle $\alpha = 20°$. In this case, the required focal length would be 10.6 mm and the corresponding *f*-number is 10.6. If we chose to allow more bounces between the lenses, we can even operate with very slow lenses. This is important, because it allows the use of diffractive lenses which can still be made very efficient at such large values of the *f*-number.

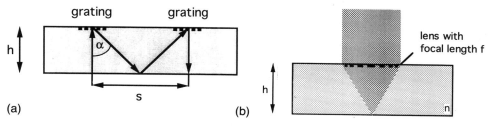

Figure 7.16 Effect of thermal changes on the stability of light propagation (a) for a pair of linear gratings, (b) for the focusing of a lens.

7.4.2 *Temperature Variations*

In this section, we consider the effect of temperature variations on the performance of a grating and a lens in a planar optical system (Jahns *et al.*, 1992a).

In the case of a grating (Fig. 7.16(a)), as it might again be used in the clock distribution scheme, the diffraction angle depends on the period p, the refractive index n and the wavelength λ

$$\sin \alpha = \frac{\lambda}{np} . \tag{7.3}$$

When the temperature T changes, the period and the refractive index change too:

$$\frac{\partial p}{\partial T} = \beta p \quad \text{and} \quad \frac{\partial n}{\partial T} = \gamma n. \tag{7.4}$$

Here, β and γ denote the linear thermal expansion coefficient of the substrate material and the temperature coefficient of the refractive index. The lateral beam position s is given as

$$s = 2h \tan \alpha. \tag{7.5}$$

Using the relationships from above, we can derive

$$\frac{\partial s}{\partial T} = 2h \tan \alpha \frac{\beta - (\beta + \gamma)}{\cos^2 \alpha} . \tag{7.6}$$

For quartz glass (SiO_2), the thermal expansion coefficient is $\beta = 5 \times 10^7/\text{K}$ and the temperature coefficient is $\gamma = 5 \times 10^{-6}/\text{K}$. As a numerical example we choose a substrate thickness of $h = 5$ mm, a propagation angle $\alpha = 20°$ and a temperature tolerance $\Delta T = 100$ K. In this case, the beam offset would be only $\Delta s = -2$ μm.

The treatment is similar for a lens (Fig. 7.16(b)). We assume that at room temperature (T_0) the focal length f of the lens is matched to the thickness h of the substrate. A temperature change affects both the focal length and the substrate thickness. We have to distinguish two cases: the diffractive lens where the focal length f_d is given as $f_d = nR^2/2\lambda$ (where R is a scaling parameter) and the reflective lens (not shown in the figure) with $f_r = -r_c/2$ (where r_c is the radius of curvature). A temperature change affects the focal lengths and the substrate thickness according to

$$\frac{\partial f_d}{\partial T} = (2\beta + \gamma)f \tag{7.7}$$

$$\frac{\partial f_r}{\partial T} = \beta f_r \tag{7.8}$$

and

$$\frac{\partial h}{\partial T} = \beta h. \tag{7.9}$$

The amount of defocusing as a function of the temperature is determined by

$$\frac{\partial(f_d - h)}{\partial T}\bigg|_{T_0} = (\beta + \gamma)f_d \tag{7.10}$$

and

$$\frac{\partial(f_r - h)}{\partial T}\bigg|_{T_0} = 0. \tag{7.11}$$

This means that in the reflective case, no thermal defocusing occurs. In the diffractive case, the amount of defocusing is still small. Using the same values as given above, we find that for a temperature change of 100 K the defocusing would be 2.8 μm.

It may be noted at this point that the thermal properties of diffractive lenses can be combined to athermalize lens systems as discussed, for example, by Behrmann and Mait in Chapter 10 of this book (Behrmann and Mait, 1997).

Although the considerations in this section are incomplete, we can conclude that planar optical systems are stable or can be made very stable with respect to various external influences.

7.5 Applications of Planar Optics

In this section we briefly consider various areas for which planar optics might be useful. The first of these are optical interconnections in computers or switching systems. A specific example might be an optoelectronic multichip module (MCM). A simplistic view of an optical MCM is shown in Fig. 7.17 where optoelectronic chips are placed on a substrate which serves as the interconnection medium. We described the various functions for such an MCM in the previous sections. A specific aspect is the thermal management in such a system. In a recent article, we presented an overall view of such a system which also takes into account possibilities for heat removal from the chips by means of an intermediate layer (Fig. 7.18), as suggested by Acklin and Jahns (1994). The intermediate layer between the chips and the substrate might be a silicon wafer with windows for the optical input and output. More details of the physical implementation are also described in Acklin and Jahns (1994).

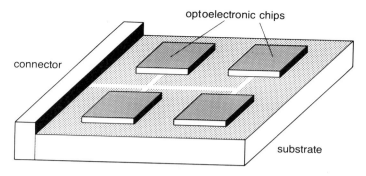

Figure 7.17 Optical multichip module using planar optics as interconnecting medium.

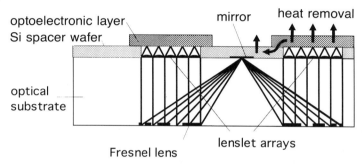

Figure 7.18 Integrated system with intermediate layer between chips and substrate. The intermediate layer serves for cooling purposes and provides a fixed distance between the chips and the lenslets on the substrates.

Another area of application where free-space optics is useful are sensors. Here it is also important to have small, rugged systems that can be implemented at low cost. A specific type of application is the confocal microscope (Wilson and Sheppard, 1984), which can be used both for imaging purposes and for sensor applications. An integrated-optic implementation based on waveguide-optics was recently suggested by Sheard *et al.* (1993). An implementation using planar optics might be of interest because of better coupling efficiencies. A scheme of an integrated free-space optical confocal microscope is shown in Fig. 7.19.

Finally, we would like to mention the use of planar optics for optical data storage as proposed recently by Shiono and Ogawa (1994). Optical pickup units are the most important devices in optical disk memory and it is advantageous to make them small and inexpensive. An integrated pickup unit can reduce the cost associated with mechanical alignment and is therefore of interest. A scheme of a planar optical pickup unit is shown in Fig. 7.20.

7.6 Conclusions

Conventional optomechanical systems suffer from being bulky and expensive. We have presented the concept of planar optics as an approach to build integrated

195

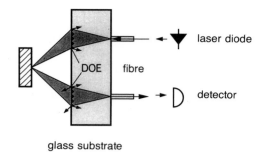

Figure 7.19 Schematic of a planar-optical confocal microscope.

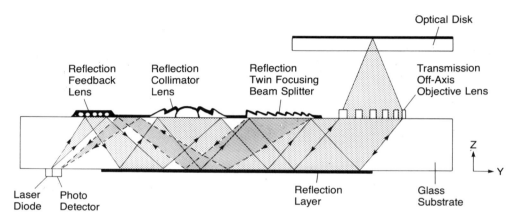

Figure 7.20 Optical pickup unit for data storage applications (after Shiono and Ogawa (1994)).

free-space optical systems without the need for costly mechanical alignment. Planar optics is based on a two-dimensional geometry which supports compatibility with known technology for the fabrication of electronic integrated circuits. We have described several experiments that show the usefulness of planar optics for interconnection, sensor, and data storage applications.

References

ACKLIN, B. and JAHNS, J. (1994) Packaging considerations for planar optical interconnection systems. *Appl. Opt.* **33**, 1391–1397.

BEHRMANN, G. P. and MAIT, J. N. (1997) Hybrid optics. In Herzig, H. P. (ed.) *Microoptics*, London: Taylor & Francis.

BRENNER, K.-H. (1991) 3D-Integration of digital optical systems. *Appl. Opt.* **30**, 25–28.

DAMMANN, H. and GÖRTLER, K. (1971) High-efficiency in-line multiple imaging by means of multiple phase holograms. *Opt. Commun.* **3**, 312–315.

GALE, M. T., ROSSI, M., KUNZ, R. E. and BONA, G. L. (1994) Laser writing and replication of continuous-relief Fresnel microlenses. In *Diffractive Optics* **11**, OSA Tech. Digest Series (Optical Society of America, Washington, D.C.), pp. 306–309.

GOODWIN, M. J., MOSELEY, A. J., KEARLEY, M. Q., MORRIS, R. C., KIRKBY, C. J. G., THOMPSON, J., GOODFELLOW, R. C. and BENNION, I. (1991) Optoelectronic components arrays for optical interconnection of circuits and subsystems. *J. Lightwave Technol.* **9**, 1639–1645.

HAMANAKA, K. (1991) Optical bus interconnection system using Selfoc lenses. *Opt. Lett.* **16**, 1222–1224.

HASE, K. R. (1987) Theoretical model for light-guiding plates in bus systems. *Arch. Elektr. Übertragungstechnik* **41**, 156.

HAUMANN, H.-J., KOBOLLA, H., SAUER, F., SCHWIDER, J., STORK, W., STREIBL, N. and VÖLKEL, R. (1990) Holographic coupling elements for optical bus system based on a light-guiding optical backplane. *Proc. SPIE* **1319**, 588–589.

HUTLEY, M. C. (1982) *Diffraction Gratings*, p. 278. New York: Academic Press.

IGA, K., OIKAWA, M., MISAWA, S., BANNO, J. and KOKUBUN, Y. (1982) Stacked planar optics: an application of the planar microlens. *Appl. Opt.* **21**, 3456–3460.

JAHNS, J. (1990) Integrated optical imaging system. *Appl. Opt.* **29**, 1998.

(1994) Tolerant design of planar optical interconnections. In *OSA Proc. Int. Opt. Design Conf.*, vol. 22, G. W. Forbes (ed.), pp. 232–235.

JAHNS, J. and ACKLIN, B. (1993) Integrated planar optical imaging system with high interconnection density. *Opt. Lett.* **18**, 1594–1596.

JAHNS, J. and HUANG, A. (1989) Planar integration of free-space optical components. *Appl. Opt.* **28**, 1602–1605.

JAHNS, J. and LEE, S. H. (eds) (1994) *Optical Computing Hardware*. Boston: Academic Press.

JAHNS, J. and MURDOCCA, M. J. (1988) Crossover networks and their optical implementation. *Appl. Opt.* **27**, 3155–3160.

JAHNS, J., LEE, Y. H., BURRUS, C. A. and JEWELL, J. L. (1992a) Optical interconnects using top-surface-emitting microlasers and planar optics. *Appl. Opt.* **31**, 592–597.

JAHNS, J., MORGAN, R. A., NGUYEN, H. N., WALKER, J. A., WALKER, S. J. and WONG, Y. M. (1992b) Hybrid integration of surface-emitting microlaser chip and planar optics substrate for interconnection applications. *IEEE Phot. Tech. Lett.* **4**, 1369–1372.

JAHNS, J., SAUER, F., TELL, B., BROWN-GOEBELER, K. F., FELDBLUM, A. Y., NIJANDER, C. R. and TOWNSEND, W. P. (1994) Parallel optical interconnections using surface-emitting microlasers and a hybrid imaging system. *Opt. Commun.* **109**, 328–337.

KOSTUK, R. K., KATO, M. and HUANG, Y. T. (1989) Reducing alignment and chromatic sensitivity of holographic optical interconnects with substrate-mode holograms. *Appl. Opt.* **28**, 4939–4944.

LEGGATT, J. S. and HUTLEY, M. (1991) Microlens arrays for interconnection of single mode fiber arrays. *El. Lett.* **27**.

LOHMANN, A. W. (1991) Image formation of dilute arrays for optical information processing. *Opt. Commun.* **86**, 365–370.

McCORMICK, F. B., TOOLEY, F. A. P., CLOONAN, T. J., SASIAN, J. M. and HINTON, H. S. (1991) Microbeam interconnections using microlens arrays. In *OSA Proc. on Phot. Switching*, H. S. Hinton and J. W. Goodman (eds), pp. 90–96.

MIDWINTER, J. E. (ed.) (1993) *Photonics in Switching*, vols I and II. London: Academic Press.

MILLER, L. F. (1969) Controlled collapse reflow chip joining. *IBM J. Res. Dev.* **13**, 239–250.

NISHIHARA, H. and SUHARA, T. (1987) Micro Fresnel lenses. In Wolf, E. (ed.) *Progress in Optics*, Vol. 24, pp. 3–37. North-Holland, Amsterdam.

PRONGUÉ, D. and HERZIG, H. P. (1989) Design and fabrication of HOE for clock distribution in integrated circuits. In *Proc. Int. Conf. Holographic Systems, Components, and Applications*, IEE Publ. No. 311, pp. 204–208.

SAUER, F. (1989) Fabrication of diffractive–reflective optical interconnects for infrared operation based on internal total reflection. *Appl. Opt.* **28**, 386–388.

SAUER, F., JAHNS, J., NIJANDER, C. R., FELDBLUM, A. Y. and TOWNSEND, W. P. (1994) Refractive–diffractive microoptics for permutation interconnects. *Opt. Eng.* **33**, 1550–1560.

SHEARD, S., SUHARA, T. and NISHIHARA, H. (1993) Integrated-optic implementation of a confocal scanning optical microscope. *J. Lightwave Techn.* **11**, 1400–1403.

SHIONO, T. and OGAWA, H. (1994) Planar-optic-disk pickup with diffractive micro-optics. *Appl. Opt.* **33**, 7350–7355.

SHIONO, T., SETSUNE, K., YAMAZAKI, O. and WASA, K. (1987) Rectangular-apertured micro-Fresnel lens arrays fabricated by electron-beam lithography. *Appl. Opt.* **26**, 587–591.

SIEGMAN, A. E. (1986) *Lasers*, ch. 19. Mill Valley, CA: University Science Books.

SONG, S. H., LEE, E. H., CAREY, C. D., SELVIAH, D. R. and MIDWINTER, J. E. (1992) Planar optical implementation of crossover interconnects. *Opt. Lett.* **17**, 1253–1255.

VELDKAMP, W. (1993) Wireless focal planes – on the road to amacronic sensors. *IEEE J. Quant. Electron.* **29**, 801–813.

VON LEHMEN, A., BANWELL, T., CHAN, W., ORENSTEIN, M., WULLERT, J., MAEDA, M., CHANG-HASNAIN, C., STOFFEL, N., FLOREZ, T. and HARBISON, J. (1991) Electronically and optically controllable vertical cavity surface emitting laser arrays for optical interconnect and signal processing applications. *Proc. SPIE* **1582**, 83–90.

WALKER, S. J. and JAHNS, J. (1992) Optical clock distribution using integrated free-space optics. *Opt. Commun.* **90**, 359–371.

WALKER, S. J., JAHNS, J., LI, L., MANSFIELD, W. M., MULGREW, P., TENNANT, D. M., ROBERTS, C. W., WEST, L. C. and AILAWADI, N. K. (1993) Design and fabrication of high-efficiency beam splitters and beam deflectors for integrated planar micro-optic systems. *Appl. Opt.* **32**, 2494–2501.

WILSON, T. and SHEPPARD, C. (1984) *Theory and Practice of Scanning Microscopy*. London: Academic Press.

8

Stacked Micro-optical Systems

W. SINGER and K. H. BRENNER

8.1 Introduction

Classical optical systems have a long history of development. One may think of the first lens systems or binoculars as 'parallel' optical systems. More recently, a complex integration of light source, beam splitter, lenses and detectors was realized in laser disc heads. Classical optical systems cover a large field ranging from low-cost systems in video cameras with approx. 10^{10} image points to expensive and high performance lens systems for photolithography with a space bandwidth product (SBP) of approx. 10^9. In the latter case, about 100 different optical elements have to be combined and aligned very precisely. Future optical products for classical imaging tasks such as lithography or for new tasks such as information processing will require even more powerful and flexible optical systems. A promising way of increasing the performance and applicability of optical systems is the micro-integration of parallel optical systems or array systems.

The general goal of micro-integration is to reduce the number of parts, or, more generally speaking, the number of degrees of freedom in the assembly. Integration provides modularity, ease of fabrication, enhanced stability and compactness of complex systems. Usually the term 'integrated optics' represents two-dimensional planar integration in waveguides. A variety of techniques is used to integrate passive optical components on a single substrate, guiding the light along predetermined paths. By including electrooptical switching mechanisms and also, more recently, by integrating semiconductors, planar integration of monolithic integrated optical circuits (OEICs) has become possible. However, the need for coupling into and out of waveguides generates interface problems. Furthermore two-dimensional planar integration permits propagation of zero- or one-dimensional optical signals only, which lessens the advantages provided by integration. The space bandwidth product for a waveguide is typically one or, for planar waveguides, in the order of 10–100. Guided-wave integrated optics does not benefit from the three-dimensionality of classical optics.

One of the main potentials of optics arises from the fact that the wavefield is three-dimensional, allowing interconnection of a large number of information

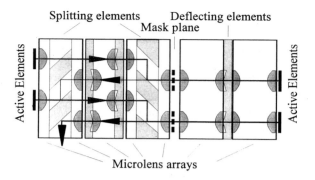

Figure 8.1 Stacked planar micro-integration of optical and optoelectronic array components. The set-up is an example of an optical adder.

channels simultaneously through space with large bandwidth and low crosstalk. In an early work K. Iga realized the potential of microlens arrays for utilizing the massive parallelism provided by optics (Iga *et al.*, 1982, 1984). He presented the concept of stacked planar optics, where array components were stacked to realize a large number of parallel but unconnected optical imaging channels. This form of integration combines the simplicity of array fabrication with the requirements of multi-lens optical systems.

A first approach to a fully interconnected and integrated three-dimensional structure utilized diffractive–reflective components to connect arrays of optical devices through free space (Brenner and Sauer, 1988). The reflective planar optics approach by J. Jahns, also covered in this book, has improved fabrication and assembly aspects by etching diffractive elements into glass substrates to provide folded three-dimensional integrated optical systems (Jahns and Huang, 1989; Jahns *et al.*, 1992; Jahns and Acklin, 1993). Thus alignment costs are reduced to a minimum at the cost of universality.

An extension of the stacked planar optics approach, which adds components for beam deflection and beam splitting to the lens arrays, was introduced by Brenner (1990, 1991). With this integration technique all types of optical systems can be realized. A complex micro-integration of optical and optoelectronic array components is shown in Fig. 8.1. The set-up is an example of an optical adder.

A simplification to stacked planar optics was introduced in Intani *et al.* (1992). Here the lateral focusing of planar microlenses is utilized. Thus the 'stack' consists of only one or two layers and only linear arrays of lenses can be utilized. Another packaging approach is based on block optics as a modular design of optical interconnection systems (Craft and Feldblum, 1992; Miyazaki *et al.*, 1994). The modular blocks are realized as separate components which can be arranged arbitrarily, but this degree of positioning freedom is gained at the expense of higher alignment effort. A recent theoretical investigation of integration of micro-optical systems was concerned with the folding of the optical modules in the different approaches, proposing packaging using massive pentagonal glass blocks (Schamaschula *et al.*, 1995). Another realization of a cascaded planar interconnection is demonstrated as a folded system by integration of roof-prism arrays, planar lens arrays, and active elements (Kakizaki *et al.*, 1995). This concept benefits from the on-axis design of stacked optical systems and the reduction of elements to be aligned in reflective planar optical systems.

Micro-optical integration using array components offers a series of advantages. Although each channel of the optical array system might be of low performance (e.g. the space bandwidth product of microlenses with a diameter of 300 μm is approximately 10^3), the array property warrants large space bandwidth products, e.g. an array of 1 cm^2 with 10^3 imaging channels provides a space bandwidth product of approximately 10^6. Stacked planar optics is well-suited for implementing micro-optical systems, especially if array optics is required. Concepts of stacked planar systems for data processing have been developed, but also simple parallel imaging systems, as required for example for large flat panel display fabrication, are based on stacked planar optics (Völkel *et al.*, 1995a and 1995b). These systems provide high flexibility and modularity in combination with low alignment costs.

Integrated stacked planar optical systems have several advantages:

- Flexible systems and system designs are possible. Single optical elements can be fabricated to form arrays and complex optical systems can be constructed from these arrays. One might even think about a 'nano-bank', the extension of a modular optical construction set into the micro-optical regime using standardized arrays of micro-optical elements.

- Due to the scaling laws of lenses, the fabrication costs of macroscopic optical components are disproportionate to component size (Lohmann, 1989). In contrast, micro-optical array components can be extended to larger arrays with no significant increase in cost.

- Integration in planar layers reduces the alignment requirements. The planar integration can be performed with the accuracy of photolithography. Currently, feature sizes of approximately 0.35 μm and alignment tolerances less than 0.05 μm are possible. The methods for aligning different optical layers can utilize the existing mask alignment technologies in VLSI semiconductor fabrication. In future systems, the alignment may be further simplified by mechanical or automatic alignment.

- Stacked planar optics is a real 3D integration of optical systems and allows the micro-integration of classical optical systems and designs. Therefore the experience of classical optical system designers and the classical optical design tools can be employed and the obstacles to applying modern technologies and micro-optics may be reduced.

- Several thousands of optical channels, e.g. for imaging or data processing, are possible. This kind of parallelism allows a large space bandwidth product in systems with small physical dimensions.

In the first section of this chapter we discuss the aspects to be considered for the realization of stacked planar micro-optical systems. Whether certain aspects are relevant or crucial depends heavily on the application. Therefore we will treat this aspect before any discussion of solutions. The next section deals with fabrication technologies for refractive micro-optical elements. In this field, several different technologies have been developed (Popovic *et al.*, 1988; Gale, 1997; Hutley, 1997). Here, we concentrate on four technologies: field assisted ion exchange in glass, the LIGA technique, thermal impressing of optical elements, and photopolymerization.

A general problem with microsystems is the alignment of the individual components. With respect to planar integration, only an alignment of the lateral position of the different substrates is necessary. Three concepts for alignment will be discussed in section 8.3: active alignment, passive alignment, e.g. with alignment grooves, and automatic alignment through flip-chip techniques. In section 8.4, we conclude with applications and examples of stacked planar integrated systems. In this section, we also discuss present problems and the need for further developments in future integrated optical systems.

8.2 Design Considerations for Micro-optical Systems

Optical interconnections are realized as a major potential to overcome the interconnection problems of electronic computers. Waveguide connections are mainly used at the level of system-to-system and board-to-board interconnections, profiting from the low loss and high data rates of optical interconnections. Free-space interconnections allow the utilization of the spatial bandwidth of optics in addition to the temporal bandwidth of waveguide optical interconnections. To be able to compete with future electronics, optical systems have to be fast, small, reliable and cheap. Therefore hybrid systems consisting of arrays of micro-optical components combined with optoelectronic smart pixels are promising candidates for future optoelectronic systems. Other applications of integrated free-space systems are in the field of optical sensors or optical storage. Here we investigate the requirements depending on these different applications.

In principle, diffractive optical elements would be well-suited for the realization of planar arrays of optical elements, since the fabrication is compatible with the conventional fabrication methods of VLSI technology (Firester et al., 1973; Stern, 1997). Present-day applications of micro-optics, e.g. optical switching units for telecommunications, utilize semiconductor lasers. These active devices, however, suffer from a wavelength instability and inhomogenety due to thermal effects and fabrication tolerances. Therefore refractive or reflective optical components are preferred. Thus, for stacked planar integration emphasis is put on refractive optical elements. Different fabrication technologies for refractive elements have been investigated. Since the technology for fabricating microlenses, for example, will be different from the technology for microprisms, a monolithic integration is currently not possible. An analysis of typical optical processing systems, however, shows that a separation of optical functions into different layers, each of a different functionality, is always possible. The layers may consist of active elements such as sources and detectors, of lenses for beam collimation and imaging, of passive filters for generating nearest neighbour interconnections, of active filters such as liquid crystal spatial light modulators (Hirabayashi et al., 1995), and of prisms for deflection. The fabrication methods of the different components will be discussed in section 8.3.

Any optical free-space system consists of a source module, the passive optical system, possibly also active elements, and a detector module. Several approaches have investigated the direct modulation of light utilizing nonlinear optical effects. However, these effects are usually very small and afford therefore large electrical or optical power. Other concepts are based on spatial light modulators by liquid crystal arrays. These devices currently suffer from long switching times. Therefore we concentrate here on optoelectronic concepts, which utilize optics only for the connection. Optoelectronic devices for switching and modulation, such as smart pixels, are future candidates for 3D integration. Examples are arrays of coupled photothyristors. These elements act in combination as detectors, sources and switching units according to the winner-takes-all principle (Kuijk *et al.*, 1991).

The source module can be constructed as a single source, as a linear source array or as a two-dimensional source array. Each source can be either an illuminated modulator or a self-luminous source. The properties of the source which are of concern here are the temporal and spatial coherence properties. The temporal coherence is directly associated with the spectral linewidth through the Wiener–Khintchine theorem. Thus the coherence length relates directly to the linewidth. The spatial coherence determines how well the source can be focused. If the full space bandwidth product is utilized in micro-optical systems, only small sources can be employed, which leads to a very high degree of spatial coherence. Therefore the degree of coherence can only be reduced by the use of thermal light sources with large linewidths and thus low temporal coherence. Incoherent illumination is required to reduce the unwanted interference effects such as speckle noise and interference fringes from partial reflections. In addition, the resolution and thus the SBP of optical systems will be reduced with coherent illumination.

The passive optical system can be divided according to functionality into components for imaging, and image or pupil division. The focusing or imaging function requires lenses. A microlens of size 300 μm is able to image a whole data array of approximately 1000 data points. However, as is known from macro-optics, there is not one lens which is optimal for all applications. One needs to distinguish between different imaging geometries. This physical limitation of classical optics is diminished by the use of micro-optics. The geometric aberrations of optical systems scale with their lateral dimensions and, therefore, aberrations are less severe in micro-optical systems (Lohmann, 1989). Nevertheless different lenses are required for different tasks. For standardization aspects it would be convenient to base the design on infinity corrected components (ICS). In such systems each lens is corrected for the case of imaging to infinity. An arbitrary scale imaging system can be realized by combining two ICSs with a given focal length ratio for scale.

For 1:1 imaging two possible implementations result. The classic 4*f*-configuration (Fig. 8.2(a)) has the advantage of a pupil plane in the plane 2*f*, where optical filters can easily be applied. However, this imaging configuration lacks the possibility of cascading several systems and suffers from vignetting. This means that an object spot, which emits into a cone which is symmetric around the optical axis, will be imaged to a cone which is not symmetric around the optical axis. Thus, in addition to the well-known fall-off of brightness towards the edges of the image, a cascading of optical systems is impeded because of the space-variance of the irradiation angles. Also, if the light needs to be coupled into a fibre, the overlap integral between the entrance fibre mode and the output asymmetric optical mode will depend on the location of the fibre. A standard method of overcoming this problem is to introduce telecentricity into the systems, as shown in Fig. 8.2(b). By adding a stop in the spatial frequency or filter plane, only a limited solid angle of the emitted source is passed through the system at the expense of optical power. The condition for telecentricity requires for the aperture stop that

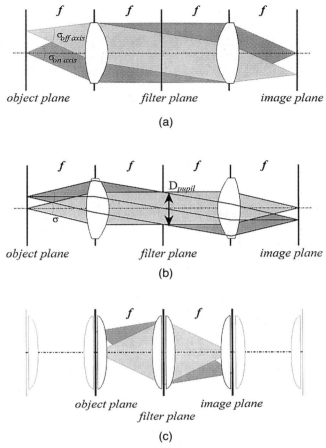

Figure 8.2 (a) 4*f*-system for 1:1 imaging utilizing ICS optics. (b) 4*f*-system for 1:1 imaging with aperture stop in the filter plane. (c) Light-pipe system for 1:1 imaging.

$D_{\text{pupil}} = D_{\text{image}}/2$. However, this system is not advantageous for array systems because crosstalk between adjacent imaging channels cannot be avoided.

An alternative method of achieving telecentricity without reducing the size of the spatial frequency plane is the so-called light pipe in Fig. 8.2(c). Here a field lens is inserted instead of the stop. The drawback of adding a third lens is, however, compensated for by the fact that the light-pipe imaging system has only half the length of the 4*f*-imaging system. As is obvious in this set-up, the full lens aperture is utilized here with telecentric imaging properties.

A different approach for 1:1 imaging with positive magnification utilizing stacked microlens arrays is used for a new type of lithography (Hugle *et al.*, 1993). In this application, imaging through parallel systems formed by stacked microlens arrays is employed for the fabrication of large flat panel displays (Fig. 8.3). Similar configurations have been applied for copying machines and displays (Anderson, 1979; Borelli *et al.*, 1991).

Image division is required whenever the whole image is to be split into two or

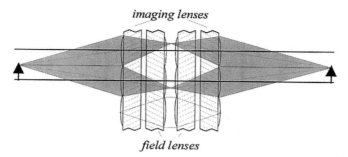

imaging lenses

field lenses

Figure 8.3   Parallel 1:1 multi-pupil imaging with microlens arrays.

more identical copies and occurs, for example, in interferometers or image processing applications (Lohman and Brenner, 1990). The inverse operation of image combination is achieved with the same components and therefore falls under the same category. Image division and combination are frequently realized by glasscube-type beam splitters which can be uncoated or coated to be polarization sensitive.

A different way of obtaining image division is pupil splitting, a function which utilizes the principle that the lateral spread of optical signals is often much larger than the angular spread. The quantity of concern here is the space bandwidth product (SBP), which determines the number of distinguishable spots that can be transferred through an optical system. The SBP thus determines the maximum possible number of interconnects and is defined as the ratio of the image area and the spot size. The image area $A_{image} = \pi D_{image}^2/4$ is the area in the image plane, which is reliably imaged through the system. It is typically smaller than or equal to the lens area. The criterion 'reliably' here must be related to the application. A fall-off of the intensity toward the edges of the output area may be tolerable for some applications but crucial for others. The definition of the spot size depends on the resolution criterion and on the coherence property of the wavefield (Born and Wolf, 1980). The definition which we will use here is based on the location of the first zeros of the diffraction spot intensity and the assumption of idealized incoherent illumination. From scalar diffraction theory, assuming a circular aperture of width D_{pupil}, this area is given by

$$A_{spot} = \pi \left(1.22 \frac{\lambda f}{D_{pupil}} \right)^2. \tag{8.1}$$

It is thus determined by the product of the focal length f and the wavelength λ. If we assume that the pupil aperture can be imaged reliably, i.e. $D_{pupil} = D_{out} = D$, the SBP is then given by

$$SBP = \frac{A_{image}}{A_{spot}} = \frac{D^4}{4(1.22\lambda f)^2}. \tag{8.2}$$

Therefore, the limitation due to the SBP is not only given by the area of the object or image plane, but also by the numerical aperture or power of a system. If the single active elements such as sources or detectors are of large size compared to the diffraction limited spot size provided by the numerical aperture of the system, the large number of degrees of freedom can be used otherwise. A convenient

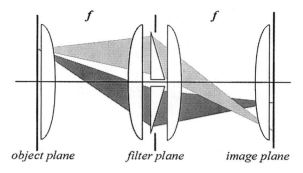

object plane   filter plane   image plane

Figure 8.4 Aperture or pupil division with a light-pipe set-up and deflecting prisms.

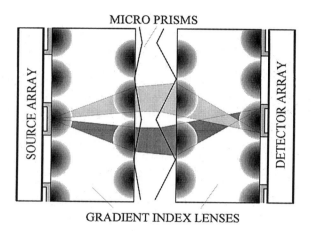

Figure 8.5 Multi-pupil imaging with microlenses and deflection with microprisms.

possibility for this is pupil division (Lohman and Brenner, 1992). Even for microlenses with a diameter of $250\,\mu$m and a focal length of typically 1 mm the number of distinguishable spots is in the order of 10^3 for visible light, as already mentioned. If such a large number of channels as provided by microlens systems is not needed, the aperture of the system can be split to perform different tasks without loss in the interconnection. Figure 8.4 shows the scheme for this set-up. If the full lens SBP is required, pupil division can also be achieved by using multiple pupil systems. A set-up similar to Fig. 8.3, utilizing lens arrays and deflecting prisms, is shown in Fig. 8.5.

8.3 Fabrication Technologies of Refractive Micro-optical Elements for Stacked Planar Integration

Several technologies have been developed for the fabrication of micro-optical components. For the application of arrays in stacked planar optics, a precise lateral structuring is required. This can be performed by lithographic-assisted fabrication

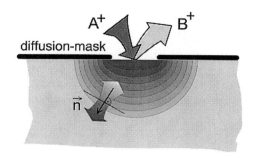

salt-melt

Figure 8.6 Elliptical index distribution due to diffusion through a mask aperture of finite diameter.

technologies. The accuracy of lithography and of step-and-repeat processing can also be exploited for the fabrication of a set of refractive optical components. Gradient-index lenses can be fabricated by ion exchange in glass, prisms by LIGA technology (<u>L</u>ithographie, <u>G</u>alvanotechnik und <u>A</u>bformung), deflecting elements and prisms by thermal impressing, and filters by photopolymerization. These technologies will be discussed shortly.

Ion exchange in glass is a very accurate and flexible method for fabricating gradient-index distributions, such as microlenses or filters in glass (Bähr *et al.*, 1992). With this process the index change is achieved by exchanging the sodium ions contained in the glass with other monovalent ions such as potassium (Tervonen, 1990), thallium (Oikawa and Iga, 1982), or silver (Bähr *et al.*, 1994). In our experiments we used silver exchange because silver is less toxic than thallium, but the index change is of the same order ($\delta n \approx 0.085$). In the first step the glass substrate is metallized with a layer of aluminium. This metal layer is structurized, forming the mask for the subsequent ion exchange process, which occurs in a melt of $AgNO_3$ at a temperature of approximately 300°C for several days (Fig. 8.6). Numerical apertures up to 0.08 can be achieved by the use of special types of glass and a proper mask design. Typical lens diameters are in the order of 300 μm with focal lengths of 1300 μm. With a special mask with additional rings the index profile can be optimized for different imaging set-ups (Singer *et al.*, 1995). Elliptical index distributions, forming astigmatic lenses (Sinzinger *et al.*, 1995), have also been fabricated via mask design. Furthermore, the thermal ion exchange can be assisted by an electrical field, accelerating and modifying the ion exchange process. With this technology, numerical apertures of up to 0.14 have been achieved within exchange times of several minutes. Figure 8.7 shows the interferogram of the transmission phase of a lens, fabricated by field-assisted ion exchange with a thermal optimization. In the latter case the ideal phase function is achieved over almost the whole aperture of the lens.

A specific advantage of gradient-index lenses are their planar surfaces. Reflection losses can be reduced to a minimum and the optical components can easily be stacked and aligned. As demonstrated already in Fig. 8.5, the glass substrates containing the microlenses can even be employed as an 'optical motherboard' for other elements such as electronic circuits, sources, detectors and

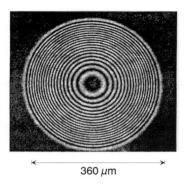

360 μm

Figure 8.7 Interferogram of a gradient-index microlens, fabricated by field-assisted ion exchange.

microprisms. Therefore ion exchange is superior to other microlens fabrication technologies, as far as stacked planar integration is concerned.

Photopolymerization is a technology for the simple fabrication of holograms. Similar processes can also be used for generating surface-relief phase structures, as required for deflecting prisms or phase filters. There are several photochemical reactions that can be initiated in polymers which result in considerable thickness changes and usually also in small index changes (Franke, 1984). Thin films of Metha-Methyl-Acrylate (MMA) are spin-coated on glass and then exposed through a photomask to UV illumination at a wavelength of 365 nm (Hg lamp). Figure 8.8 demonstrates this process. The exposed regions polymerize in a vapour of free MMA molecules, resulting in a thickness change which is proportional to the exposure. Final heating of the resist is necessary to remove unreacted MMA or photosensitive ketal molecules. The maximum thickness variation achieved in our experiments to date has been 100%, using layers of 20 μm thickness. Figure 8.9 shows a result for a 3 × 3 microprism deflector (Bagordo *et al.*, 1993; Rohrbach and Brenner, 1995).

To achieve even larger deflection angles, higher deflecting prisms than are obtainable with photopolymerization are required. This can be achieved by the thermal impressing of optical structures in PMMA. Figure 8.10 shows the principle for the fabrication. A metal or glass stamp with a polished surface at a desired prism angle is heated to a temperature slightly above the glass temperature of PMMA. The bevel is then pressed into the PMMA layer. In a step-and-repeat process under computer control a PMMA substrate can be structured with prisms at different positions and angles with a lateral accuracy of about 2 μm – which is sufficient for this task.

The deep-lithography or LIGA process was developed at the Kernforschungszentrum Karlsruhe, for microstructuring PMMA with synchrotron radiation (Rogner *et al.*, 1992; Brenner *et al.*, 1993). To this end the PMMA substrate is covered with a metal mask and exposed to coherent synchrotron radiation. A similar process for the fabrication of refractive components was developed with proton lithography (Brenner *et al.*, 1990a; Frank *et al.*, 1991). The regions exposed to the radiation contain shorter molecules, broken chains from the PMMA, that can be removed with a suitable solvent. With this technique

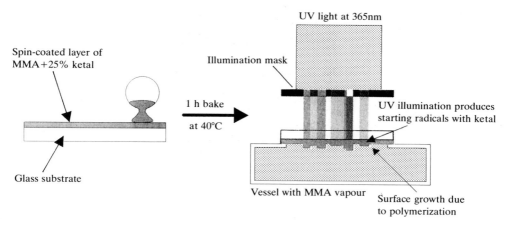

UV light at 365nm

Spin-coated layer of
MMA+25% ketal

Illumination mask

1 h bake
at 40°C

UV illumination produces
starting radicals with ketal

Glass substrate

Vessel with MMA vapour

Surface growth due
to polymerization

Figure 8.8 Set-up for photopolymerization.

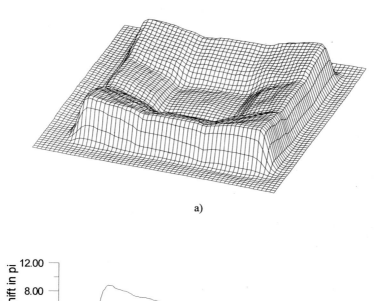

a)

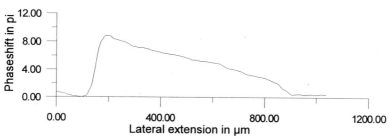

b)

Figure 8.9 (a) 3-D surface plot of a 3×3 prism array. (b) Linescan of the phase of
one of the prisms.

209

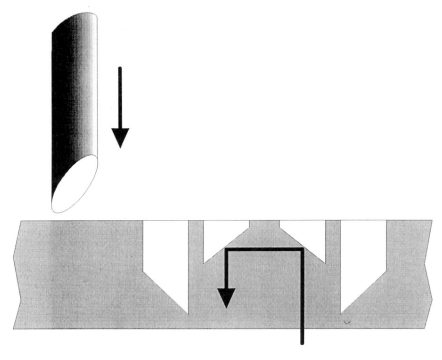

Figure 8.10 Thermal impressing for the fabrication of microprism arrays.

three-dimensional PMMA structures with feature size of 5 μm and up to 700 μm height can be fabricated (Fig. 8.11). The surface roughness can be as small as 50 nm, making it a useful technology for three-dimensional micro-integration. The application of LIGA is just as good for the fabrication of the optical components itself as for packaging units. The plastic microstructures achieved may already be the end product of manufacture. But there are further process steps to fabricate replicas by electroforming and, with a view to mass-production, plastic moulding (Gale, 1997).

8.4 Packaging of Micro-optical Systems

One of the most significant differences between a new technology such as photonics and a well-established technology such as silicon electronics is packaging. Packaging involves much more than just placing components in a package. It covers the whole range of interface specifications that will simplify the combination of different components. In the context of this proposal there are three types of interface: the mechanical interface, the optical interface and the electrical interface.

Optical components are usually a critical issue in product development since every component has three positional and three rotational degrees of freedom. In assembly each component has to be aligned according to these degrees of freedom. The goal of packaging here is to remove these degrees of freedom. The

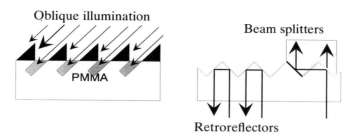

Figure 8.11   Microprism fabrication by the LIGA technique.

specific advantage of array optics is the reduction of lateral degrees of freedom by application of arranged arrays. The lateral alignment of the single elements is not required since the lateral position is controlled by the lithographic fabrication technologies. Thus the number of elements to align is comparable to macro-optics. The alignment of the optical elements to the optical axis is anticipated with planar arrays, because the planar layers can easily be stacked, possibly with distance layers. Thus only the lateral position of different layers is critical. For this alignment, different concepts have been proposed. The first ideas are based on a mechanical alignment, e.g. on alignment grooves. However, for most applications passive alignment will not be sufficient, therefore an active alignment will be required. For mass-production, alignment technologies from semiconductor integration can be employed, such as the different bonding techniques.

In the optics industry alignment has been simplified by providing mounts that only allow a movement of the components in one direction. For optical processing systems with components in the millimetre range this goal can be achieved by defining more complex blocks which are prefabricated and which can only be mounted on prespecified positions. Examples of these building blocks are image splitters, image combiners, imaging systems, light sources, and mechanical mounts for handling. In order to be able to freely combine these building blocks it is necessary to set up mechanical standards with respect to the geometric dimensions and with respect to the shape of the interface planes. Mechanical alignment of standardized optical layers in building blocks can help the applicant to realize micro-optical set-ups in a short time and with moderate accuracy. For mechanical alignment, either a separate alignment block is necessary in combination with alignment structures in the layers to be aligned (Fig. 8.12), or the alignment structures are integrated with the optical layers, e.g. by moulding techniques (Fig. 8.13). The latter technique will be well-suited for 'plug-and-play' applications in research and education. Such a concept has been described by Jahns and Huang (1989) for so-called flat optics.

However, for complex applications and optimized designs, mechanical alignment will not in general be sufficient. For these cases, active alignment will be required. A specific problem with the alignment of stacked planar optics is the alignment through thick substrates. This can be performed either by commercial mask-aligners or by special alignment masks such as Fresnel zone plates. The well-defined focus of the micro-Fresnel lens can be used as a mark for further alignment, as shown in Fig. 8.14. The accuracy of this technique depends only on the accuracy of the illumination angle perpendicular to the layers.

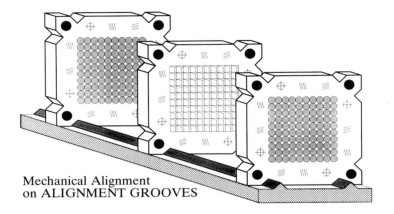

Figure 8.12 Mechanical alignment on alignment grooves, fabricated by wet chemical etching in silicon.

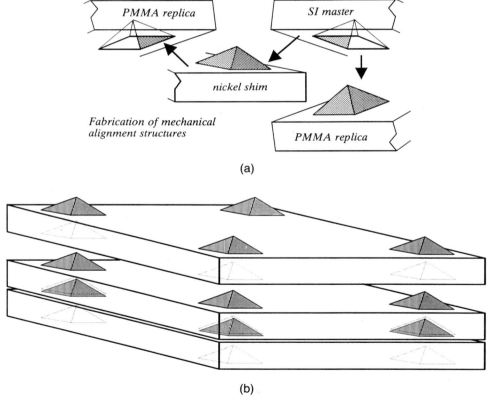

(a)

(b)

Figure 8.13 (a) Fabrication of mechanical alignment structures (naps and grooves) for plug-and-play applications. (b) Application of mechanical alignment structures for stacked optical systems, the optical 'LEGO™'.

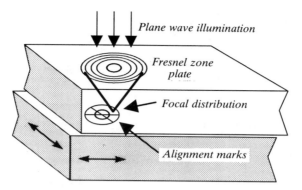

Figure 8.14 Active alignment through thick substrates with Fresnel zone plates.

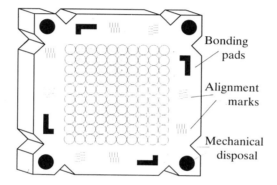

Figure 8.15 Integration of several alignment techniques.

Active or mechanical alignment may be time-consuming or expensive, reducing the speed of stacked planar optics. For mass fabrication therefore an automatic alignment is provided. This can be performed by the well-known flip-chip bonding technique, by structuring the layers with metallic bonding pads for auto-alignment (Fig. 8.15) (Jahns *et al.*, 1992). After deposition of solder on the bonding pads and heating of the composition, the layers will be dragged by the surface tension with respect to the assembly. Bonding pads can also be combined with other alignment techniques (Fig. 8.15) or electronic interfaces in an optical motherboard.

Several standards already exist for electric interfaces, which whenever possible should be maintained in order to be compatible with existing electronics. However, the optoelectronic components should also match the optical and mechanical interface specifications. A promising aspect in planar integration is the application of glass substrates as optical motherboards. Similar concepts have been developed with folded optics applications. The complexity but simplicity of the concept is outlined in Fig. 8.16.

8.5 Applications and Examples

A large class of optical systems can be constructed by combining gradient-index microlens arrays, impressed microprisms and mask arrays with LIGA structures.

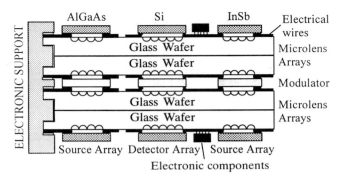

Figure 8.16 Optical motherboard for the integration of electronic and optoelectronic components on micro-optical layers.

Typical optical systems are constructed as 4*f*-systems or light-pipes which have been described previously. Even if the Fourier transform property is not required, e.g. for spatial filtering, such a system is still preferred because of standardization aspects and the capability of cascading. The input plane typically consists of a source array, which can be composed of a modulator array and a single source, or can be constructed as for example an LED array. The output plane includes a detector array or an interface to a fibre bundle. In complex set-ups such as image processing units coupled arrays of photothyristors can also be employed. A first cascading of such devices utilizing refractive micro-optical elements has been reported in Brenner *et al.* (1992).

One specific advantage of stacked planar optics is its flexibility. The available optical and optoelectronic components allow a micro-integration of simple but very effective set-ups, as well as the construction of complex optical systems, e.g. for data processing. For short-term applications less complex systems can also be envisaged. Specific areas of application are:

- Stacked planar optics for display technology. In certain applications, stacking of complex systems will not be required, but the array property of the planar integrated optical components is employed. An application for microlens arrays is, for example, beam shaping for LED arrays (Fig. 8.17) or direction-sensitive coding for displays (Fig. 8.18). The latter example can be employed for security reasons as well as for enhancing the brightness of displays. A further simple concept of stacked planar optics for imaging, utilizing only microlenses, has already been shown in Fig. 8.3. The set-up consists in the limit of only three planar microlens arrays. However, this set-up will possibly trigger tremendous developments in flat panel display fabrication (Dändliker *et al.*, 1994).

- Micro-optical sensor. A wide range of optical measuring and sensing techniques have been applied on an industrial scale. These systems are typically based on bulk optics and by three-dimensional integration they can be reduced in size and improved in terms of stability. Microinterferometers for measuring distances or physical parameters such as temperature, pressure, distance etc. are examples of this type. Micro-optical integration also allows *in situ* measurements in medical applications, for example. Array optics offers the implementation of parallel measuring devices or of redundancy.

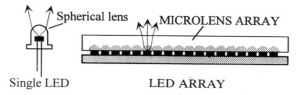

Figure 8.17 Beam-shaping for LED arrays with planar microlens arrays.

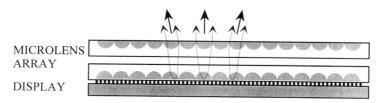

Figure 8.18 Direction-sensitive display for various applications.

- Parallel pick-up systems. For optical disk applications the high interconnection density of imaging can improve the rate of data transfer. With integrated planar pick-up systems based on microlenses and microprisms several hundreds of pixels can be read simultaneously.

- Optical interconnections. Three-dimensional micro-integration can provide suitable technology for different levels of communication. At the system-to-system level integrated connectors for parallel fibre interconnect can be built (Fig. 8.19) (Sasaki *et al.*, 1992). For flexible connectors the alignment again profits from the planar surfaces. However, a connection to a stacked planar optical system which benefits from the space bandwidth product of every lens is not straightforward. In this case, an intermediate set-up of cascaded image divisions has to reduce the high density of data points to be coupled into the fibres with larger physical dimensions.

  At the board-to-board level optical bus connectors are more feasible (Hamanaka, 1991). Stacked planar optics can form the backplane of electronic or optical boards. At the chip-to-chip level a scheme similar to the generic micro-optical set-up in Fig. 8.16 can be employed. The chips can be fixed directly to the optical components, forming the optical motherboard. In this application, direct communication between electronic processing units will be possible with approximately 10^5 data channels, actually restricted only by the detector and source dimensions. The electrical support of the electronic integrated circuits can be performed conventionally.

- Optical data processing units. Different concepts for systems for optical data processing have been developed (Brenner *et al.*, 1990b; Haney, 1992). Demonstrators have already been built with macro-optics (Eckert *et al.*, 1992; McCormick *et al.*, 1993). By integration of modulators, prisms and microlenses, complex micro-optical systems such as data processing units can be implemented. Figure 8.20 shows an example of a general modulator design, employing aperture stops or electrooptical modulators as filters in the space and frequency domains.

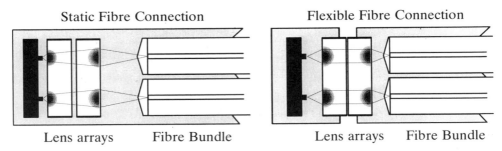

Figure 8.19 Static and flexible fibre connectors with planar microlens arrays.

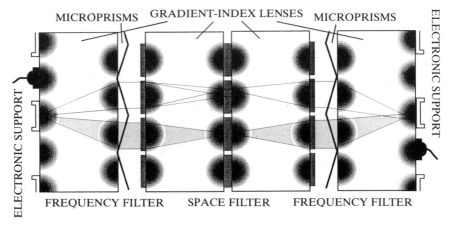

Figure 8.20 Space and frequency filtering with active modulators of stops.

The concept for an optical adder has already been shown in Fig. 8.1. A first demonstration of a micro-optical integrated interlace device is shown in Fig. 8.21 (Moisel and Brenner, 1994). Microprisms were directly aligned on a gradient-index microlens array by embossing and replication (Fig. 8.22). In the image plane, the two object planes are interlaced according to the prism angles. In the demonstrator, two different masks and illumination by a halogen lamp were applied instead of source arrays. Because of the use of refractive components only, different wavelengths could be employed. Figure 8.23 shows the measurement of the overlay of two masks, one with rectangular structures, the other with circular structures. Not visible in this figure is the green colour of the circles and the red colour of the squares, obtained by colour filters in the data planes.

8.6 Conclusions

The planar integration of micro-optical components enables the realization of standard optical applications such as sensors as well as the micro-integration of complex optical systems. Stacked planar optics is a real three-dimensional integration of optical systems. Stacked integration of planar optical elements in combination with optoelectronic devices enables therefore the realization of complex optical systems for data processing as well as simply the enhancement

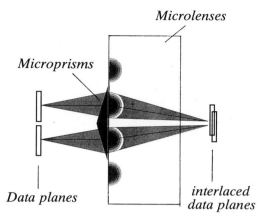

Figure 8.21 Integrated interlace device consisting of microprisms and planar microlenses.

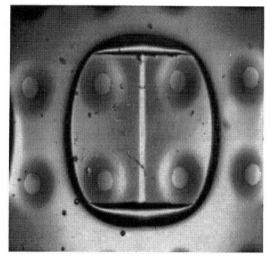

Figure 8.22 Replicated microprism on a planar microlens array.

of displays. The optical components are used on-axis in a similar way to classical macroscopic applications. As macro-optical set-ups can readily be miniaturized for practical applications, available knowledge of macro-optical systems can be transferred directly to complex micro-optical systems. An additional benefit for micro-optical arrangements in planar layers is the reduction in the degrees of freedom due to the lithographic fabrication of optical arrays and the ease of alignment in the assembly.

While the folded-planar integration benefits from the integration of all optical components in one planar layer, three-dimensional integration with stacked planar optics is more flexible and provides larger space-bandwidth products. However, this increased flexibility and parallelism of stacked planar optics has to be paid for by the more sophisticated fabrication technologies of the refractive optical components as well as by the alignment requirements of the single layers in the

Figure 8.23 Interlace experiment: left channel, right channel and interlaced images.

composition. On the other hand, by arranging in one layer only components of one type, the fabrication process can be kept simple and the components can be applied in a more flexible way. The microlenses can be fabricated by thermal ion exchange in glass. Filter structures can be achieved with the same technology, but also with photopolymerization. Prisms made from LIGA masters can easily be replicated and combined with the lenses and filters to form stacked optical systems of great complexity.

Concepts have been proposed for aligning stacks of layers of the different components. Examples are the active alignment by alignment marks. As the expense of active alignment is relatively high in complex systems, the next step may be to carry out the alignment by naps and grooves if possible, which are mounted on the different layers. A third step should be a self-aligning technique, such as the flip-chip technologies.

However for special applications stacked planar optics will be too costly and, especially for mass-fabricated sensors or small systems, planar integration in folded systems may be preferable. The benefits of stacked planar optics are in their exhaustive array properties and in their flexibility. If neither flexibility, nor the array property is required, other packaging technologies are superior. However, these kinds of set-ups with low integration density will easily be fabricable by stacking of planar components, if standardized array components are available. One specific advantage of stacked planar optics will be the possibility of constructing optical systems with components from the board. For this reason, a standardization of micro-optical array components is required. In the near future a full range of different microlenses, microprisms, beam splitters and modulators with adapted specifications should be obtainable. Furthermore, active elements such as photothyristor arrays or integrated LED arrays are required. The implementation of optimized systems will heavily depend on the availability of standardized components and packaging units. Then plug-and-play set-ups with micro-optics will increase the scientific experience and the commercial interest in integrated optical applications of all kinds.

Acknowledgements

Research in a complex field such as micro-optics is only possible with enthusiastic scientists. We therefore want to thank especially our colleagues J. Moisel, Ch.

Passon, J. Bähr, Th. Merklein, S. Sinzinger, M. Testorf, A. Rohrbach and W. Eckert. We further acknowledge fruitful discussions with H. P. Herzig and R. Völkel.

References

ANDERSON, R. H. (1979) Close-up imaging of documents and displays with lens arrays. *Appl. Opt.* **18**, 477–484.

BAGORDO, G., BRENNER, K. H., MERKLEIN, T. and ROHRBACH, A. (1993) Fabrication of micro-optic elements by UV-initiated polymerization. In *Integrated Optics and Micro-Optics in Polymers*, Teubner 'Texte zur Physik' Bd. **27**, 177–192.

BÄHR, J., BRENNER, K. H., SINZINGER, S., SPICK, T. and TESTORF, M. (1994) Index-distributed planar microlenses for three-dimensional micro-optics fabricated by silver–sodium ion exchange in BGG35 substrates. *Appl. Opt.* **33**, 5919–5924.

BÄHR, J., BRENNER, K. H., MOISEL, J., SINGER, W., SINZINGER, S., SPICK, T. and TESTORF, M. (1992) Diffusion elements in glass: comparison and optimization of the diffusion response in different substrates. *ICO Topical meeting on Optical Computing*, Proceedings SPIE Vol. **1806**, Minsk 1992, pp. 234–242.

BORELLI, N. F., BELLMAN, R. H., DURBIN, J. A. and LAMA, W. (1991) Imaging and radiometric properties of microlens arrays. *Appl. Opt.* **30**, 3633–3642.

BORN, M. and WOLF, E. (1980) *Principles of Optics*, 6th edn. New York: Pergamon.

BRENNER, K. H. (1990) Optical implementations of symbolic substitution. In Athale, R. A. (ed.) *Digital Optical Computing*, Proc. Soc. Photo-Opt. Instrum. Eng., Vol. CR35, 190–196.

(1991) 3D-integration of digital optical systems. In *Optical Computing 1991*, Technical Digest Series Vol. 6, Salt Lake City, pp. 25–28.

BRENNER, K. H. and SAUER, F. (1988) Diffractive–reflective optical interconnections. *Appl. Opt.* **27**, 4251–4254.

BRENNER, K. H., FRANK, M., KUFNER, M. and KUFNER, S. (1990a) H⁺-lithography for 3-D integration of optical circuits. *Appl. Opt.* **29**, 3723–3724.

BRENNER., K. H., KUFNER, M. and KUFNER, S. (1990b) Highly parallel arithmetic algorithms for a digital optical processor using symbolic substitution logic. *Appl. Opt.* **29**, 1610–1618.

BRENNER, K. H., ECKERT, W., KUFNER, S., KUIJK, M., HEREMANS, P., MOISEL, J. and SINZINGER, S. (1992) Cascading of two pnpn-Photothyristor arrays in a microoptical system: an experimental demonstration. *OSA Meeting on Diffractive Optics*. Vol. 9, New Orleans 1992, pp. 3–8.

BRENNER, K. H., KUFNER, M., KUFNER, S., MOISEL, J., MÜLLER, A., SINZINGER, S., TESTORF, M., GÖTTERT, J. and MOHR, J. (1993) Application of three-dimensional microoptical components formed by lithography, electroforming and plastic molding. *Appl. Opt.* **32**, 6464–6469.

CRAFT, N. C. and FELDBLUM, A. Y. (1992) Optical interconnects based on arrays of surface-emitting lasers and lenslets. *Appl. Opt.* **31**, 1735–1739.

DÄNDLIKER, R., GRAY, S., CLUBE, F., HERZIG, H. P. and VÖLKEL, R. (1994) Non-conventional techniques for optical lithography. In *MNE Micro- and Nano-Engineering 94*, Davos 1994.

ECKERT, W., PASSON, C. and BRENNER, K. H. (1992) Demonstration of an optical pipeline adder based on systolic arrays. *ICO Topical meeting on Optical Computing*, Proceedings SPIE Vol. **1806**, Minsk 1992, pp. 590–599.

FIRESTER, A. H., HOFFMAN, D. M., JAMES, E. A. and HELLER, M. E. (1973) Fabrication of planar phase elements. *Opt. Commun.* **8**, 160–162.

FRANK, M., KUFNER, M., KUFNER, S. and TESTORF, M. (1991) Microlenses in polymethyl methacrylate with high relative aperture. *Appl. Opt.* **30**, 2666–2667.

FRANKE, H. (1984) Optical recording of refractive-index patterns in doped poly-(methyl methacrylate) films. *Appl. Opt.* **23**, 2729.

GALE, M. T. (1997) Direct writing of continuous relief micro-optics. In Herzig, H. P. (ed.) *Micro-optics*, pp. 87–126. London: Taylor & Francis.

(1997) Replication. In Herzig, H. P. (ed.) *Micro-optics*, pp. 153–177. London: Taylor & Francis.

HAMANAKA, K. (1991) Optical bus interconnection system using Selfoc lenses. *Opt. Lett.* **16**, 1222–1224.

HANEY, M. W. (1992) Pipelined optoelectronic free-space permutation network. *Opt. Lett.* **17**, 282–284.

HIRABAYASHI, K., YAMAMOTO, T. and YAMAGUCHI, M. (1995) Free-space optical interconnections with liquid-crystal microprism arrays. *Appl. Opt.* **34**, 2571–2580.

HUGLE, W. B. *et al.* (1993) Lens array photolithography, US patent application, #08/114,732 (August 1993).

HUTLEY, M. (1997) Refractive lenslet arrays. In Herzig, H. P. (ed.) *Micro-optics*, pp. 127–152. London: Taylor & Francis.

IGA, K., KOKUBUN, Y. and OIKAWA, M. (1984) *Fundamentals of microoptics*. New York: Academic Press.

IGA, K., OIKAWA, M., MISAWA, S., BANNO, J. and KOKUBUN, Y. (1982) Stacked planar optics: an application of the planar microlens. *Appl. Opt.* **21**, 3456–3460.

INTANI, D., BABA, T. and IGA, K. (1992) Planar microlens relay optics utilizing lateral focusing. *Appl. Opt.* **31**, 5255–5258.

JAHNS, J. and ACKLIN, B. (1993) Integrated planar optical imaging system with high interconnection density. *Opt. Lett.* **18**, 1594–1596.

JAHNS, J. and HUANG, A. (1989) Planar integration of free-space optical components. *Appl. Opt.* **28**, 1602–1605.

JAHNS, J., LEE, Y. H., BURRUS, Ch. A. and JEWELL, J. L. (1992) Optical interconnects using top-surface-emitting microlasers and planar optics. *Appl. Opt.* **31**, 592–597.

KAKIZAKI, S., HORAN, P. and HEGARTY, J. (1995) Surface-normal cascaded planar interconnection with easy alignment. *Opt. Lett.* **20**, 2294–2296.

KUIJK, K., HEREMANS, P. and BORGHS, G. (1991) Array to array transcription of optical information by means of surface emitting thyristors. *IEDM Techn. Digest 91*, 433.

LOHMAN, G. and BRENNER, K. H. (1990) Morphological optical image processor. In *Proc. ICO 15, Optics in Complex Systems*, Garmisch-Partenkirchen 1990, p. 161.

LOHMAN, G. E. and BRENNER, K.-H. (1992) Space invariance in optical computing systems. *Optik* **89**, 123–134.

LOHMANN, A. W. (1989) Scaling laws of lens systems. *Appl. Opt.* **28**, 4996–4998.

MCCORMICK, F. B., CLOONAN, T. J., TOOLEY, F. A. P., LENTINE, A. L., SASIAN, J. M., BRUBAKER, J. L., MORRISON, R. L., WALKER, S. L., CRISCI, R. J., NOVOTNY, R. A., HINTERLONG, S. J., HINTON, H. S. and KERBIS, E. (1993) Six-stage digital free-space optical switching network using symmetric self-electro-optic-effect devices. *Appl. Opt.* **32**, 5153–5171.

MIYAZAKI, D., TANIDA, J. and ICHIOKA, Y. (1994) Reflective block optics for packaging of optical computing systems. *Opt. Lett.* **19**, 1281–1283.

MOISEL, J. and BRENNER, K. H. (1994) Demonstration of a 3D integrated refractive microsystem. In *Proc. International Conference on Optical Computing*, Inst. Phys. Conf. Ser. No. 139, Edinburgh 1994, pp. 259–262.

OIKAWA, M. and IGA, K. (1982) Distributed-index planar microlens. *Appl. Opt.* **21**, 1052–1056.

POPOVIC, Z. D., SPRANGUE, R. A. and CONELL, G. A. N. (1988) Technique for monolithic fabrication of microlens arrays. *Appl. Opt.* **27**, 1281–1284.

ROGNER, A., EICHER, J., MÜNCHMEYER, D., PETERS, R.-P. and MOHR, J. (1992) The LIGA technique – what are the new opportunities. *J. Micromech. Microeng.* **2**, 133–140.

ROHRBACH, A. and BRENNER, K. H. (1995) Surface-relief phase structures generated by light-initiated polymerization. *Appl. Opt.* **34**, 4747–4754.

SASAKI, A., BABA, T. and IGA, K. (1992) Focusing characteristics of convex-shaped distributed-index microlens. *Jpn. J. Appl. Phys.* **32**, 1611–1617.

SCHAMSCHULA, M. P., REARDON, P., CAULFIELD, H. J. and HESTER, C. F. (1995) Regular geometries for folded optical modules. *Appl. Opt.* **34**, 816–827.

SINGER, W., TESTORF, M. and BRENNER, K. H. (1995) Gradient-index microlenses: numerical investigation of different spherical index profiles with the wave propagation method. *Appl. Opt.* **34**, 2165–2171.

SINZINGER, S., BRENNER, K. H., MOISEL, J., SPICK, Y. and TESTORF, M. (1995) Astigmatic gradient index elements for laser-diode collimation and beam shaping. *Appl. Opt.* **34**, 6626–6632.

STERN, M. (1997) Binary optics fabrication. In Herzig, H. P. (ed.) *Micro-optics*, pp. 53–85. London: Taylor & Francis.

TERVONEN, A. (1990) A general model for fabrication processes of channel waveguides by ion exchange. *J. Appl. Phys.* **67**(6), 2746–2752.

VÖLKEL, R., HERZIG, H. P., NUSSBAUM, PH., SINGER, W., WEIBLE, K. J., DÄNDLIKER, R. and HUGLE, W. B. (1995a) Microlens lithography: A new fabrication method for large displays. In *SID-conference Asia Display 1995*, Hamamatsu, Japan 1995, pp. 713–716.

VÖLKEL, R. NUSSBAUM, PH., WEIBLE, K. J., HERZIG, H. P., DÄNDLIKER, R., HASELBECK, S., EISNER, M. and SCHWIDER, J. (1995b) Fabrication of non-conventional microlens arrays. In *Microlens Arrays*, EOS Topical Meetings Digest, Vol. 5, pp. 116–120.

9

Laser Beam Shaping

J. R. LEGER

9.1 Introduction

The unique properties of lasers have made them valuable tools in applications ranging from medicine to metrology. In applications such as holography and coherent optical communication, high spectral and spatial coherence are of primary importance. This is achieved by designing the laser to operate in a single longitudinal and single transverse mode. Other applications such as optical recording and playback simply require high spatial coherence from a single transverse mode. The high spatial coherence (accompanied by good phase uniformity) permits focusing of the beam to a spot size that is limited entirely by diffraction from the focusing optic. Still other applications such as material processing, welding and laser pumping place a premium on power, often at the expense of spatial coherence. Lasers for these applications may operate in several spatial modes simultaneously, and the ability to concentrate their output light is limited by the radiance theorem (Boyd, 1983).

The near- and far-field spatial light distribution from a given laser is usually governed by the design of the laser resonator. Many gas and solid-state lasers utilize conventional stable spherical mirror cavities. These cavities can be operated in a single spatial mode resulting in a Gaussian-shaped near and far field, or can be made to lase in many spatial modes simultaneously. Others use unstable resonators (Siegman, 1965 and 1986) to achieve high-power operation in a single spatial mode with large diameter beams. However, the near-field mode profile is not always desirable due to a central obscuration. Variable reflectivity mirror unstable resonators can sometimes be used to circumvent this problem (Lavigne *et al.*, 1985). High-power semiconductor and CO_2 lasers often have single-aperture power limitations due to thermal and optical damage constraints. For this reason, high-power versions of these lasers are often designed as one- or two-dimensional arrays of lower-power lasers.

In many cases, the spatial characteristics of a particular laser beam or array of beams are not ideally suited to the application, and it is desirable to modify these characteristics. This chapter explores several optical techniques for laser beam

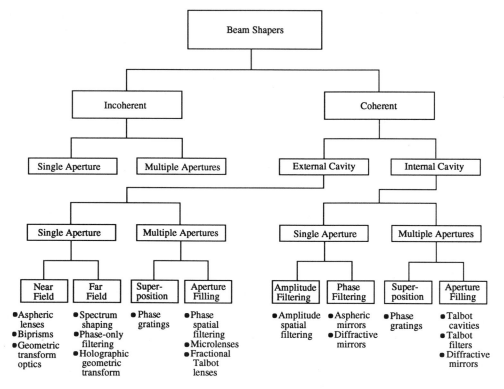

Figure 9.1 Outline of beam-shaping techniques.

shaping, where 'shaping' is meant to imply simple modifications of beam geometry, intensity profile, divergence properties, and phase front. Conversion of laser beams into more complex patterns is described in Chapters 1 and 11. The spectral properties of the beam are generally not considered, and the techniques described work equally well for a single and multiple longitudinal mode laser.

Figure 9.1 shows a general overview of beam-shaping techniques. The chapter concentrates mainly on techniques that require spatially coherent sources. Section 9.2 reviews the benefits of beam shaping from the standpoint of applications and laser performance. In section 9.3, we review external-cavity beam-shaping techniques applied to single- and multiple-aperture laser systems. Near-field and far-field shaping is considered. Section 9.4 explores beam shaping inside laser resonators. Again, we present different techniques for single- and multiple-aperture systems. We will see that intracavity beam shaping not only produces desirable output profiles, but can also improve the performance of the laser system itself.

9.2 Benefits of Laser Beam Shaping

9.2.1 *Matching the Beam Shape to the Application*

The shaping requirements of a specific application can be categorized as near field or far field, and whether the intensity or complex field is important. Perhaps the simplest near-field application that requires intensity shaping is near-field scene illumination. Illuminating a scene with a quasi-uniform beam improves the signal-to-noise level across the entire scene and reduces the dynamic range requirements of the detector. A second illumination task is encountered in materials processing such as laser annealing of semiconductors. In this case, the intensity flatness requirements can be very strict on both the macroscopic and microscopic level (Poate and Mayer, 1982). Both these applications are concerned simply with the intensity profile of the beam, and the beam phase is arbitrary. Some basic examples of intensity beam shaping for near-field illumination are contained in section 9.3.

Examples of applications that require near fields with uniform phase as well as intensity are optical holography, laser doppler velocimetry, interferometry, MTF testing, and optical information processing. In these cases, a single-spatial-mode laser source must be used. The optical systems that are required for these applications are generally more complex, and are reviewed in sections 9.3 and 9.4.

If the laser beam needs to be projected far from the aperture, or if it is focused in the focal plane of a lens, the far-field performance is of greatest concern. Applications that would benefit from a uniform far-field profile include optical recording and printing, and detector-array-based laser radar (Veldkamp and Kastner, 1982). A final application that requires mode shaping is coupling of light into fibres and waveguides. In this case, the coupling efficiency is a function of the overlap between the incident light and the waveguide mode, and it is important to obtain the correct light intensity and phase at the waveguide for efficient coupling.

Laser arrays require special considerations for beam shaping. If the lasers in the array are mutually coherent, the far-field complex amplitude is simply given by the Fourier transform of the field at the laser array. The Strehl ratio can be defined as the intensity of the array at the centre of the far field divided by the same intensity from a uniformly filled aperture of the same size and total power. It is easy to show (Leger, 1993) that the Strehl ratio of a coherent uniform-phase array is approximately given by the ratio of the emitting aperture area to the total area of the array (called the array fill factor). The light from an underfilled array is dispersed throughout the far field into grating lobes in much the same way that light from a diffraction grating forms discrete diffraction orders. Thus, if the objective is to concentrate light in the far field, the light from the individual laser beams must be converted into a uniform beam with an appropriate optical system.

Arrays of mutually incoherent sources present a slightly different problem. In many applications, the light must be concentrated into a small spatial region while simultaneously maintaining a small divergence. For example, when coupling light from an array of diode lasers into a multimode fibre optic, the light must be concentrated in the core of the fibre while the divergence must be contained within the fibre numerical aperture. Reducing the size of the array with conventional optics simultaneously increases the angular divergence, leaving the étendue (the product of the total array area and the solid-angle divergence) unchanged. However, if micro-optics are used to expand the effective size of each emitter in the array individually, the divergence from each emitter is reduced while the overall array size is unchanged. Thus, the beam-shaped array can be demagnified while still staying within the numerical aperture of the fibre. This permits more laser apertures to be coupled to the core of the fibre. As in the coherent case, there may be large advantages to shaping an array of beams to increase the fill factor. The limitations imposed by the radiance theorem and design strategies for incoherent arrays are considered in Leger and Goltsos (1992).

9.2.2 *Laser System Improvements with Beam Shaping*

The performance of a laser is influenced by the interaction between the lasing mode and the gain medium. Beam shaping of this mode inside the laser can lead to several performance enhancements. If the laser medium is pumped with a particular profile, a laser mode with the same size and profile will best extract the gain. For example, longitudinal pumping of a solid-state laser with a fibre-bundle-coupled semiconductor laser gives rise to a circular pump region that is approximately uniform across the top in the image plane of the fibre bundle. A circular flat-top laser mode should be able to extract more energy from the gain medium than a conventional Gaussian mode. A second advantage enjoyed by a flat-top mode is reduced spatial hole burning. Gaussian fundamental modes extract large amounts of gain from the centre of the medium but are incapable of extracting significant gain at its edge because of the fall-off of mode intensity. This allows gain to build up in the cavity that can only be accessed by higher-order modes, promoting multi-spatial-mode operation. The uniform intensity of the flat-top mode can theoretically access all the gain in the medium, and should not be susceptible to spatial hole burning. Thus, it is easier to ensure single-spatial-mode operation.

A third potential advantage of flat-top modes when applied to solid-state lasers is higher heat extraction efficiency. The diameter of the laser crystal must be chosen to pass the fundamental mode with negligible loss. In a conventional spherical resonator, the Gaussian beam has long tails that must be passed by the crystal, resulting in oversized crystal diameters. If the same mode is converted to a flat-top beam with negligible tails, the diameter of the crystal can be reduced without affecting the loss. This narrower-diameter crystal will be easier to cool, potentially reducing the complexity of the laser system.

A fourth advantage of intracavity mode shaping is a potential increase in mode discrimination ability. In a stable spherical mirror laser resonator, an aperture is usually used to provide selective loss to the higher-order modes. However, since the fundamental Gaussian has energy in the tails, the aperture cannot be overly restrictive without introducing significant loss to the fundamental mode. In addition, the higher-order Hermite–Gaussian modes are separated only slightly from the fundamental, and this separation becomes smaller as the Fresnel number of the resonator is increased. On the other hand, if the fundamental mode is designed to have all its energy in a well-defined area, an aperture can pass virtually the entire mode without introducing loss. Simultaneouly, the cavity parameters can be chosen such that the higher-order modes are much larger than the fundamental mode at the aperture, resulting in increased loss and ensuring high modal discrimination. We will discuss this in detail in section 9.4.

Multiple-aperture lasers can also benefit from mode shaping in several different ways. For a laser array to be coherent (meaning that the phases of all the laser apertures are locked together), the lasers must be coupled by some mechanism to generate a single spatial mode that extends across the entire array. There are many techniques for achieving this coupling. One of the simplest that is appropriate for one- and two-dimensional laser arrays is a Talbot cavity. This technique employs the Talbot self-imaging effect (Talbot, 1836; Rayleigh, 1881) to provide the reimaging of the mode in the laser resonator. The loss to each of the array modes is a function of the array fill factor, with increased loss for a reduced fill factor. Consequently, there is an optimal fill factor for balancing the fundamental mode loss against higher-order mode discrimination. An intracavity optical system can be used to shape each beam to achieve this optimum fill factor.

A second use for array beam shaping is to reduce the divergence of laser apertures in a Talbot cavity. The Talbot effect is based on a paraxial diffraction theory, and non-paraxial propagation leads to aberrations. Again, micro-optical elements can be employed to reduce the divergence of each aperture, and ensure paraxial propagation.

Finally, sophisticated diffractive mirrors can be used to shape the profile of the array mode, ensuring minimum fundamental mode loss, maximum modal discrimination, and reduced spatial hole burning.

9.3 Coherent External Resonator Techniques

A spatially coherent light field can be represented by a complex function $a(x, y) \exp[j\phi(x, y)]$, where $a(x, y)$ represents the amplitude and $\phi(x, y)$ the phase of the electric field. Beam shaping consists of converting this field into a more desirable amplitude and phase $b(x, y) \exp[j\theta(x, y)]$. In principle, this can be performed straightforwardly by fabricating an optic with the complex transmittance

$$t(x, y) = \frac{b(x, y)}{a(x, y)} \exp[j(\theta(x, y) - \phi(x, y))]. \tag{9.1}$$

In the simplest beam-shaping problems, the original intensity $|a(x, y)|^2$ is not modified and the transmittance of the optical element is phase-only. Lenses, axicons, lenslet arrays, etc. are all of this simple type. The first example of section 9.3.1 shows an anamorphic microlens that removes astigmatism from a tapered semiconductor amplifier.

Changing the intensity profile $|a(x, y)|^2$ is generally more difficult. Since $|t(x, y)| \leq 1$, modification of the intensity profile by a simple application of equation (9.1) requires absorption and subsequent loss of power. A more power-efficient method utilizes a phase plate followed by free-space propagation. The second part of section 9.3.1 describes optical systems that modify the near-field intensity of a laser beam. In the third part, we show examples of optical systems for modifying the far-field intensity. Finally, arrays of identically spaced sources can take advantage of special methods to increase the array fill factor. We consider these cases in section 9.3.2.

9.3.1 *Single-aperture Techniques*

Refractive Aspheric Microlenses

The first optical element to be considered is an anamorphic refractive microlens fabricated by the mass transport technique (Liau *et al.*, 1989; Swensen *et al.*, 1995).

227

Unlike many simpler fabrication methods, mass transport permits the microlens to be accurately shaped to an arbitrary figure. This flexibility makes the technique particularly useful for demanding applications such as high-numerical-aperture collimation optics and anamorphic systems. Diode lasers and laser amplifiers are particularly suited to this technology. Edge-emitting lasers generally have elliptical output apertures, dissimilar transverse and lateral divergences, and often contain astigmatism at the output facet. Tapered-waveguide amplifiers and unstable resonators (Welch *et al.*, 1992; Kintzer *et al.*, 1993) have even greater beam asymmetry (e.g. 200:1) and astigmatism. However, free-space diffraction causes the narrow beam direction to expand more quickly than the wide beam; at a specific propagation distance the two widths become equal, resulting in an approximately circular beam. A single aspheric surface can then be used to remove the remaining phase curvature and collimate the beam. Since the distance between the laser and this circular-beam plane is typically only a few hundred microns, microlenses are required to perform the beam collimation.

The mass transport process starts by dry etching a binary pattern into a semiconductor such as gallium phosphide or indium phosphide. Figure 9.2(a) shows a pattern designed to produce a highly anamorphic lens for circularizing and collimating light from a tapered unstable-resonator diode laser (Liau *et al.*, 1994). The thickness of the transported lens is controlled by the width of the etched feature. Upon heating to just below the melting point of the semiconductor, surface diffusion causes the mass to redistribute over a local region, and a smooth aspheric surface results. Figure 9.2(b) shows a profilometer trace along the two directions of the anamorphic lens. The focal lengths are chosen specifically to remove the astigmatism present in the unstable-resonator diode laser. In addition, the surface shape is optimized to eliminate aberrations present in conventional spherical surfaces. The *f*-number transverse to the laser junction is *f*/0.9. This lens has been shown to produce nearly circular beams with near-diffraction-limited performance from a highly elliptical source.

Near-field Beam Shaping

A variety of optical devices have been proposed for shaping a laser beam in the near field. In some applications, the intensity of the shaped beam is of prime importance, and the beam phase can be arbitrary. Beam formers of this type are very simple. A beam former proposed for materials processing converts a Gaussian fundamental laser mode into a quasi-uniform intensity by using a pair of biprisms (Lacombat *et al.*, 1980). The first biprism splits the Gaussian beam in the middle along one of the transverse axes, and redirects the two halves along opposite angles. The beams are brought together again by mirrors, and are made to overlap so that the uniformity is improved. A second biprism and set of mirrors along the other axis performs the same function in the orthogonal direction. The four overlapping beams have greatly improved uniformity. However, the mutual coherence of the beams produces interference patterns. In some cases, the frequency of the interference fringes may be sufficiently high to be unimportant. In others, the interference pattern must be eliminated. This can be accomplished along one axis by using a half-wave plate to rotate the polarization of one of the beams; a phase modulator can be used along the other axis to sweep the fringes across the image and average out their effect. A four-surface prism has

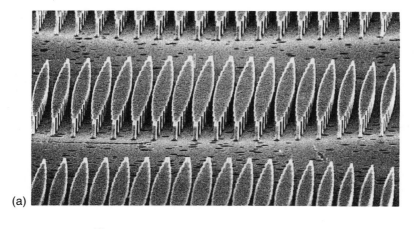

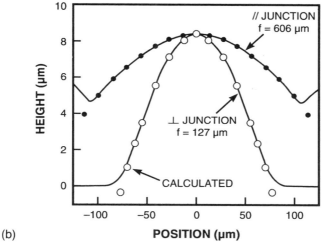

(a)

(b)

Figure 9.2 Fabrication and performance of an anamorphic refractive microlens by mass transport. (a) Initial etch pattern. (b) Surface figure after transport. (Liau *et al.*, 1994)

been used to perform the same shaping in a single element (Kawamura *et al.*, 1983).

Near-field beam shaping that requires control of the phase front is often performed by two phase elements separated in space. The first phase element redirects the light to form the desired intensity profile at the second phase element. The second element can then be used to establish the desired phase across the beam. A geometric transformation is derived between the coordinates of the first plane and the second using the principle of energy balance (Han *et al.*, 1983). For example, if a Gaussian beam with intensity $i(r) = \exp(-2r^2/r_0^2)$ is incident on the first optical surface, the energy contained in a circle of radius r is given by

$$E(r) = 2\pi \int_0^r i(r)r\,dr = \frac{\pi r_0^2}{2}[1 - \exp(-2r^2/r_0^2)].$$

(9.2)

229

To convert the Gaussian into a flat-top mode, this light energy can be equated to the energy at a second optical surface within a circle of radius R from a uniform light beam with constant intensity σ:

$$2\pi \int_0^R i(r)r\,dr = \sigma\pi R^2. \tag{9.3}$$

Combining equations (9.2) and (9.3) we have a relationship between r and R:

$$r = \left[\frac{r_0^2}{2} \ln\left(\frac{1}{1 - 2\sigma R^2/r_0^2} \right) \right]^{1/2}. \tag{9.4}$$

An aspheric refractive lens figure can be specified at the first surface to perform this coordinate transformation (Rhodes and Shealy, 1980). The second phase-correcting surface can then be determined by equal path length considerations.

It is possible to approximate the required transformation by using four spherical elements and balancing the spherical aberration (Shafer, 1982). Alternatively, a holographic coordinate transformation method (Bryngdahl, 1974a, 1974b) can be used to generate circular flat-top beams (Han *et al.*, 1983; Roberts, 1989). Finally, this method has been extended to include rectangularly shaped uniform profiles (Aleksoff *et al.*, 1991).

Far-field Beam Shaping

Beam shaping in the far field generally requires the preparation of a beam whose Fourier transform is of the desired shape. As with near-field beam shaping, there are some applications where the phase of the far-field distribution is unimportant. In this case, the phase of the near-field beam can be used as a free variable, and the shape of the far-field intensity can be influenced by proper near-field phase selection. This procedure is sometimes called spectrum shaping. A straightforward method of selecting the optimum phase distribution is given by the Gerchberg–Saxton algorithm for iterative phase retrieval (Gerchberg and Saxton, 1972). In this method, the intensity of the desired far-field wavefront and the intensity of the near-field illumination are used as constraints, and the phase is selected to optimize these two constraints in an iterative manner. After a sufficient number of iterations, the process converges to a solution that approximately satisfies both constraints. The near-field phase function is then used to fabricate a diffractive optical element. Figure 9.3(a) shows the phase function that converts a Gaussian near-field intensity distribution into a flat-top far-field intensity distribution. A computer-generated hologram of this phase function was fabricated and illuminated by a Gaussian laser beam (Lee, 1981). The resulting far-field pattern obtained in the focal plane of a lens is shown in Fig. 9.3(b).

Many applications require the phase of the far field to be uniform in addition to the intensity. For example, a laser radar system may require a uniform far-field intensity with a minimum divergence from a fixed telescope aperture size. Phase variations in the far field require an undesirable increase in aperture size. The same argument holds for concentrating light at the focal plane of a lens. To produce a uniform rectangularly illuminated far-field distribution with uniform phase, the near-field amplitude distribution must approximate a $\text{sinc}(ax)\text{sinc}(by)$ function, where $\text{sinc}(x) = \sin(\pi x)/(\pi x)$. Similarly, a circularly illuminated far-field distribu-

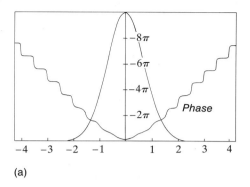

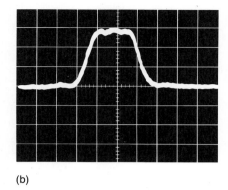

(a)   (b)

Figure 9.3 (a) Phase function produced by iterative phase retrieval algorithm. The illuminating Gaussian is also shown for scale. (b) Far field produced by a computer-generated hologram with phase function of (a) and Gaussian illumination. (Lee, 1981)

tion requires a $J_1(2\pi r)/r$ near-field distribution, where J is a Bessel function of the first kind. One straightforward method of converting a Gaussian beam into a one-dimensional sinc function is to design a phase plate with the phase transitions of a sinc function (Veldkamp and Kastner, 1982). Thus, the plate simply consists of a coarse square-wave phase grating with a central phase zone that is twice the width of the other zones. The effect of illuminating this plate with a Gaussian distribution is to approximate a sinc function. Since the Gaussian and the sinc are similar in shape in the area of the central lobe of the sinc function, the approximation is surprisingly good, and reasonably flat far fields can be achieved. The location of the phase transitions with respect to the Gaussian beam width and the modulation depth of the grating can be optimized for best far-field shape and efficiency.

Although this phase-only method produces a reasonably flat main lobe, there is excess energy contained in the side lobes. In some applications, this may be detrimental. For a better approximation to a one-dimensional sinc function, a second interlaced square-wave grating has been included in the plate (Veldkamp, 1982). The second grating was chosen to have a much higher frequency to diffract light out of the region of interest. The grating modulation depth was chosen to be identical to the original sinc function pattern, and the amount of attenuation was controlled by varying the grating duty cycle. Figure 9.4(a) shows a picture of the mask used to fabricate the phase grating. The interlaced structure can be clearly seen. An interlaced beam shaper was fabricated and tested with a CO_2 laser; the far-field response is shown in Fig. 9.4(b). The total efficiency of the beam shaper was measured to be 74%.

A single optical element is able to convert a Gaussian beam into a one-dimensional sinc function or a two-dimensional $J_1(2\pi r)/r$ function with fairly high efficiency because of the similarity in shape between the functions. This is not the case for two-dimensional rectangular far-field illumination, where the sinc(ax)sinc(by) function has its energy concentrated along the x and y axes. In this case, the above method is very inefficient and a more complex arrangement is required. One method of producing an accurate two-dimensional sinc function

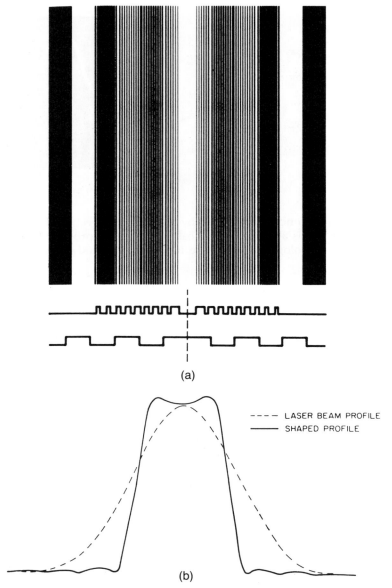

(a)

LASER BEAM PROFILE
SHAPED PROFILE

(b)

Figure 9.4 (a) Mask used to generate the phase plate. (b) Far-field distribution from a CO_2 laser with and without the beam shaper. (Veldkamp, 1982)

is via the geometric transformation optics used for near-field beam shaping (Eismann *et al.*, 1989). The method uses two phase elements separated by a finite distance (Fig. 9.5). The first element redirects the light into the desired sinc function intensity distribution, and the second establishes the correct phase. The initial design of the first element uses the stationary phase method (Born and Wolf, 1970) to determine an approximate phase distribution for converting a Gaussian beam into a sinc function shape. This phase distribution was used as a

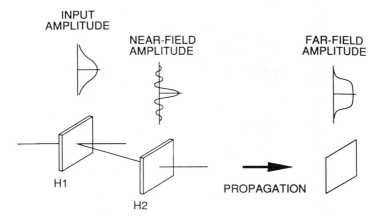

Figure 9.5 Dual hologram method for uniform far-field illumination. (Eismann *et al.*, 1989)

starting-point for an iterative phase retrieval solution. The phase retrieval method was found to improve the results remarkably. The second optical element, produced holographically, cancelled the phase distribution introduced by the first mapping and encoded the proper sign transitions onto the sinc function.

9.3.2 *Multiple-aperture Techniques*

Many single laser systems are inherently limited in power by physical constraints. For example, a single semiconductor laser may be limited by catastrophic facet damage as well as thermal considerations at high power levels. For this reason, some high-power laser systems contain arrays of smaller single lasers. There are two basic requirements to maximize the source radiance. First, the lasers must be phase-locked to achieve mutual coherence across the array. Second, the light distribution across the array must be uniform in amplitude and phase. External optical techniques for establishing coherence across two-dimensional arrays are reviewed by Leger (1993, 1994). In this section, we review some of the methods for combining light from a coherent array into a uniform plane wave.

The two basic methods of combining coherent light are shown in Fig. 9.6 using semiconductor lasers for illustration. The first, called superposition, collimates each of the beams and directs them to cross at a common location. Thus, the beams occupy the same space but not the same direction. An optical element is placed in this area of overlap to redirect the distribution into a common propagation angle. The second method, called aperture filling, uses an optical element to spread out the light across the array so that the array fill factor (the percentage of the array containing the laser apertures) approaches unity. Both these methods are capable of producing a light field whose radiance is the sum of the individual laser radiances.

Superposition with Gratings

Consider the optical set-up of Fig. 9.6(a). Light from each aperture is incident on a common plane, and an interference pattern is formed. We would like the light field after the optical element to be as uniform as possible in amplitude and

233

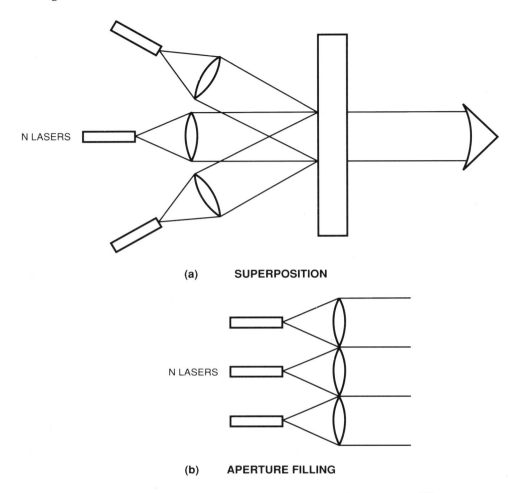

(a)   **SUPERPOSITION**

(b)   **APERTURE FILLING**

Figure 9.6 Laser beam combination. (a) Superposition. (b) Aperture filling.

phase to maximize its Strehl ratio (Leger, 1993). The phase of the interference pattern can be cancelled by a complex-conjugate phase element; the amplitude can only be attenuated by a passive element, resulting in an undesirable loss of power. One can adjust the phases of the individual lasing apertures, however, to ensure that the interference pattern is as uniform as possible. When this element is used inside an optical cavity to phase-lock the lasers (Leger *et al.*, 1986), the lasers will automatically adjust to the proper phase state for a given optical phase element. However, there are restrictions placed on the design of the element. Specifically, the transmittance of the element must be symmetric and real (phase values of 0 or π only). This implies that the phase state of the laser array must be selected to be symmetric and real as well. For laser arrays of reasonable size, it is possible to exhaustively examine all possible phase states to determine the optimum (Leger *et al.*, 1987b). Typically, the uniform phase state has the poorest performance; best performance is obtained from a phase state with a mixture of 0 and π phases. For example, the optimum phase state for 16 lasers is given by

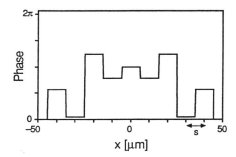
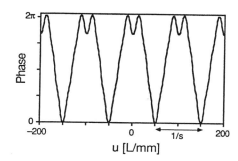

Figure 9.7 (a) Plot of phase plate profile placed over the nine emitting apertures of a diode laser array. (b) Plot of phase plate profile placed in the focal plane of the lens in Fig. 9.6(a). (Hertzig *et al.*, 1989)

$\{0, \pi, 0, \pi, \pi, 0, 0, 0, 0, 0, 0, \pi, \pi, 0, \pi, 0\}$. This state couples 84.1% of the light into an on-axis order.

If a binary phase state is still used but the symmetry restriction is removed, much higher coupling efficiencies can be obtained. For example, a seven-bit Barker code (Barker, 1953) can be used to define the phase state of a seven-element laser array as $\{\pi, \pi, \pi, 0, 0, \pi, 0\}$. The phase-correcting element is no longer binary, but the coupling efficiency is increased to 95% (Leger *et al.*, 1987b).

Finally, if we assume complete freedom in the choice of phase state, even better results can be obtained. An optimization procedure was developed in Herzig *et al.* (1989) to minimize the residual amplitude ripple at the optical element. An optimized phase state was determined, and is illustrated in Fig. 9.7(a). The array consisted of nine equally spaced emitters phase-locked in the in-phase mode. Figure 9.7(b) shows the predicted structure of the phase plate placed in the focal plane of the lens in Fig. 9.6(a). The total coupling efficiency of this structure is theoretically predicted to be 99.3%.

Evanescently-coupled diode laser arrays usually prefer to lase in an oscillating phase mode $(0, \pi, 0, \pi, \ldots)$. For this case, the above system can be reoptimized to convert this mode into a uniform field (Ehbets *et al.*, 1993). For an array consisting of 10 elements, the theoretical coupling efficiency is 97.3%. An experimental verification of this system was shown using a diode laser array.

Aperture Filling by Phase Spatial Filtering

The spatial filtering system shown in Fig. 9.8 can be used to perform aperture filling (Swanson *et al.*, 1987). The operating principle is the same as that of a Zernike phase contrast microscope for converting phase modulation into intensity modulation. In the present case, however, the system is used in reverse to convert a non-uniform intensity into a non-uniform phase. The phase-shifting element in the back focal plane of the lens modifies the phase of the zero-order Fourier component but does not affect the higher-order terms. It can be shown (Swanson *et al.*, 1987) that a binary amplitude distribution in the laser array plane can be converted into a uniform intensity distribution with a binary phase in the output plane as long as the fill factor is $\geq 25\%$. (Arrays with smaller fill factors can be used, but the coupling efficiency will be reduced.) A phase grating is then placed

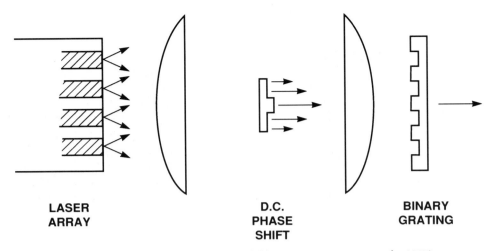

LASER   D.C.   BINARY
ARRAY   PHASE   GRATING
                      SHIFT

Figure 9.8 Optical system for phase spatial filtering. (Swanson *et al.*, 1987)

in the output plane to cancel the binary phase and convert the beam into a uniform plane wave. The optical device also works with non-binary distributions with a small reduction in efficiency. Gaussian beams have been shown to produce Strehl ratios of 0.9–0.97. Figure 9.9 shows a simple experiment that converts a line source array with a 25% fill factor into a filled array. The far field from each of these patterns is also shown. The far field of the original array contains several side lobes, and the power in the main lobe is equal to the fill factor (25%). The far field of the transformed array consists of essentially a single lobe, containing 92% of the total power. This optical system has also been demonstrated with a Y-coupled diode laser array (Leger *et al.*, 1987a) and a CO_2 laser array (Holz *et al.*, 1987).

Aperture Filling with Microlenses

Conceptually, one of the simplest techniques for increasing the array fill factor is to expand each beam until the individual light cones fill the field. For some laser arrays (e.g. CO_2 lasers), this may require an array of lenses to perform the expansion. For others such as edge-emitting diode laser arrays, the divergence from each single-mode waveguide is usually sufficient so that no beam expansion optics are required. At the point where the light cones meet, a simple lens element can be used to collimate each individual laser, resulting in a nearly uniform intensity. The collimating lens array may have strict requirements, however. First, the periodicity of the lens array must match the periodicity of the laser array. For diode laser arrays, this periodicity can be as small as 10 μm. Second, the fill factor of the lenses must be high so that the resulting wavefront is as continuous as possible. This often requires the lenses to be on a common substrate with close packing. Third, the numerical aperture of the lenses must match the numerical aperture of the lasers in the array. For edge-emitting diode lasers, numerical apertures as high as 0.5 may be required. Fourth, the lenses may be required to remove residual aberration from the individual lasers. Finally, the optical

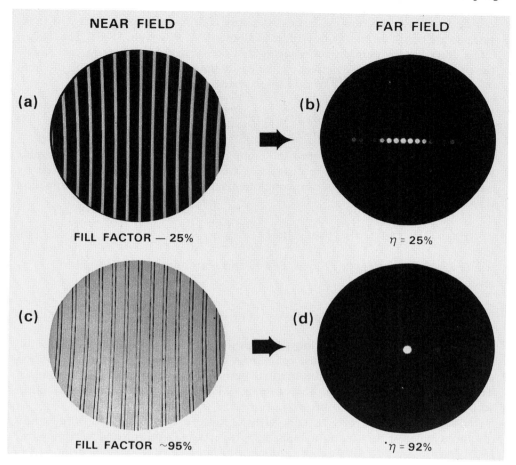

NEAR FIELD   **FAR FIELD**

(a)   FILL FACTOR — 25%   (b) $\eta = 25\%$

(c)   FILL FACTOR ~95%   (d) $\eta = 92\%$

Figure 9.9 Demonstration of aperture filling by phase spatial filtering. (a) Original coherent line sources with 25% fill factor. (b) Far-field intensity pattern of (a). (c) Image of line sources after phase spatial filtering. (d) Far-field intensity pattern of (c). (Swanson *et al.*, 1987)

aberration introduced by the lenses must be low so that the resulting wavefront will not be distorted. These requirements present a significant challenge to current micro-optical fabrication techniques. Both diffractive optic lens arrays (Leger *et al.*, 1988b) and mass transport lens arrays (Diadiuk *et al.*, 1989) have been successfully employed to increase the fill factor of coherent arrays. Chapter 5 of this book details several current methods for fabricating microlenses and microlens arrays.

Fractional Talbot Effect Aperture Filling

If the individual beams in Fig. 9.6(b) are allowed to propagate far enough to overlap before encountering an optical element, they start to form interference patterns. Interference from a coherent array has several unusual properties that can be used to create unique laser resonators and beam shapers. Many of these systems are based on the Talbot effect (Talbot, 1836; Rayleigh, 1881). The Talbot effect describes the periodic self-imaging that occurs in coherent arrays that satisfy the paraxial condition. The effect can be easily understood by expressing the field of an array $a(x, y)$ in terms of a Fourier series:

$$a(x, y, z = 0) = \sum_{m=-\infty}^{\infty} \sum_{n=-\infty}^{\infty} b_{mn} \exp\left[j2\pi\left(\frac{mx + ny}{d} \right) \right] \tag{9.5}$$

where d is the separation between the emitters in the array and b_{mn} are complex weights. The diffracted field a distance z from the array can be computed by multiplying each of the Fourier components by the Fresnel transfer function (Goodman, 1968):

$$H(m, n, z) = \exp\left(j2\pi \frac{z}{\lambda} \right) \exp\left[-j\pi\lambda z \left(\frac{m^2 + n^2}{d^2} \right) \right] \tag{9.6}$$

where λ is the wavelength of light, and m and n are integers. At values of $z = z_T$ given by

$$z_T = 2\frac{d^2}{\lambda} \tag{9.7}$$

the transfer function of equation (9.6) results in a constant phase term for all integer values of m and n. Thus, apart from this constant phase, the distribution at $z = 2nd^2/\lambda$ is identical to the original periodic object, where n is an integer. These self-images are termed Talbot images. The Talbot resonators described in section 9.4.2 are based on this effect.

Besides these basic self-imaging planes, there are many other propagation distances that give rise to unusual imaging properties (Winthrop and Worthington, 1965; Packross *et al.*, 1986). For example, the fractional Talbot planes at $z_T/2n$, $n \geq 1$ contain n equally spaced copies of the original aperture. These copies are registered with the original aperture for even n and shifted by $1/2n$ period for odd n. A second set of multiple images is formed at the planes corresponding to $z_T/(2n - 1)$, $n \geq 1$, where in this case $(2n - 1)$ equally spaced and properly registered images result. In each case, the phase is constant across a single copy of the aperture, but changes from copy to copy within one period.

The fractional Talbot effect can be used to convert an array of uniformly illuminated apertures into a uniform beam (Cassarly and Finlan, 1989; Liu and Liying, 1989; Kachurin *et al.*, 1991). In general, an array of square apertures with a fill factor of f (in one dimension) is converted into a uniform intensity distribution at

$$z_0 = \frac{z_T}{2N} = \frac{d^2}{N\lambda} \tag{9.8}$$

where $N = 1/f$. At this location, the fractional Talbot effect produces N copies of the aperture, filling the array. The phase across a single period consists of N equally spaced constant phase sections given by (Leger and Swanson, 1990)

$$\phi = \frac{\pi l^2}{N} \tag{9.9}$$

where I is an integer for even N and a half-integer for odd N ranging from $-N/2$ to $(N/2) - 1$. By placing a phase plate with the complex conjugate distribution of equation (9.9) at z_0, the resulting uniform intensity beam can be made uniform in phase as well. This method also works with Gaussian beams or other distributions. The efficiency is somewhat reduced, however, due to the non-uniform nature of the Gaussian and the interference from the various copies of the Gaussian beams. Variations on this method have been demonstrated to increase the fill factor of diode laser arrays (D'Amato *et al.*, 1989; Waarts *et al.*, 1992) and CO_2 laser arrays (Lescroart *et al.*, 1995).

9.4 Beam Shaping Inside Laser Resonators

The potential advantages of beam shaping inside laser resonators have been pointed out previously. In this section, we describe intracavity mode shaping techniques that are appropriate for both single-aperture lasers and multiple-aperture laser systems (laser arrays). We will explore various methods of producing a mode profile with desirable characteristics, and determine whether the design can enhance the modal discrimination of the cavity.

9.4.1 *Single-aperture Techniques*

The modes of a laser resonator are described by a set of mutually orthogonal two-dimensional field functions. Each mode has the property that its field must reproduce itself after propagating one round-trip in the cavity. However, intracavity apertures and finite-size mirrors attenuate the field, and the propagation length changes its phase. Thus, each mode is modified by a specific complex constant or eigenvalue. The general modal integral equation of a laser resonator can be written as

$$\int K(\overline{x}, \overline{x}') U_\nu(\overline{x}') d^2\overline{x}' = \gamma_\nu U_\nu(\overline{x}) \tag{9.10}$$

where the integral kernel $K(\overline{x}, \overline{x}')$ describes the round-trip propagation in the cavity including diffraction, mirror surfaces and apertures; $U_\nu(\overline{x})$ are the eigenfunctions or modes of the equation, and γ_ν their corresponding eigenvalues. The squared magnitude of an eigenvalue associated with a particular mode gives the round-trip attenuation of that mode due to diffraction. We will explore techniques for modifying the mode shape (eigenfunction) and the mode losses (eigenvalues) of single-aperture resonators.

Amplitude Spatial Filtering

The most common resonator structure consists of two spherical mirrors spaced to satisfy the stability condition (Yariv, 1991). If the mirrors and the aperture are sufficiently large, this resonator structure produces a fundamental mode described by a two-dimensional Gaussian function with a spherical wavefront. In rectangular coordinates, a higher-order mode can be expressed in terms of the fundamental by multiplying the fundamental mode by a two-dimensional Hermite polynomial of a given order. As a Gaussian beam propagates, it simply changes the radius of curvature of the spherical wavefront, and the width of the Gaussian intensity. Thus, if the size and location of the beam waist are chosen correctly, the phase fronts of the mode at each mirror match the radius of curvature of the spherical mirror, and the mirror effectively returns the complex conjugate beam. The complex conjugate beam retraces the path and regenerates the original Gaussian beam, thus establishing itself as a mode of the cavity.

The above cavity can be modified by introducing a spatial filter to produce selective absorption, or by modifying the surface figure of the mirrors. In either case, the shape of the modes, as well as their attenuation, can be changed. The simplest filter consists of an aperture placed between the two mirrors. This aperture provides a larger absorption to the higher-order modes than the fundamental, and tends to promote mode discrimination. With sufficiently small apertures, the mode shape is also distorted from a pure Gaussian.

The mode shape can be controlled by using more sophisticated apertures. In the experiment shown in Fig. 9.10(a), the laser cavity consists of two planar mirrors, a lens of focal length f, and two spatial filters (Kermene *et al.*, 1992). The planar mirrors are placed in the front and back focal planes of the lens. In the absence of the filters, the Fourier transform of the light distribution at mirror M_1 is produced at mirror M_2. Upon reflection and propagation back to M_1, this distribution is transformed back to an inverted copy of the original distribution. Thus, this cavity is degenerate to all symmetric modes. In order to select a specific mode, spatial filtering is applied in both mirror planes. The first spatial filter T_1 consists of a square aperture of width a. The second filter T_2 contains a grid of thin opaque strips that are spaced to occur at the zeros of the function $\text{sinc}(xa/\lambda f)\,\text{sinc}(ya/\lambda f)$, where λ is the focal length of the light. A uniform square mode at M_1 will have nulls at the opaque strips in M_2 and therefore have reduced loss. This mode will be established as the fundamental mode of the cavity, and all other distributions will be attenuated to a larger degree by one of the two amplitude spatial filters. The intensity distribution at the output of mirror M_1 is shown in Fig. 9.10(b) from an Nd:YAG laser ($\lambda = 1.06\,\mu$m) with this cavity configuration. The size of the beam is 1.93 mm, and the length of the cavity is 800 mm. The far-field intensity was observed to be close to a $\text{sinc}^2(x, y)$ distribution, demonstrating that the laser was operating in a single spatial mode and that the phase of this mode was uniform.

Aspheric Mirrors

An alternative approach to mode shaping in a laser resonator is to modify the phase of the cavity mirrors (Bélanger and Paré, 1991; Leger and Li, 1992) and introduce intracavity phase plates (Leger *et al.*, 1994a). A schematic diagram of a generalized resonator is shown in Fig. 9.11, where the end mirror, output mirror and phase plate contain aspheric reflectances and transmittances. These aspheric elements have been fabricated using diamond turning (Bélanger *et al.*, 1992) and diffractive optics (Leger *et al.*, 1994b). If we are allowed total design freedom in selecting the phase transmittance of the mirrors (as is the case for diffractive optic mirrors), it is possible to tailor the fundamental mode to have an arbitrary intensity and phase profile. Consider a laser cavity consisting simply of two mirrors with arbitrary phase reflectances spaced by a distance l (no intracavity phase plate).

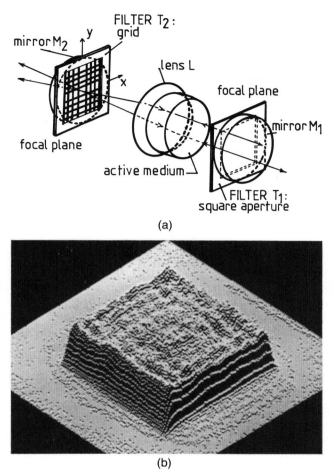

(a)

(b)

Figure 9.10 Intracavity mode shaping with amplitude spatial filters. (a) Laser cavity containing spatial filters at both planar mirrors. (b) Intensity distribution at mirror M_1. (Kermene *et al.*, 1992)

Let the desired amplitude and phase of the mode just past the first mirror be described by the complex function $a(x, y)$. This can be expressed equivalently in terms of its angular plane wave spectrum $A(u, v)$ as

$$a(x, y) = \int_{-\infty}^{\infty} \int_{-\infty}^{\infty} A(u, v) \exp[j2\pi(xu + yv)] \, du \, dv \tag{9.11}$$

where u and v are spatial frequencies.

The distribution at the second mirror is given by

$$b(x', y') = \int_{-\infty}^{\infty} \int_{-\infty}^{\infty} A(u, v) \exp[j2\pi(xu + yv)]$$

$$\times \exp[jkl\sqrt{1 - (\lambda u)^2 - (\lambda v)^2}] \, du \, dv. \tag{9.12}$$

241

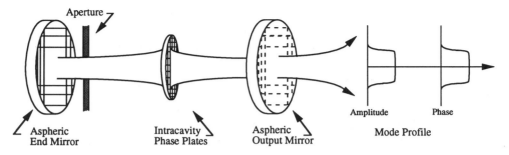

Figure 9.11 Generalized aspheric mirror laser cavity. (Leger *et al.*, 1995a)

If we construct the second mirror with a reflectance $t_2(x', y')$ of

$$t_2(x', y') = \frac{b^*(x', y')}{b(x', y')} \tag{9.13}$$

where * indicates the complex conjugate, the return wave is given by

$$b(x', y') t_2(x', y') = b^*(x', y')$$

$$= \int_{-\infty}^{\infty} \int_{-\infty}^{\infty} A^*(u, v) \exp[-j2\pi(xu + yv)]$$

$$\times \exp[-jkl\sqrt{1 - (\lambda u)^2 - (\lambda v)^2}] \, du \, dv. \tag{9.14}$$

Propagation back to the first mirror results in

$$\int_{-\infty}^{\infty} \int_{-\infty}^{\infty} A^*(u, v) \exp[-j2\pi(xu + yv)] \, du \, dv = a^*(x, y). \tag{9.15}$$

If now the first mirror reflectance is chosen to be

$$t_1(x, y) = \frac{a(x, y)}{a^*(x, y)} \tag{9.16}$$

the original distribution $a(x, y)$ has reproduced itself after one round-trip in the laser cavity, thereby establishing itself as a mode of the system. This phase-conjugate cavity is reminiscent of resonators based on Brillouin scattering or four-wave mixing (Auyeung *et al.*, 1979). Note, however, that the aspheric mirror phases are fixed, so this low-loss imaging only occurs for the desired fundamental mode.

Since the reflectances of the two mirrors in equations (9.13) and (9.16) are phase-only, they do not absorb any light. By making these elements sufficiently large, diffractive losses are kept to a minimum and the loss to the fundamental mode is very small. However, the losses to the higher-order modes are also small, and the cavity has very little modal discrimination. To ensure single-mode operation, it is usually necessary to provide a loss mechanism such as an absorbing aperture (shown in Fig. 9.11) to increase the loss of the higher-order modes. The selection of the cavity parameters that optimize the mode discrimination is reviewed below.

Optimization of cavity parameters The design of a particular laser cavity depends upon the desired fundamental mode. We will concentrate on a square super-Gaussian distribution of order n with an amplitude distribution given by

$$a(x, y) = \exp\left[-\left(\frac{x}{\omega_0}\right)^n \right] \exp\left[-\left(\frac{y}{\omega_0}\right)^n \right]. \tag{9.17}$$

For $n \gg 1$, the super-Gaussian approximates a flat-top distribution, and has all the potential advantages described in section 9.2. In addition, the flat-top distribution makes spatial filtering particularly effective since the aperture can be placed very close to the super-Gaussian edge without significantly attenuating the mode. Note also that since the super-Gaussian $a(x, y)$ is a real function, the mirror transmittance given by equation (9.16) is unity, and a simple plane mirror can be used. In this case, only a single aspheric mirror need be fabricated.

One of the principal considerations of resonator design is the loss associated with each mode. In general, we would like to minimize the loss to the fundamental mode and maximize the loss to all higher-order modes. The optimization of the cavity length (Paré and Bélanger, 1992; Leger *et al.*, 1995a) can be determined by calculating the losses of the first two modes for a variety of different cavity lengths. In general, it has been determined that maximum modal discrimination of a super-Gaussian fundamental mode occurs when the cavity length is roughly given by the Rayleigh range $z_0 = \pi\omega_0^2/\lambda$ defined by a conventional Gaussian with beam waist ω_0. A second important parameter is the size of the aperture at the flat mirror. When the size of the aperture is given by $2\omega_0$, a super-Gaussian fundamental mode uniformly illuminates the aperture with very little clipping (assuming $n \gg 1$), whereas the higher-order modes (with substantial light outside the aperture) are efficiently filtered out. The fundamental mode thus has negligible diffractive losses, whereas the loss to all the higher-order modes can be very large.

Experimental laser systems A CO_2 laser has been designed with a flat output mirror and an aspheric end mirror (Bélanger *et al.*, 1992). The set-up is shown in Fig. 9.12. Experiments were performed using fourth and sixth-order super-Gaussians as the fundamental mode amplitude. The circularly symmetric super-Gaussian amplitude $a(r)$ was defined as

$$a(r) = \exp\left[-\left(\frac{r}{\omega_0}\right)^n \right] \tag{9.18}$$

where ω_0 is the beam waist at the $1/e$ amplitude point and n is the order number. Because of the circular symmetry and the large wavelength ($\lambda = 10.6\,\mu\text{m}$), the mirror could be fabricated by diamond turning. In these experiments, an increase of up to 50% of monomode energy extraction was measured compared with that of a semiconfocal resonator of the same dimension.

For more complex mirror designs, diffractive optical elements have been used to exhibit the aspheric mirror response (Leger *et al.*, 1994b). The calculated mirror profile is converted into a diffractive optic by subtracting off integer multiples of the lasing wavelength and quantizing the resulting relief structure into discrete phase levels (see Chapter 1). A diffractive aspheric end mirror was calculated to produce a square super-Gaussian intensity. An Nd:YAG laser was constructed,

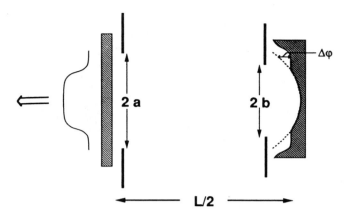

Figure 9.12 CO_2 laser resonator for producing a sixth-order super-Gaussian fundamental mode. Paré and Bélanger, 1992)

containing this diffractive end mirror, a flat output coupling mirror with square aperture, and a flashlamp-pumped Nd:YAG crystal. The resulting laser mode (measured inside the cavity with a sampling beam splitter) is shown in Fig. 9.13. The rms variation across the top of the mode is 1.5% of the peak value. The fundamental mode was fitted to a generalized two-dimensional super-Gaussian, and found to correspond closely to a 14th-order super-Gaussian of width $2\omega_0 = 1.3$ mm. Figure 9.13(b) shows the far-field pattern of this mode. As expected, it looks similar to the squared Fourier transform of a rectangular function, $|\sin(\pi df)/(\pi df)|^2$, where d is the width of the rectangle and f is the spatial frequency. The Strehl ratio of the far field was measured to be 94.5%, corresponding to an rms phase aberration across the near-field pattern of approximately 0.04 waves.

The higher-order cavity modes were studied by placing wire spatial filters in the nodes of the higher-order modes (Chen *et al.*, 1995). These wires were selected to produce high fundamental mode loss and negligible loss to higher-order modes. By proper wire placement, the lasing thresholds of the first three spatial modes (TEM$_{00}$, TEM$_{10}$ and TEM$_{11}$) were measured as a function of the square aperture size at the flat output mirror. These data are shown in Fig. 9.14 (discrete points) compared to the theoretical predictions (continuous lines) for a 14th-order super-Gaussian. The round-trip power loss of a specific mode due to diffraction is given by equation (9.10) as $|\gamma_\nu|^2$. The laser gain required to overcome this loss (called the modal threshold gain) can then be defined as

$$G_\nu = \frac{1}{|\gamma_\nu|^2}. \tag{9.19}$$

G_ν is a good measure of modal discrimination and fundamental mode loss, where a value of unity corresponds to a lossless cavity. There is very little fundamental mode loss for a 1.3 mm aperture (threshold gain is approximately unity). However, the threshold gain of the second-lowest-order mode (TEM$_{10}$) is greater than 3.0, corresponding to a large loss (66%).

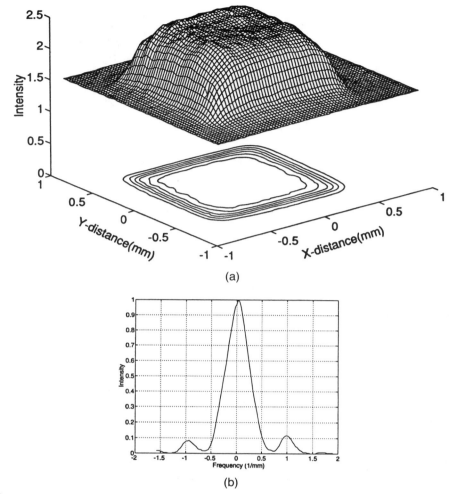

(a)

(b)

Figure 9.13 Experimental measurements of diffractive aspheric mirror Nd:YAG laser. (a) Near-field pattern. (b) Far-field pattern. (Chen *et al.*, 1995)

Diffractive aspheric mirrors have also been used as external cavities for wide-stripe diode lasers (Mowry and Leger, 1995). In this case, anamorphic optics are used to collimate the light emanating from the laser waveguide. Since the planar waveguide is single-mode transverse to the junction, the external mirror need only control the mode in one dimension. A $600\,\mu m$ wide diode laser was antireflection-coated on both output facets, and an external cavity was constructed consisting of one plane output mirror and one diffractive aspheric mirror. The aspheric mirror was designed to produce a $500\,\mu m$ wide one-dimensional super-Gaussian of order eight in the direction parallel to the junction. The $600\,\mu m$ gain region served as an effective aperture for higher-order mode discrimination, and no additional apertures were used. The near field was close to the desired super-Gaussian shape, and the far field was diffraction-limited with a divergence

245

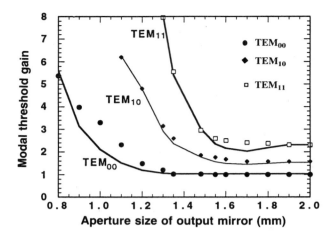

Figure 9.14 Measurement of modal threshold gain for the three lowest-order spatial modes of a diffractive mirror laser cavity. (Chen *et al.*, 1995)

of 0.12 mrad. The resulting laser system was able to produce up to 2.8 watts of optical power in this diffraction-limited mode.

Intracavity Phase Plates

The simple two-mirror cavity was shown to establish an arbitrary real mode profile with a single aspheric mirror and an arbitrary complex profile with two aspheric mirrors. When simple mode profiles such as super-Gaussians are desired, however, large discrimination between spatial modes occurs only when the cavity length is of the order of a Rayleigh range. Thus, for large beam diameters, these methods can result in long cavity lengths. The addition of intracavity elements can significantly increase the mode discrimination and permit utilization of shorter cavities and larger modes (Leger *et al.*, 1994a). A cavity consisting of a simple phase plate and a single diffractive aspheric mirror is shown in Fig. 9.15. The design of the aspheric mirror proceeds as before. The mode at the output mirror can be selected to be any real function, but the diffraction pattern incident upon the diffractive aspheric mirror is modified by the phase plate. The aspheric mirror is designed to return the phase conjugate of this pattern. As the light passes through the phase plate on the return trip, it recreates the desired mode. Thus, if a laser gain medium is placed between the phase plate and the flat end mirror, the mode in the gain medium can be simple (e.g. super-Gaussian) and can extract power from the medium in an optimum fashion.

The presence of the phase plate can increase the discrimination of the cavity dramatically. One simple phase plate consists of a grating with amplitude transmittance

$$t(x) = \exp[jm \sin(2\pi f_g x + \phi)] \tag{9.20}$$

where m is the modulation index of the grating, f_g is the grating frequency, and ϕ is the grating phase shift. Without the phase plate, a 25 cm long cavity designed to produce a 20th-order super-Gaussian with a beam size of 2 mm has very poor

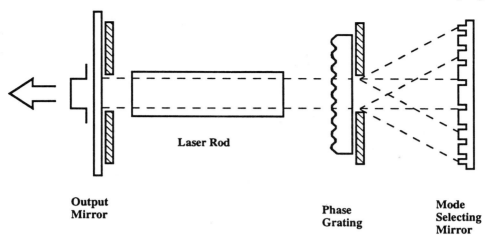

Output Mirror   **Laser Rod**   **Phase Grating**   **Mode Selecting Mirror**

Figure 9.15 Laser cavity containing an intracavity phase plate. (Leger *et al.*, 1994a)

modal discrimination (second-order modal threshold gain $G_2 = 1.006$). However, by choosing the proper grating frequency f_g, the threshold gain can be increased to a value of $G_2 = 2.5$ (corresponding to a loss of 60% to the second-order mode). The phase of the grating also influences the modal discrimination, and, in this case, provides a maximum discrimination when $\phi = 0$. A phase plate consisting of a two-dimensional version of equation (9.20) has been shown to exhibit high modal discrimination inside an Nd:YAG laser resonator (Leger *et al.*, 1994a).

The mode discrimination ability of the sinusoidal phase plate is reduced for higher grating frequencies, and has virtually no effect when the grating frequency is sufficiently high to allow the grating orders to separate before reaching the diffractive aspheric mirror. A different behaviour is seen for a random-phase plate, where there are no discrete diffraction orders (Leger *et al.*, 1995a). In this case the modal threshold gain continues to increase along with the spatial frequency bandwidth of the plate.

9.4.2 *Multiple-aperture Techniques*

In this section, we consider arrays of single-mode lasing apertures inside a common resonator structure. First, we review the method of coupled modes for analyzing the array performance. The method is derived for diode lasers, but can be generalized to apply to any laser array. Next, we describe the modal performance of a Talbot cavity, and describe some of the spatial filtering methods that have been employed to control the modes and increase the modal discrimination. Finally, the method of aspheric mirrors introduced in the previous section is generalized for operation with arrays.

Coupled Mode Theory

We consider an array of single-mode diode laser apertures, each with a Gaussian profile. A one-dimensional formulation is shown for simplicity, but can be readily extended to two dimensions. The system consists of an array of waveguides with a rear facet amplitude reflectivity of r_0. The front facet is assumed to have a perfect antireflection coating, and the external mirror located a distance in front of the array is coated with amplitude reflectivity r. Coupling between the waveguides results from free-space diffraction between the array and the mirror and any phase modulation that is recorded on the mirror. Although in principle it is possible to solve the eigenvalue problem of equation (9.10) to determine the modal performance of this system, coupled mode theory offers a simplified mathematical tool when the paraxial approximation is valid. We assume that the array consists of N independent lasers, where the spatial mode of the nth laser is given by $e_n(x)$. The amplitude distribution of the array is then given by

$$a(x) = \sum_{n=1}^{N} b_n e_n(x) \tag{9.21}$$

where b_n is the complex amplitude of the nth laser mode. If we assume a constant known shape for each laser mode, then $a(x)$ is uniquely determined by a vector of amplitude coefficients

$$\boldsymbol{b}^t = \{b_1, b_2, \ldots, b_N\}. \tag{9.22}$$

Coupled mode theory (Butler *et al.*, 1984; Kapon *et al.*, 1984) relates the coefficient vector $\boldsymbol{b}$ to a coupling matrix $\mathbf{K}$ via an eigenvalue equation

$$[r_0 r \exp(\mathrm{i}2\sigma L)\mathbf{K} - \mathbf{I}]\boldsymbol{b} = 0 \tag{9.23}$$

where $\mathbf{I}$ is the identity matrix, σ is the complex-valued propagation constant, L is the length of the diode lasers, and $\mathbf{K}$ is an $N \times N$ matrix of coefficients k_{mn} describing the coupling between the mth and nth waveguides (Yariv, 1991).

The distributed modal threshold gain g_ν is defined in terms of the gain coefficient (in cm^{-1}) required to achieve threshold:

$$g_\nu = \frac{1}{L}\ln\left(\frac{1}{r_0 r}\right) + \frac{1}{L}\ln\left(\frac{1}{|\gamma_\nu|}\right) \tag{9.24}$$

where γ_ν is the νth eigenvalue of equation (9.23). The first term is mode-independent, and consists simply of the loss due to the rear facet r_0 and the finite reflectivity of the external mirror r. The second term is a function of mode number, and can be used to increase mode discrimination and enhance single mode performance.

Talbot Cavities

The Talbot effect described in section 9.3.2 can be employed inside a resonator to produce a low-loss array mode of coupled lasers (Antyukhov *et al.*, 1986; Golubentsev *et al.*, 1987; Leger *et al.*, 1988a; D'Amato *et al.*, 1989; Piché *et al.*, 1990; Hornby *et al.*, 1993; Sanders *et al.*, 1994). Talbot imaging theory provides an approximate explanation of the resonator behaviour for an array mode consisting of a large number of in-phase apertures. If the distance between the laser apertures and the flat end mirror is chosen to be $z_\mathrm{T}/2$ (where $z_\mathrm{T} = 2d^2/\lambda$ is the Talbot distance defined in section 9.3.2 and d is the spacing between waveguides), the light from the laser array is self-imaged via the Talbot effect after one round-trip, and couples back into the waveguides with little loss. However, a completely mutually incoherent array will not form a Talbot image and the return light will be spread across the array. The waveguide apertures act as spatial filters, coupling only a small fraction of the return light back into the guides and raising the lasing threshold of the incoherent state. Thus, from this simple theory it is apparent that the mutually coherent in-phase array mode is preferred over a totally incoherent mode.

A more complete analysis is required to determine the actual modal performance of the Talbot cavity. Coupled mode theory can be employed to calculate the N eigenmodes ψ_ν and associated diffractive losses (Mehuys *et al.*, 1991) of an N-laser cavity. The k_{mn} components of the coupling matrix $\mathbf{K}$ are computed by propagating a Gaussian beam from the mth waveguide and calculating the amount coupled into the nth waveguide. The eigenvectors and eigenvalues of equation (9.23) are then calculated. It is clear that the coupling of a uniformly spaced array is space invariant, so the matrix $\mathbf{K}$ is Toeplitz in form (all diagonal elements are equal). The eigenvectors of a Toeplitz matrix are known to be given by sinusoids (Grenander and Szego, 1958), such that the resulting array modes can be written as

$$\psi_\nu(x) = \sum_{n=1}^{N} \sin(n\theta_\nu) e_n(x) \tag{9.25}$$

$$\theta_\nu = \frac{\nu\pi}{N+1}, \quad \nu = 1, \ldots, N. \tag{9.26}$$

The eigenvalues (and corresponding losses) associated with each mode are a function of the cavity length, the fill factor (ratio of aperture size to aperture spacing), and the number of elements in the array. Figure 9.16 shows the loss associated with the in-phase ($\nu = 1$) and oscillating-phase ($\nu = N$) modes as a function of cavity length z_M for an eight-element array with a fill factor of 0.16. r_0 and r in equation (9.24) are assumed to be unity and $1/\sqrt{e}$ respectively. For the laser length $L = 250\,\mu m$, this corresponds to a mode-independent loss of $20\,cm^{-1}$. Low loss to the fundamental mode occurs when the Talbot imaging condition is satisfied, or

$$2z_M = nz_T \tag{9.27}$$

where $2z_M$ is the round-trip propagation distance, z_T is the Talbot distance given in equation (9.7), and n is an integer. It is clear from the figure that a low-loss image of the oscillating-phase mode is also formed at this cavity length. The loss to both these modes is entirely a result of edge effects from the finite array. For finite-size arrays, there is a small difference between losses of the two modes (Golubentsev *et al.*, 1993; Leger *et al.*, 1994c), but this difference rapidly approaches zero for large N. The maximum modal discrimination coupled with minimum fundamental mode loss occurs when $2z_M = z_T/2$. In this case, the oscillating-phase mode is the fundamental mode. Since this configuration leads to a single spatial supermode, many laser arrays have been designed with a half-Talbot round-trip propagation distance (Leger and Griswold, 1990; Waarts *et al.*, 1992). The oscillating-phase mode can then be converted back into an in-plane mode by an external phase plate (Heidel *et al.*, 1986).

The modal discrimination of a Talbot cavity decreases with increasing array size.

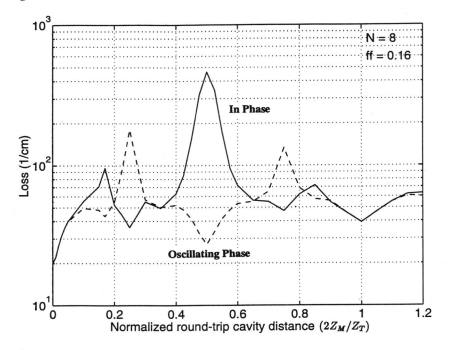

Figure 9.16 Diffractive losses of a Talbot cavity for the in-phase and oscillating-phase array modes as a function of round-trip cavity distance.

Laser arrays may therefore tend to lase in several spatial array modes simultaneously. The discrimination can be increased by reducing the fill factor (Mehuys *et al.*, 1991). However, reduced fill factors lead to larger edge effects and increased fundamental mode loss. This effect limits the ability to achieve high modal discrimination with large laser arrays. In diode laser arrays, a second limit is imposed by the paraxial condition. Since the Talbot effect is based on the paraxial form of free-space propagation, the lasing apertures must be large enough to ensure this condition is valid (Leger and Swanson, 1990).

Mode Selection with Intracavity Spatial Filters

Fourier (Leger, 1989; Aleksandrov *et al.*, 1990) and Fresnel (D'Amato *et al.*, 1989; Golubentsev *et al.*, 1990; Leger and Griswold, 1990) spatial filters can be placed inside Talbot cavities to provide loss to the undesirable modes. Fresnel spatial filters can be used to break the quasi-degeneracy at a round-trip distance of z_T by placing an absorbing grid at the flat end mirror. At $z_T/2$, the oscillating-phase mode produces a proper image of the mode, whereas the in-phase mode produces a shifted image. By placing the absorbing grid registered with the original source apertures, the oscillating-phase mode will be attenuated, whereas the in-phase mode will pass through the transparent areas between the absorbing stripes. These techniques have also been used in waveguides for one-dimensional spatial filtering (Mawst *et al.*, 1989; Van Eijk *et al.*, 1991).

Mode Tailoring with Diffractive Mirrors

By using diffractive optics to implement the aspheric mirror technique of section 9.4.1, arbitrary fundamental mode shapes can be realized. Therefore, these same techniques are readily applicable to laser arrays simply by defining the fundamental mode to be an array. A desirable fundamental mode for many arrays consists of an equally spaced series of Gaussian beams, defined by

$$a(x, y) = \sum_{n=-N/2}^{N/2} \exp\left[-\frac{(x - nd)^2 + y^2}{\omega_0^2} \right] \tag{9.28}$$

where ω_0 is the beam waist of a single Gaussian beam in the array and d is the array spacing. This mode shape differs from a simple Talbot cavity mode because each of the Gaussians is designed to have equal intensity. As mentioned in section 9.2.2, this has the advantage of utilizing all the available gain in the array and thus helps prevent spatial hole burning. In addition, the mode is chosen to be in-phase, so the on-axis far-field diffraction pattern will be maximized.

It is possible to design a diffractive mirror cavity of any length with the fundamental mode $a(x, y)$ of equation (9.28). The diffraction pattern of the desired mode is calculated at the end mirror, and a diffractive element is designed to reflect the complex conjugate of this distribution. The resultant field returned to the array reproduces $a(x, y)$ and the light couples back into the array waveguides with no loss. This is distinctly different from the Talbot cavity, where edge effects always produce some loss. The discrimination against higher-order modes can be evaluated by coupled-mode theory (Leger *et al.*, 1995b). Unlike the Talbot cavity, this resonator is space variant, and each of the coupling terms k_{mn} must be calculated separately to determine the matrix **K** completely. Because **K** is not Toeplitz in form, the eigenvectors no longer consist of sampled sinusoids, but rather take on more complex distributions.

The effect of the diffractive mirror cavity length on higher-order mode discrimination is shown in Fig. 9.17. As before, the output coupler amplitude reflectivity $r = 1/\sqrt{e}$, corresponding to a 20 cm^{-1} mode-independent loss from the 250 μm long lasers. For each external cavity length, a new diffractive mirror profile is computed to produce the fundamental mode of equation (9.28) with negligible diffractive loss. The total threshold gain of this mode is thus a constant 20 cm^{-1} and is not shown on the figure. The total threshold gain of the next-highest-order mode shown in the figure is a strong function of cavity length. For the fill factor of 0.12 chosen in this simulation, the cavity length with the highest modal discrimination (highest second-order mode loss) corresponds to a round-trip cavity distance of $2z_M = z_T$. This is very different from a Talbot cavity, where a round-trip cavity distance of z_T produces negligible mode discrimination.

The modal discrimination of the diffractive mirror cavity is reduced by increasing the fill factor or the array size, similar to a conventional Talbot cavity. However, unlike the Talbot cavity, the discrimination is also reduced by using too small a fill factor or array size. The decrease in discrimination due to large arrays can be compensated for by using the appropriate fill factors. Since small fill factors do not increase the fundamental mode loss, the diffractive mirror cavity can be designed to provide substantial mode discrimination for large arrays in a low-loss cavity.

An experimental verification of this cavity was shown using a diode laser array

251

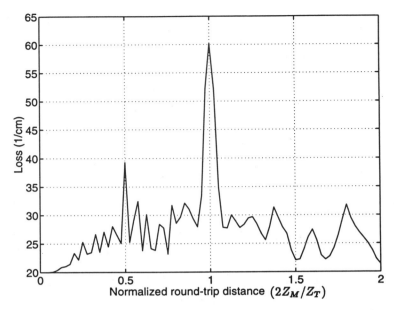

Figure 9.17 Distributed modal threshold gain of the second-order mode as a function of round-trip cavity length for a diffractive mirror cavity designed for an eight-laser array. (Leger *et al.*, 1995b)

(Leger and Mowry, 1993). A diffractive optical element was designed to excite a mode consisting of eight uniformly spaced Gaussians. One facet of a diode laser array was antireflection coated, and the other was given a high-reflection coating. Anamorphic optics were used to couple the light out of the array. The diffractive mirror was placed one-quarter Talbot distance behind an image of the AR-coated facet, corresponding to a round-trip propagation distance of $z_T/2$. This distance produced the optimum modal discrimination for the particular array fill factor. Figure 9.18 shows the far-field diffraction pattern from this cavity. The diffraction pattern consists of many 'grating lobes' due to the low fill factor of the array. This pattern can be converted into a single on-axis peak using the aperture-filling optics described in section 9.3. The central diffraction peak is maximum, with symmetric peaks on both sides. This is characteristic of an in-phase mode. Note that a conventional Talbot cavity of this length should lase in the oscillating-phase mode.

9.5 Conclusions

In this chapter, we have discussed a variety of laser beam-shaping techniques that can be applied to coherent fields. Shaping methods were divided into two categories: those that are applied to an existing laser beam (external shaping) and those that are an integral part of the laser cavity (internal cavity shaping). The primary advantage of external shaping is that it can be applied to any existing laser. Internal cavity shaping, on the other hand, can often be used to enhance the

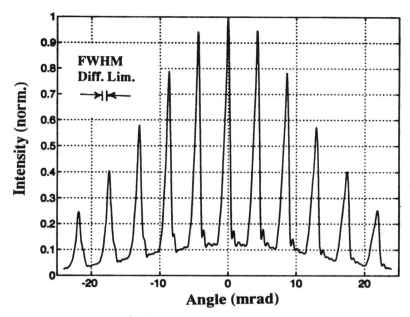

Figure 9.18 Experimental far-field pattern from an eight-element diode laser array in a diffractive-mirror external resonator. (Leger and Mowry, 1993)

performance of the laser as well as provide the correct beam shape. Both methods allow the designer to optimize the beam shape to a specific application.

Single-aperture external beam shapers were described that convert Gaussian beams into flat-top distributions for both near- and far-field applications. In some cases, the phase of the light field was unimportant, and the shaper simply created a uniform intensity. Other shaper designs resulted in a uniform intensity and phase. Laser arrays were seen to have special properties that could be exploited to construct unique external beam shapers. Internal beam-shaping systems contained intracavity amplitude and phase elements to excite specific desired modes and discriminate against higher-order modes. Applications to diode lasers, CO_2 lasers, and solid-state lasers were shown.

Most of the beam shapers described require specialized optics and advanced fabrication techniques. Continued improvements in fabrication technology should make these beam-shaping methods increasingly more effective and common in future laser systems.

References

ALEKSANDROV, A. G., ANGELUTS, A. A., VASIL'TSOV, V. V., ZELENOV, E. V. and KURUSHIN, E. A. (1990) Phase locking of radiators in a multichannel laser by a spatial filter. *Kvantovaya Elektron.* **17**, 1462–1463.

ALEKSOFF, C. C., ELLIS, K. K. and NEAGLE, B. D. (1991) Holographic conversion of a Gaussian beam to a near-field uniform beam. *Opt. Eng.* **30**, 537–543.

ANTYUKHOV, V. V., GLOVA, A. F., KACHURIN, O. R., LEBEDEV, F. V., LIKHANSKII,

V. V., NAPARTOVICH, A. P. and PIS'MENNYI, V. D. (1986) Effective phase locking of an array of lasers. *JETP Lett.* **44**, 78–81.

AUYEUNG, J., FEKETE, D., YARIV, A. and PEPPER, D. M. (1979) A theoretical and experimental investigation of the modes of optical resonators with phase-conjugate mirrors. *IEEE J. Quantum Electron.* **QE-15**, 1180–1188.

BARKER, R. H. (1953) In Jackson, W. (ed.) *Communication Theory,* pp. 273–287. London: Butterworths.

BÉLANGER, P. A. and PARÉ, C. (1991) Optical resonators using graded-phase mirrors. *Opt. Lett.* **16**, 1057–1059.

BÉLANGER, P. A., LACHANCE, R. L. and PARÉ, C. (1992) Super-Gaussian output from a CO_2 laser by using a graded-phase mirror resonator. *Opt. Lett.* **17**, 739–741.

BORN, M. and WOLF, E. (1970) *Principles of Optics,* 4th edn, p. 752. Oxford: Pergamon Press.

BOYD, R. W. (1983) *Radiometry and the Detection of Optical Radiation.* New York: Wiley.

BRYNGDAHL, O. (1974a) Optical map transformations. *Opt. Commun.* **10**, 164–168.
 (1974b) Geometrical transformations in optics. *J. Opt. Soc. Am.* **64**, 1092–1099.

BUTLER, J. K., ACKLEY, D. E. and BOTEZ, D. (1984) Coupled-mode analysis of phase-locked injection laser arrays. *Appl. Phys. Lett.* **44**, 293–295.

CASSARLY, W. and FINLAN, J. M. (1989) Paper PD2, OSA Annual Meeting, Orlando.

CHEN, D., WANG, Z. and LEGER, J. R. (1995) Measurements of the modal properties of a diffractive-optic graded-phase resonator. *Opt. Lett.* **20**, 663–665.

D'AMATO, F. X., SIEVERT, E. T. and ROYCHOUDHURI, C. (1989) Coherent operation of an array of diode lasers using a spatial filter in a Talbot cavity. *Appl. Phys. Lett.* **55**, 816–818.

DIADIUK, V., LIAU, Z. L., WALPOLE, J. N., CAUNT, J. W. and WILLIAMSON, R. C. (1989) External-cavity coherent operation of InGaAsP buried-heterostructure laser array. *Appl. Phys. Lett.* **55**, 2161–2163.

EHBETS, P., HERZIG, H. P., DÄNDLIKER, R., REGNAULT, P. and KJELBERG, I. (1993) Beam shaping of high-power laser diode arrays by continuous surface-relief elements. *J. Mod. Opt.* **40**, 637–645.

EISMANN, M. T., TAI, A. M. and CEDERQUIST, J. N. (1989) Iterative design of a holographic beamformer. *Appl. Opt.* **28**, 2641–2650.

GERCHBERG, R. W. and SAXTON, W. O. (1972) A practical algorithm for determination of phase from image and diffraction plane pictures. *Optik* **35**, 237–246.

GOLUBENTSEV, A. A., LIKHANSKII, V. V. and NAPARTOVICH, A. P. (1987) Theory of phase locking of an array of lasers. *Sov. Phys. JETP* **66**, 676–682.
 (1993) Mode structure in a laser set with close component coupling. In *High-Power Multibeam Lasers and Their Phase Locking,* SPIE Vol. 2109, pp. 219–228. Bellingham: SPIE.

GOLUBENTSEV, A. A., KACHURIN, O. R., LEBEDEV, F. V. and NAPARTOVICH, A. P. (1990) Use of a spatial filter for phase locking of laser array. *Sov. J. Quantum Electron.* **20**, 934–938.

GOODMAN, J. W. (1968) *Introduction to Fourier Optics.* New York: McGraw-Hill.

GRENANDER, U. and SZEGO, G. (1958) *Toeplitz Forms and Their Applications.* Berkeley: University of California Press.

HAN, C.-Y., ISHII, Y. and MURATA, K. (1983) Reshaping collimated laser beams with Gaussian profile to uniform profiles. *Appl. Opt.* **22**, 3644–3647.

HEIDEL, J. R., RICE, R. R. and APPELMAN, H. R. (1986) Use of a phase corrector plate to generate a single-lobed phased array far field pattern. *IEEE J. Quantum Electron.* **QE-22**, 749–752.

HERZIG, H. P., DÄNDLIKER, R. and TEIJIDO, J. M. (1989) Beam shaping for high power laser diode array by holographic optical elements. In *Holographic Systems, Components and Applications,* pp. 133–137, Conf. Pub. No. 311. London: Institute of Electrical

Engineers.

HOLZ, M., LEGER, J. R. and SWANSON, G. J. (1987) Lossless sidelobe suppresion for a phase-locked CO_2 waveguide laser. *Conf. on Lasers and Electro-optics: Digest of technical papers*, 26 April–1 May, Baltimore, Maryland, p. 356. New York: OSA.

HORNBY, A. M., BAKER, H. J., COLLEY, A. D. and HALL, D. R. (1993) Phase-locking of linear arrays of CO_2 waveguide lasers by the waveguide-confined Talbot effect. *Appl. Phys. Lett.* **63**, 2591–2593.

KACHURIN, O. R., LEBEDEV, F. V., NAPARTOVICH, A. P. and KHLYNOV, M. E. (1991) Control of the angular distribution of the radiation emitted by phase-locked laser arrays. *Sov. J. Quantum Electron.* **21**, 351–354.

KAPON, E., KATZ, J. and YARIV, A. (1984) Supermode analysis of phase-locked arrays of semiconductor lasers. *Opt. Lett.* **10**, 125–127.

KAWAMURA, Y., ITAGAKI, Y., TOYODA, K. and NAMBA, S. (1983) A simple optical device for generating square flat-top intensity irradiation from a Gaussian laser beam. *Opt. Commun.* **48**, 44–46.

KERMENE, V., SAVIOT, A., VAMPOUILLE, M., COLOMBEAU, B. and FROEHLY, C. (1992) Flattening of the spatial laser beam profile with low losses and minimal beam divergence. *Opt. Lett.* **17**, 859–861.

KINTZER, E. S., WALPOLE, J. N., CHIN, S. R., WANG, C. A. and MISSAGGIA, L. J. (1993) High-power strained-layer amplifiers and lasers with tapered gain regions. *IEEE Photon. Technol. Lett.* **5**, 605–608.

LACOMBAT, M., DUBROEUCQ, G. M., MASSIN, J. and BRÉVIGNON, M. (1980) Laser projection printing. *Solid State Tech.* **23**, 115–121.

LAVIGNE, P., MCCARTHY, N. and DEMERS, J.-G. (1985) Design and characterization of complementary Gaussian reflectivity mirrors. *Appl. Opt.* **24**, 2581–2586.

LEE, W. H. (1981) Method for converting a Gaussian laser beam into a uniform beam. *Opt. Commun.* **36**, 469–471.

LEGER, J. R. (1989) Lateral mode control of AlGaAs laser array in a Talbot cavity. *Appl. Phys. Lett.* **55**, 334–336.

— (1993) External methods of phase locking and coherent beam addition of diode lasers. In Evans, G. A. and Hammer, J. M. (eds) *Surface Emitting Semiconductor Lasers and Arrays*, pp. 379–433. Boston: Academic Press.

— (1994) Microoptical components applied to incoherent and coherent laser arrays. In Botez, D. and Scifres, D. R. (eds) *Diode Laser Arrays*, pp. 123–179. Cambridge: Cambridge University Press.

LEGER, J. R. and GOLTSOS, W. C. (1992) Geometrical transformation of linear diode-laser arrays for longitudinal pumping of solid-state lasers. *IEEE J. Quantum Electron.* **28**, 1088–1100.

LEGER, J. R. and GRISWOLD, M. P. (1990) Binary-optics miniature Talbot cavities for laser beam addition. *Appl. Phys. Lett.* **56**, 4–6.

LEGER, J. R. and LI, X. (1992) Modal properties of an external diode-laser-array cavity with phase conjugate optics. *Bull. Am. Phys. Soc.* **37**, 1212.

LEGER, J. R. and MOWRY, G. (1993) External diode-laser-array cavity with mode selecting mirror. *Appl. Phys. Lett.* **63**, 2884–2886.

LEGER, J. R. and SWANSON, G. J. (1990) Efficient array illuminator using binary-optics phase plates at fractional-Talbot planes. *Opt. Lett.* **15**, 288–290.

LEGER, J. R., CHEN, D. and DAI, K. (1994a) High modal discrimination in a Nd:YAG laser resonator with internal phase gratings. *Opt. Lett.* **19**, 1976–1978.

LEGER, J. R., CHEN, D. and MOWRY, G. (1995a) Design and performance of diffractive optics custom laser resonators. *Appl. Opt.* **34**, 2498–2509.

LEGER, J. R., CHEN, D. and WANG, Z. (1994b) Diffractive optical element for mode shaping of a Nd:YAG laser. *Opt. Lett.* **19**, 108–110.

LEGER, J. R., MOWRY, G. and CHEN, D. (1994c) Modal analysis of a Talbot cavity. *Appl.*

Phys. Lett. **64**, 2937–2939.

LEGER, J. R., MOWRY, G. and LI, X. (1995b) Modal properties of an external diode-laser array cavity with mode-selecting mirrors. *Appl. Opt.* **34**, 4302–4311.

LEGER, J. R., SCOTT, M. L. and VELDKAMP, W. B. (1988a) Coherent addition of AlGaAs lasers using microlenses and diffractive coupling. *Appl. Phys. Lett.* **52**, 1771–1773.

LEGER, J. R., SWANSON, G. J. and HOLZ, M. (1987a) Efficient side-lobe suppression of laser diode arrays. *Appl. Phys. Lett.* **50**, 1044–1046.

LEGER, J., SWANSON, G. J. and VELDKAMP, W. B. (1986) Coherent beam addition of GaAlAs lasers by binary gratings. *Appl. Phys. Lett.* **48**, 888–890.

LEGER, J. R., SWANSON, G. J. and VELDKAMP, W. B. (1987b) Coherent laser addition using binary phase gratings. *Appl. Opt.* **26**, 4391–4399.

LEGER, J. R., SCOTT, M. L., BUNDMAN, P. and GRISWOLD, M. P. (1988b) Astigmatic wavefront correction of a gain-guided laser diode array using diffractive microlenses. *Proc. SPIE* **884**, 82–89.

LESCROART, G., MULLER, R. and BOURDET, G. L. (1995) Phase filter design for efficient side lobe suppression in phase coupled CO_2 waveguide laser array. *Opt. Commun.* **115**, 233–240.

LIAU, Z. L., DIADIUK, V., WALPOLE, J. N. and MULL, D. E. (1989) Gallium phosphide microlenses by mass transport. *Appl. Phys. Lett.* **55**, 97–99.

LIAU, Z. L., WALPOLE, J. N., MULL, D. E., DENNIS, C. L. and MISSAGGIA, L. J. (1994) Accurate fabrication of anamorphic microlenses and efficient collimation of tapered unstable-resonator diode lasers. *Appl. Phys. Lett.* **64**, 3368–3370.

LIU, L. and LIYING, Z. (1989) Aperture filling of phase-locked arrays by phase correction of self-imaging. *Chinese J. of Lasers* **16**, 37.

MAWST, L. J., BOTEZ, D., ROTH, T. J., SIMMONS, W. W., PETERSON, G., JANSEN, M., WILCOX, J. Z. and YANG, J. J. (1989) Phase-locked array of antiguided lasers with monolithic spatial filter. *Electron. Lett.* **25**, 365–366.

MEHUYS, D., STREIFER, W., WAARTS, R. and WELCH, D. F. (1991) Modal analysis of linear Talbot-cavity semiconductor lasers. *Opt. Lett.* **16**, 823–825.

MOWRY, G. and LEGER, J. R. (1995) Large-area, single-transverse-mode semiconductor laser with diffraction-limited super-Gaussian output. *Appl. Phys. Lett.* **66**, 1614–1616.

PACKROSS, B., ESCHBACH, R. and BRYNGDAHL, O. (1986) Image synthesis using self imaging. *Opt. Commun.* **56**, 394–398.

PARÉ, C. and BÉLANGER, P. A. (1992) Custom laser resonators using graded-phase mirrors. *IEEE J. Quantum Electron.* **28**,. 355–362.

PICHÉ, M., GODIN, P. and BÉLANGER, P. A. (1990) Self-imaging laser resonators using the Talbot effect. In *Optical Resonators*, SPIE Vol. 1224, pp. 256–264. Bellingham: SPIE.

POATE, J. M. and MAYER, J. W. (1982) *Laser Annealing of Semiconductors*. New York: Academic Press.

LORD RAYLEIGH (1881) *Philos. Mag.* **11**, 196–205.

RHODES, P. W. and SHEALY, D. L. (1980) Refractive optical systems for irradiance redistribution of collimated radiation: their design and analysis. *Appl. Opt.* **19**, 3545–3553.

ROBERTS, N. C. (1989) Beam shaping by holographic filters. *Appl. Opt.* **28**, 31–32.

SANDERS, S., WAARTS, R., NAM, D., WELCH, D., SCIFRES, D., EHLERT, J., CASSARLY, W., FINLAN, J. M. and FLOOD, K. (1994) High power coherent two-dimensional semiconductor laser array. *Appl. Phys. Lett.* **64**, 1478–1480.

SHAFER, D. (1982) Gaussian to flat-top intensity distribution lens. *Opt. Laser Technol.* **14**, 159–160.

SIEGMAN, A. E. (1965) Unstable optical resonators for laser application. *Proc. IEEE* **53**, 277–287.

(1986) *Lasers*. Mill Valley, CA: University Science Books.

SWANSON, G. J., LEGER, J. R. and HOLZ, M. (1987) Aperture filling of phase-locked laser arrays. *Opt. Lett.* **12**, 245–247.

SWENSEN, J. S., FIELDS, R. A. and ABRAHAM, M. H. (1995) Enhanced mass-transport smoothing of $f/0.7$ GaP microlenses by use of sealed ampoules. *Appl. Phys. Lett.* **66**, 1304–1306.

TALBOT, W. H. F. (1836) Facts relating to optical science No. IV. *Philos. Mag.* **9**, 401–407.

VAN EIJK, P. D., REGLAT, M., VASSILIEFF, G., KRIJNEN, G. J. M., DRIESSEN, A. and MOUTHAAN, A. J. (1991) Analysis of the modal behavior of an antiguide diode laser array with Talbot filter. *J. Lightwave Technol.* **9**, 629–634.

VELDKAMP, W. B. (1982) Laser beam profile shaping with interlaced binary diffraction gratings. *Appl. Opt.* **21**, 3209–3212.

VELDKAMP, W. B. and KASTNER, C. J. (1982) Beam profile shaping for laser radars that use detector arrays. *Appl. Opt.* **21**, 345–356.

WAARTS, R. G., NAM, D. W., WELCH, D. F., MEHUYS, D., CASSARLY, W., EHLERT, J. C., FINLAN, J. M. and FLOOD, K. M. (1992) Semiconductor laser array in an external Talbot cavity. *Proc. SPIE* **1634**, 288–298.

WELCH, D. F., PARKE, R., MEHUYS, D., HARDY, A., LANG, R., O'BRIEN, S. and SCIFRES, D. (1992) 1.1 W CW diffraction-limited operation of a monolithically integrated flared-amplifier master oscillator power amplifier. *Electron. Lett.* **28**, 2011.

WINTHROP, J. T. and WORTHINGTON, C. R. (1965) Theory of Fresnel images. I. Plane periodic objects in monochromatic light. *J. Opt. Soc. Am.* **55**, 373.

YARIV, A. (1991) *Optical Electronics*, 4th edn, Chap. 13. Philadelphia: Holt, Rinehart and Winston.

10

Hybrid (Refractive/Diffractive) Optics

G. P. BEHRMANN and J. N. MAIT

10.1 Introduction

Advances in the design, fabrication and analysis of diffractive optical elements (DOEs) have made them a viable alternative to refractive elements for optical systems in interconnects, data storage and infrared imaging (Lee, 1993). Because of their unique properties, it is desirable in some instances to combine DOEs with refractive and reflective components in order to achieve special functions or desired levels of performance. From these applications a new field of optical design, hybrid optics, has emerged.

As the name implies, a hybrid optic consists of a diffractive surface that is etched, micromachined or embossed onto the surface of either a refractive or reflective optical component. This one-piece construction offers savings in both mass and volume. In fact, hybrid design techniques give the optical designer more flexibility in materials selection because the diffractive surface can be used to modify the bulk optical properties, e.g. dispersion and thermal behaviour, of refractive materials.

The applications of hybrid optical elements are limited only by the imagination of the designer (Sasian and Chipman, 1993; Shafer, 1993). However, most hybrid designs attempt to solve traditional problems that have been previously addressed with all-refractive solutions that require the specification of several lens materials and surfaces. Therefore, in this chapter, we emphasize these problems, the most common of which is achromatization. Achromatic lens systems are those that can operate over a wide wavelength band. In section 10.2, we discuss the design of achromatic hybrids and indicate some of their limitations. This is followed in section 10.3 with a discussion of design techniques for athermalization. An athermalized lens system is one that exhibits stability over a wide variation in temperature. In addition, we describe the requirements for hybrid lens systems that are both athermal and achromatic. In section 10.4, we discuss how a diffractive surface can be used to reduce the spherical aberration of a refractive lens. A hybrid approach also has advantages in the design of array generators and in section 10.5 we present designs of this nature. Finally, we end in section 10.6 with some concluding remarks.

10.2 Achromatization

When the illumination of an optical system is polychromatic, it is necessary to consider system performance over the entire wavelength band. An optical system's sensitivity to wavelength is measured by its level of chromatic aberration. All-reflective optical systems exhibit no chromatic aberrations and are, therefore, best used in the presence of polychromatic illumination. However, because the optical path is folded by each optical element, all-reflective systems are impractical to design in many cases and one is forced to use refractive materials whose optical properties vary with wavelength.

In this section we introduce the concept of dispersion and how it gives rise to chromatic aberrations in optical systems. We present techniques for designing basic achromatic lenses, including those that make use of hybrid optics, and discuss the advantages and disadvantages of a hybrid achromat.

10.2.1 *Dispersion of a Refractive Lens*

For objects located at infinity, the lens power $\phi(\lambda)$ and focal length $f(\lambda)$ of a refractive lens can be determined from the lensmaker's formula

$$\phi(\lambda) = \frac{1}{f(\lambda)} = [n(\lambda) - 1]\left(\frac{1}{R_1} - \frac{1}{R_2}\right). \tag{10.1}$$

Here $n(\lambda)$ is the index of refraction of the lens material at wavelength λ and $1/R_1$ and $1/R_2$ are the respective curvatures of the front and back surfaces of the lens. For a BK7 convex-plano lens, i.e. $R_2 = \infty$, with a paraxial focal length of 100 mm at $\lambda = 587.6$ nm, $R_1 = 51.68$ mm (Fig. 10.1(a)).

The variation in refractive index with wavelength is known as the dispersion of a material. For broadband optical systems, use of a highly dispersive material can significantly degrade image quality. The dispersion of a refractive lens material that is suitable for use in the visible spectrum is specified by its Abbe number V_r:

$$V_r = \frac{1}{t_r} = \frac{n_{d-1}}{n_F - n_C} \tag{10.2}$$

where t is the variation in power with wavelength

$$t = \frac{\Delta\phi(\lambda)}{\phi(\lambda)}. \tag{10.3}$$

The terms n_F, n_d and n_C correspond to the index of refraction at the wavelengths 486.1 nm (the hydrogen F-line), 587.6 nm (the helium d-line), and 656.3 nm (the hydrogen C-line), respectively. For broadband applications in other parts of the spectrum, e.g. the 8–12 μm region in the infrared, V_r is determined by the indices of the short, middle and long wavelengths that describe the band of interest. Note that the larger the value of V_r, the less dispersive the material. In Table 10.1, refractive index data and Abbe numbers are listed for several common glasses.

If the input to a refractive lens is not monochromatic, but consists of a continuous band of wavelengths, a continuum of foci is produced along the optic

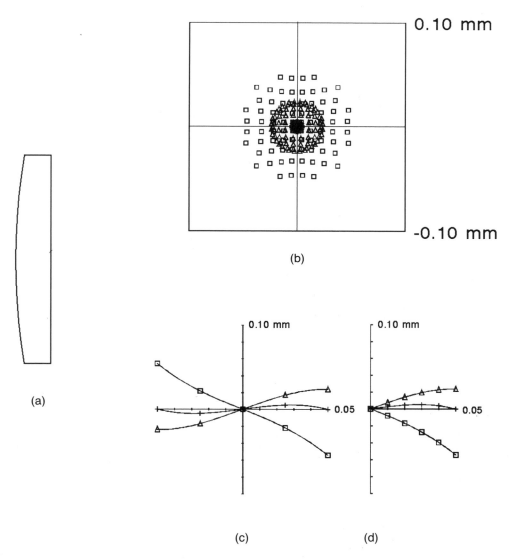

Figure 10.1 Performance of an *f*/10, 100 mm focal length BK7 singlet at *f*(λ_d). (a) Cross-section of lens. (b) On-axis spot diagram. Ray-intercept plots in (c) tangential and (d) sagittal planes. $\lambda_F = 486.1$ nm is indicated by $\square$, $\lambda_d = 587.6$ nm by +, and $\lambda_C = 656.3$ nm, by $\triangle$.

axis. This continuum of foci is known as primary axial chromatic (PAC) aberration and is measured by the distance between the foci produced by the wavelengths λ_F and λ_C (Smith, 1990)

$$d_{PAC} = f(\lambda_F) - f(\lambda_C).$$ (10.4)

If the BK7 singlet considered above is an *f*/10 lens, $d_{PAC} = -1.56$ mm. The effect of chromatic aberration manifests itself as a bright monochromatic focal spot surrounded by a polychromatic halo, which is illustrated in the spot diagram of

261

Table 10.1 Refractive index data and Abbe numbers for common optical glasses

Material	n_F	n_d	n_C	V_r
BK7	1.522 83	1.516 80	1.514 32	64.17
SF11	1.806 45	1.784 72	1.775 99	25.76
TiF4	1.595 21	1.584 06	1.579 45	37.04
TiF6	1.630 70	1.616 50	1.610 80	30.97
LaSF9	1.868 96	1.850 26	1.842 59	32.23

the singlet at $f(\lambda_d)$ in Fig. 10.1(b) and in the ray intercept plot of Fig. 10.1(c). Here, light rays at wavelength λ_d, denoted by the plus signs, are in focus, but are surrounded by the defocused rays at wavelengths λ_F and λ_C, denoted by squares and triangles, respectively.

10.2.2 Achromatic Doublets

By selecting lens materials that have different levels of dispersion, one can design a doublet that has reduced chromatic aberration. For thin lenses in close contact, the individual powers for achromatization can be determined from the equations

$$1 = \frac{\phi_1}{\phi} + \frac{\phi_2}{\phi} \tag{10.5a}$$

$$0 = \frac{1}{V_1}\frac{\phi_1}{\phi} + \frac{1}{V_2}\frac{\phi_2}{\phi}. \tag{10.5b}$$

Here, ϕ is the power of the doublet and V_1 and V_2 are the Abbe numbers of the lens materials. For a 100 mm focal length cemented achromat constructed from BK7 and SF11, the powers of the individual lenses are $\phi_{BK7} = 0.0167 \text{ mm}^{-1}$ and $\phi_{SF11} = -0.0067 \text{ mm}^{-1}$.

To compensate for the addition of the colour-correcting SF11 component, the power in the BK7 element is greater than the total power of the achromat. Recall from equation (10.1) that the more powerful a lens, the greater the curvature and hence the greater the required amount of material. Thus, the large positive power required indicates the difficulty associated with making achromats that have short focal lengths.

The use of equations (10.5a) and (10.5b) in lens design reduces, but does not eliminate, chromatic aberration. Elimination of PAC ensures only that the long and short wavelengths share a common focus. The distribution of foci generated by the other wavelengths still has finite extent and a residual amount of chromatic aberration remains. This residual aberration, measured by the distance between the common focus of $f(\lambda_F)$ and $f(\lambda_C)$ and the focus $f(\lambda_d)$, is known as secondary axial chromatic (SAC) aberration or secondary spectrum, which, for a thin achromatic doublet, is (Smith, 1990)

$$d_{SAC} = f(\lambda_F) - f(\lambda_d) = f\frac{P_2 - P_1}{V_1 - V_2} \tag{10.6a}$$

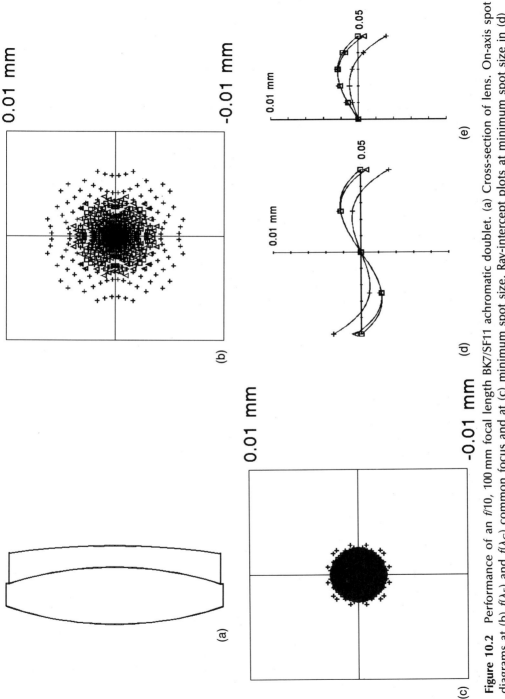

Figure 10.2 Performance of an *f*/10, 100 mm focal length BK7/SF11 achromatic doublet. (a) Cross-section of lens. On-axis spot diagrams at (b) $f(\lambda_F)$ and $f(\lambda_C)$ common focus and at (c) minimum spot size. Ray-intercept plots at minimum spot size in (d) tangential and (e) sagittal planes. Wavelength labelling is as in Fig. 10.1.

where P is the relative partial dispersion:

$$P = \frac{n_F - n_d}{n_F - n_C}.$$ (10.6b)

For an $f/10$ BK7/SF11 achromat that has a 100 mm focal length, $d_{SAC} = 0.013$ mm, a value that is acceptable for most imaging applications. A secondary spectrum can be observed in the spot diagram of Fig. 10.2(b) where the wavelengths λ_F and λ_C share a common focus, but are surrounded by the defocused rays of λ_d. It should be noted that this location does not produce the minimum possible spot size. The best focus is produced at an intermediate position between $f(\lambda_d)$ and the common focus of $f(\lambda_F)$ and $f(\lambda_C)$. This distinction is illustrated in Figs 10.2(b) and (c).

10.2.3 Dispersion of a Diffractive Lens

Up to this point, we have considered some of the important issues related to the design of refractive achromats. However, the maturity of diffractive optics technology now makes it possible to consider DOEs for the correction of chromatic aberration (Stone and George, 1988). Traditional analysis and design techniques can be applied, so long as the dispersion of a DOE is described in terms similar to those used for refractive lens materials.

A diffractive lens is a grating structure that when illuminated by a plane wave produces a focal point. Its operation is based on the principles of the Fresnel zone plate, whose focal length is determined by the width of its circular zone radii and the wavelength of illumination (Hecht, 1987). As shown in Fig. 10.3, the zone radii r_m are defined so that the distance from the edge of each zone m to the focal point f_0 is a multiple of the design wavelength λ_0. In the paraxial region, where $r_m \ll (f_0/m)^2$, for a given value of m, the focal length f_0 is

$$f_0 = \frac{n_i r_m^2}{2m\lambda_0}$$ (10.7)

where n_i is the refractive index of the image space. Although in many instances the image space is air ($n_i = 1$), in integrated and planar optical systems, all components are fabricated in and on a single substrate (Jahns *et al.*, 1992). The optical path lengths therefore need to be determined in this medium.

The focal length of a diffractive lens as a function of wavelength follows from equation (10.7):

$$f(\lambda) = \frac{1}{\phi(\lambda)} = f_0 \frac{\lambda_0}{\lambda}.$$ (10.8)

The dispersion and effective Abbe number V_d for a diffractive lens can be determined from equation (10.8) (Stone and George, 1988):

$$V_d = \frac{1}{t_d} = \frac{\lambda_d}{\lambda_F - \lambda_C}.$$ (10.9)

In the visible range $V_d = -3.452$ and in the 8–12 μm infrared region $V_d = -2.5$. It is important to note that V_d is a function only of the wavelengths that describe

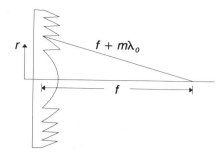

Figure 10.3 Cross-sectional view of a surface-relief diffractive lens.

the band and is independent of the lens material properties. In addition, V_d is negative, which implies that the focal lengths of long wavelengths are shorter than the focal lengths of short wavelengths, a characteristic that is opposite to that of refractive lens materials.

10.2.4 Hybrid Achromats

When the dispersion of the diffractive lens is expressed in terms of an Abbe number, it is possible to use equations (10.5a) and (10.5b) to solve for the refractive and diffractive powers that result in a single material achromat. The refractive and diffractive lens powers for a 100 mm focal length achromatic lens made from BK7 are $\phi_r = 0.0095$ mm^{-1} and $\phi_d = 0.0005$ mm^{-1}, respectively.

Techniques to fabricate hybrid achromats are mature and hybrid achromats are now available from many commercial sources. In fact, hybrid achromatic lenses are an integral part of several military and commercial systems. For example, hybrid achromats have been used in the design of lightweight forward-looking-infrared rifle sights (Chen and Anderson, 1993) and, in optical data storage applications, hybrid achromats have been used to overcome the difficulties associated with laser mode hopping (Kubalak and Morris, 1992).

The elimination of a second lens material is an obvious advantage of the hybrid achromat. However, there are other advantages worth noting. For example, in the hybrid achromat both the refractive and diffractive powers are positive and less than the total power of the lens. Thus, hybrid achromats can be made with weaker surfaces than all-refractive achromats and, since a weak refractive surface has a small curvature, less material is required to make the lens. In addition, small curvatures aid in the reduction of monochromatic aberrations such as spherical aberration (Kingslake, 1978).

Further, because the dispersion in the diffractive element is comparatively large, only a small amount of diffractive power is needed for achromatization. Weak diffractive surfaces, i.e. those with long focal lengths, have relatively large feature sizes and are easily manufacturable. For example, if the BK7 hybrid operates at $f/10$, the width of the smallest zone is approximately 11.8 μm, a feature size that is easily produced by most DOE fabrication techniques (Swanson, 1989).

The performance of an $f/10$ BK7 hybrid achromat at minimum spot size is illustrated in Fig. 10.4. As was the case for the refractive achromat, the hybrid

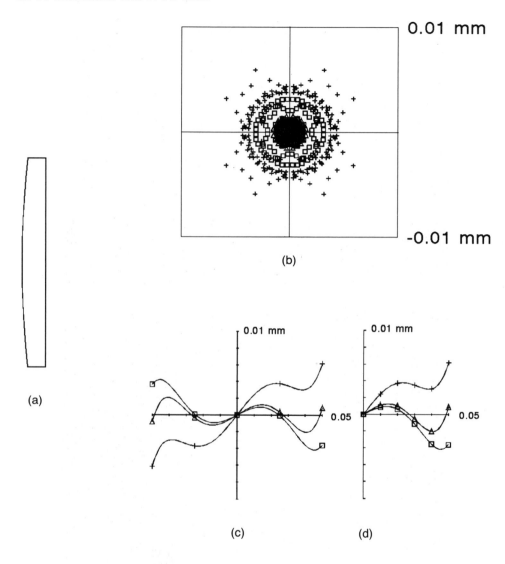

Figure 10.4 Performance of an f/10, 100 mm focal length BK7 hybrid achromatic doublet at minimum spot size. (a) Cross-section of lens. (b) On-axis spot diagram. Ray-intercept plots in (c) tangential and (d) sagittal planes. Wavelength labelling is as in Fig. 10.1.

achromat suffers from SAC. The relative partial dispersion of the diffractive lens is

$$P_d = \frac{\lambda_F - \lambda_d}{\lambda_F - \lambda_C} \tag{10.10}$$

which can be used in equation (10.6a) to calculate the d_{SAC} for the hybrid. Unfortunately, hybrid achromats in the visible wavelength range tend to have

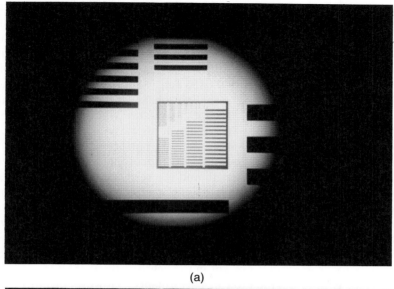

(a)

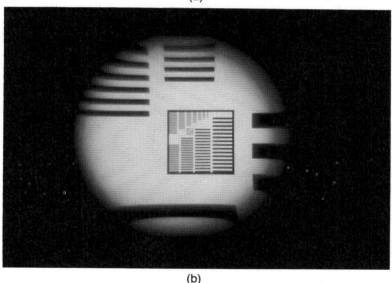

(b)

Figure 10.5 Imaging performance of a hybrid achromat. White light images produced by (a) an *f*/1.8, 20.7 mm focal length LaNF21 achromatic hybrid and (b) the refractive plano-convex lens used in the manufacture of the hybrid. (©European Space Agency.)

higher levels of SAC than their refractive counterparts. For example, for the BK7 hybrid $d_{SAC} = -0.166$ mm, which is more than a factor of ten greater than that for the BK7/SF11 refractive achromat. However, in the infrared regions of 3–5 μm and 8–12 μm, hybrid achromats and achromatic doublets can be designed to have essentially the same levels of SAC (Wood, 1992).

The white light image produced by a hybrid achromat fabricated by the

European Space Agency is presented in Fig. 10.5(a) (Czichy, 1993). The $f/1.8$ 20.7 mm lens was designed to be achromatic from 587.57 nm to 643.86 nm, with 632.81 nm as the nominal operating wavelength. It was fabricated by cementing a plano-convex LaFN21 refractive lens to a planar substrate of the same material upon which had been etched the diffractive lens. The nominal powers of the refractive and diffractive lenses are $0.0456 \, \text{mm}^{-1}$ and $0.0027 \, \text{mm}^{-1}$, respectively. The diffractive lens was fabricated as an eight-phase-level element with a minimum feature of $3 \, \mu\text{m}$. For comparison, Fig. 10.5(b) is the white light image produced by the plano-convex lens alone. The effect of aberrations is apparent.

A second residual aberration that can limit the performance of hybrid achromats is spherochromatism (Davidson *et al.*, 1993). Spherochromatism is the variation in spherical aberration with wavelength and is a result of the extremely dispersive nature of DOEs. (Spherical aberration itself is a monochromatic aberration related to the variation in focus with lens radius.) One can compare spherochromatism in the achromatic doublet and the hybrid achromat by inspecting the ray intercept plots of Figs 10.2(d) and (e) and 10.4(c) and (d). Spherochromatism is evidenced by the differing amounts of deviation from a straight line for each wavelength. Although spherochromatism is present in the achromatic doublet (see Figs 10.2(d) and (e)), there is little degradation in performance. However, from Figs 10.4(c) and (d), note that, as lens radius increases, the variation in spherical aberration with wavelength becomes more pronounced. Thus, spherochromatism is a critical consideration for low f-number lenses (Fritz and Cox, 1989).

For DOEs, diffraction efficiency is also an important consideration. Because the grating periods are large in comparison to the wavelengths of illumination, most hybrid achromats can be analyzed with scalar diffraction theory. Scalar theory predicts that a Fresnel zone plate fabricated with continuous phase diffracts 100% of the incident energy at a single wavelength λ_0 into a single diffracted order (Buralli and Morris, 1992b; Londoño and Clark, 1992). If one assumes that a diffractive surface is 100% efficient for λ_0 at order m, the diffraction efficiency at other wavelengths is (Swanson, 1989)

$$\eta(\lambda) = \left(\frac{\sin\{\pi[(\lambda_0/\lambda) - m]\}}{\pi[(\lambda_0/\lambda) - m]} \right)^2. \tag{10.11}$$

As illustrated in Fig. 10.6, when illuminated at other wavelengths, a DOE diffracts energy into other orders. If light from other orders is allowed to propagate through the system, it produces background energy at the image plane, which reduces image contrast and, hence, the modulation transfer function (MTF) of the system.

As defined by Buralli and Morris (1992b), the polychromatic integrated efficiency

$$\eta_{\text{int,poly}} = \frac{\int_{\lambda_{\min}}^{\lambda_{\max}} \eta(\lambda) \, d\lambda}{\lambda_{\min} - \lambda_{\max}} \tag{10.12}$$

scales the system MTF that results from conventional aberration analysis and is therefore a useful figure of merit for evaluating the effects of diffraction efficiency on the image quality of broadband systems. Here, $\lambda_{\min}$ and $\lambda_{\max}$ are the upper and lower bounds of the wavelength range. An approximate value of $\eta_{\text{int,poly}}$ for

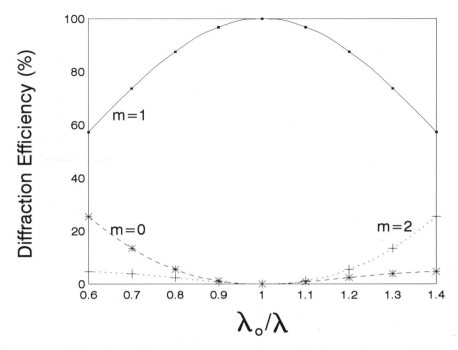

Figure 10.6 Diffraction efficiency versus λ_0/λ for zeroth, first and second diffracted orders of a diffractive lens. Lens is optimized for diffraction into first order.

$m = 1$ can be derived if $\eta(\lambda)$ is expanded in a power series and integrated term by term:

$$\eta_{\text{int,poly}} \cong 1 + \frac{\pi^2}{3\lambda_0}(\lambda_{\min} + \lambda_{\max} - \lambda_0) - \frac{\pi^2}{9\lambda_0^2}(\lambda_{\min}^2 + \lambda_{\min}\lambda_{\max} + \lambda_{\max}^2). \tag{10.13}$$

Table 10.2 lists the values of $\eta_{\text{int,poly}}$ for the visible region, 0.4–0.7 μm, and the two infrared regions, 3–5 μm and 8–12 μm. Note that in all three regions the integrated efficiency exceeds 90% and is maximum at 95% in the 8–12 μm band. Although the diffraction efficiencies are high, it is important to remember that the final MTF is also influenced by aberrations and the limits of fabrication.

10.3 Athermalization

Military and aerospace applications of DOEs have made DOE performance in harsh environments a critical issue. In these applications, ambient air temperatures can vary greatly and, in some cases, systems are expected to perform over an 80°C temperature range. At the same time, commercial demands, such as bringing optical fibres into the home, also require performance over a significant temperature range.

Recent work on the thermal and material properties of DOEs and their effects on the performance of diffractive optical systems has led to the development of DOE design techniques that allow for temperature compensation (Hudyma and

Table 10.2 Approximate values for integrated efficiency for three common wavelength bands

Wavelength band (μm)			
λ_{min}	λ_0	λ_{max}	$\eta_{int,poly}$ (%)
0.40	0.55	0.70	91.8
3.00	4.00	5.00	93.1
8.00	10.00	12.00	95.6

Kampe, 1992; Behrmann and Bowen, 1993; Londoño et al., 1993). In this section, we discuss methods to analyze the thermal behaviour of optical systems and also discuss design techniques for passive athermalization with DOEs.

10.3.1 Opto-thermal Expansion Coefficient

When a system operates over a wide temperature range, several important issues must be addressed. Consider that, as the system represented in Fig. 10.7 is heated or cooled, not only do the lens materials and mounting materials physically expand and contract but the refractive indices of the lens and image space also change. A useful parameter for evaluating the thermal behaviour of optical systems is the opto-thermal expansion coefficient x_f, which has been used in the design of all-refractive passively athermal lens systems (Jamieson, 1981; Rogers, 1991).

The opto-thermal expansion coefficient is defined as the normalized change in focal length due to a change in lens temperature (Jamieson, 1981). Its derivation for a refractive lens follows from equation (10.1):

$$x_{f,r} = \frac{1}{f}\frac{df}{dT} = \alpha_g - \frac{1}{n-1}\left(\frac{dn}{dT}\right).$$

(10.14a)

The change in focal length is

$$\Delta f = f x_{f,r} \Delta T.$$

(10.14b)

Here, n is the refractive index of the lens, dn/dT is the change in index of the lens material relative to the change in index of the image space, and α_g is the thermal expansion coefficient of the lens material. Note that the opto-thermal expansion coefficient is independent of the lens shape and is dependent only on material properties. However, recall from section 10.2 that n is also a function of wavelength and, therefore, thermal behaviour is wavelength-dependent.

For a complex optical system, the opto-thermal expansion coefficient is a function of the thermal properties of all lenses and airspaces. To evaluate the effects of temperature, it is necessary either to determine the opto-thermal expansion coefficient for the entire system or to evaluate the system on a piecewise basis.

To analyze the thermal behaviour of diffractive systems using techniques that have been developed for refractive optical systems, we must derive an expression for the opto-thermal expansion coefficient of a diffractive lens. From equation

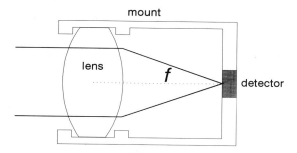

Figure 10.7 A simple optical system.

(10.7), in the paraxial region the focal length of a diffractive lens can be expressed in terms of its zone spacings. As the temperature changes, the zone spacings expand and contract and the zone radii r_m can, to the first order, be expressed as a function of temperature by

$$r_m(\Delta T) = r_m(1 + \alpha_g \Delta T).$$ (10.15)

The combination of equations (10.7) and (10.15) yields the opto-thermal expansion coefficient for a diffractive lens

$$x_{f,d} = \frac{1}{f}\frac{df}{dT} = 2\alpha_g + \frac{1}{n_i}\frac{dn_i}{dT}.$$ (10.16)

A comparison of equations (10.14a) and (10.16) reveals that there are fundamental differences in the thermal behaviour of refractive and diffractive lenses. For example, the change in focal length of a diffractive lens is a function only of α_g. It is not a function of thermally induced changes in the refractive index of the lens material. Although the diffraction efficiency of a diffractive lens is affected by temperature changes in the refractive index of the lens, for most materials the effects are negligible (Behrmann and Bowen, 1993).

Table 10.3 lists refractive and diffractive opto-thermal expansion coefficients for several optical materials. The image space medium is assumed to be air. The opto-thermal expansion coefficients for refractive lenses cover a wide range of values, both positive and negative; this property is attributed to the fact that dn/dT can be positive or negative and depends on the lens material. On the other hand, since $x_{f,d}$ is dominated by α_g, the diffractive opto-thermal expansion coefficients are all positive and cover a smaller range.

Two commonly used materials in this list are germanium, frequently used in infrared military applications, and acrylic, a popular plastic material for commercial products. When compared with common glass materials, these materials have extremely high values for $x_{f,r}$. It has, therefore, been difficult to design refractive infrared and plastic visible optical systems to operate over wide temperatures. However, the low diffractive opto-thermal expansion coefficients for germanium and acrylic indicate that insensitivity to temperature changes may be easier to achieve when these materials are used for diffractive lenses (Goto *et al.*, 1989; Gan, 1990).

Note that athermalization does not require the optical system to have a low

Table 10.3 Refractive and diffractive opto-thermal expansion coefficients for common optical materials

Material	$x_{f,r}$ (ppm/°C)	$x_{f,d}$ (ppm/°C)
BK7	0.98	13.89
SF11	−10.35	11.29
TiF4	7.66	17.80
TiF6	20.94	27.80
FK52	27.64	27.89
acrylic	315.00	129.00
polycarbonate	246.00	131.00
fused silica	−21.10	1.10
Ge	−123.76	12.20
ZnSe	−26.42	15.40
ZnS	−34.82	13.80

opto-thermal expansion coefficient; rather, the opto-thermal expansion coefficient of the system should be matched to the thermal expansion of the mounting material. Thus, the lens and mounting materials must be chosen so that the change in image position with temperature corresponds to the change in position of the focal plane. Designs of this nature are discussed in the following section.

10.3.2 *Hybrid Athermats*

By taking advantage of the unique thermal properties of refractive and diffractive lenses, it is possible to design athermalized hybrid lenses, referred to as athermats, in a manner similar to that used to design achromatic hybrids. Athermalized hybrid lenses offer improved thermal behaviour with savings in mass and volume.

In addition, as discussed in the previous section, DOEs make it possible to use materials such as germanium and acrylic that previously were difficult to use over wide temperature ranges. Consider, for example, acrylic. Acrylic offers several advantages for use as a lens material; not only is it inexpensive, lightweight and easy to shape, but also there are established techniques for generating aspheric diffractive surfaces on acrylic (Muranaka *et al.*, 1988; Clark and Londoño, 1989). Unfortunately, a refractive acrylic lens has a high opto-thermal expansion coefficient, 315 ppm/°C, the effects of which are highlighted in the following example. (The unit ppm indicates parts per million, $\times 10^{-6}$.)

Assume that the system in Fig. 10.7 consists of an f/5 convex-plano acrylic lens with a 100 mm focal length. The lens is all-refractive and the mounting material is aluminium ($\alpha_m = 23$ ppm/°C). The wavelength of illumination is 632.8 nm. If perfect alignment is assumed at 25°C and the entire system is heated to 50°C, then the focal point and detector assume new positions. The new focus is determined from equation (10.14b) and the detector displacement is given by

$$\Delta L = L_0 \alpha_m \Delta T \tag{10.17}$$

where $L_0 = 100$ mm for this example. The distance between the focal point and detector position is the net defocus, which in this case is 730 μm. At the illumination wavelength 632.8 nm, the net defocus adds approximately six waves of optical path difference peak-to-valley (OPD P–V) to the wavefront. Spot diagrams generated by the acrylic singlet at the detector before and after the temperature change are presented in Figs 10.8(a) and (b). Note that the spot size increases by a factor of 3. The spot diagrams also exhibit evidence of spherical aberration, the reduction of which we consider in section 10.4. At present we are concerned only with compensating thermal behaviour.

Because $x_{f,r}$ and $x_{f,d}$ for acrylic are different, a hybrid element can be designed such that the net opto-thermal expansion coefficient matches the coefficient of expansion of the mounting material. Although traditional athermal doublets are made from two different lens materials (Jamieson, 1981), if the diffractive lens is fabricated on the surface of the refractive element, the hybrid has the effective mass and volume of a singlet. For the hybrid athermat, the total lens power ϕ is the sum of the individual powers:

$$1 = \frac{\phi_r}{\phi} + \frac{\phi_d}{\phi} \tag{10.18a}$$

where ϕ is the net optical power and ϕ_r and ϕ_d are the respective refractive and diffractive powers of the hybrid. The net opto-thermal expansion coefficient is given by

$$x_f = x_{f,r}\frac{\phi_r}{\phi} + x_{f,d}\frac{\phi_d}{\phi}. \tag{10.18b}$$

Solution of equations (10.18a) and (10.18b) yields the refractive and diffractive powers that satisfy the focal length and thermal requirements of the lens. The refractive and diffractive powers necessary to athermalize the acrylic lens to an aluminium mount are $\phi_r = -0.0058$ mm^{-1} and $\phi_d = 0.0158$ mm^{-1}. The performance of this lens is illustrated in the spot diagrams of Figs 10.8(c) and (d). Over the temperature range 25–50°C the change in the spot size is approximately only 0.1%. However, it is critical to note that the diffractive power in the athermalized hybrid is high, which implies that its minimum feature is small. For this example, the width of the outer zones of the diffractive lens is only on the order of 4 μm and does not present a fabrication difficulty.

The measured amount of defocus over a 50°C change in temperature for a hybrid athermat fabricated by the Army Research Laboratory is presented in Fig. 10.9 (Behrmann *et al.*, 1994). The *f*/4.3 110 mm lens was designed to be athermal to aluminium and was fabricated by etching a binary-phase diffractive lens on the planar surface of a plano-concave fused silica lens. The nominal powers of the refractive and diffractive lenses are -0.0090 mm^{-1} and 0.0180 mm^{-1}, respectively.

It is also important to consider the chromatic aberration introduced when the hybrid athermat is illuminated at wavelengths other than the design wavelength. As discussed in section 10.2, the Abbe number is a useful figure of merit for evaluating the wavelength sensitivity of an optical material. The effective Abbe number V_{eff} for any doublet is

$$V_{eff} = \frac{1}{t_{eff}} = \phi\left(\frac{\phi_1}{V_1} + \frac{\phi_2}{V_2}\right)^{-1}. \tag{10.19}$$

273

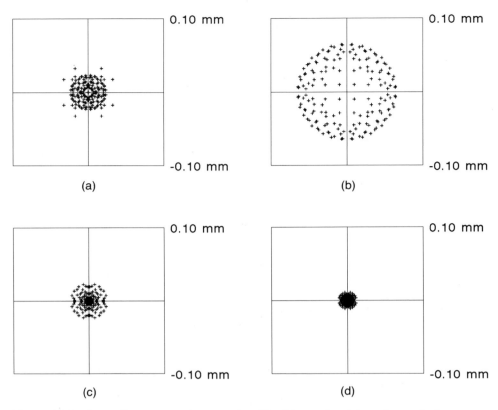

Figure 10.8 Spot diagrams at detector for *f*/5, 100 mm focal length acrylic lenses. On-axis spot diagrams for a refractive acrylic singlet at (a) *T* = 25°C and (b) *T* = 50°C. (c)–(d) as in (a)–(b), except that the lens is a hybrid acrylic athermat.

In the visible spectrum $V_r = 52$ and $V_d = -3.45$ for acrylic. Thus, for the athermat designed in the previous section $V_{eff} = -2.13$, which implies that the chromatic aberrations of the lens are severe. In fact, if the bandwidth is defined as the width of the spectrum over which OPD P–V is 0.25 waves, this hybrid athermat has a useful bandwidth of only ±0.1 nm, which limits it to applications whose illumination is monochromatic and stable.

10.3.3 *Hybrid Athermal Achromats*

The athermalization design procedure is similar to that used to achromatize refractive and diffractive surfaces. For achromatization, power is distributed in the refractive and diffractive surfaces based on the dispersion properties of the lens materials and in athermalization, power is distributed based on their thermal properties. However, as the example in the previous section indicates, lens design for broadband applications must consider chromatic performance in conjunction with thermal behaviour.

One approach to designing a hybrid doublet that is both achromatic and

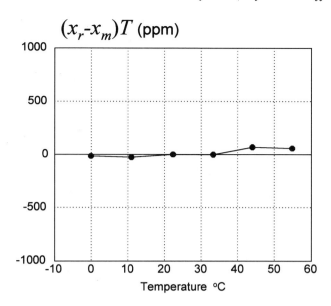

Figure 10.9. Measured normalized defocus due to changes in temperature for a hybrid fused silica doublet that is athermal to aluminium.

athermal to its mount is to use a mounting material whose thermal expansion coefficient matches the opto-thermal expansion coefficient of a hybrid achromat, which from equations (10.5a) and (10.5b) and equation (10.18b) is

$$x_f = \frac{V_r x_{f,r} - V_d x_{f,d}}{V_r - V_d}.$$
(10.20)

Note that x_f is a function only of the refractive and diffractive properties of the lens material. In the visible spectrum TiF6 ($x_f = 21.6$ ppm/°C) provides achromatization and near athermalization for aluminium ($\alpha = 23$ ppm/°C) and TiF4 ($x_f = 8.52$ ppm/°C) is a good choice for a titanium mount ($\alpha = 8.82$ ppm/°C). However, because of their large negative opto-thermal expansion coefficients, the lens materials most commonly used in the 8–12 μm spectrum, germanium (Ge), zinc selenide (ZnSe), and zinc sulphide (ZnS), require mounting materials that contract when heated. Properties of these materials are listed in Tables 10.3 and 10.4.

To arbitrarily select an opto-thermal expansion coefficient, correct chromatic aberration, and attain some degree of flexibility in materials selection requires a triplet that is described by the following linear system of equations:

$$1 = \frac{\phi_1}{\phi} + \frac{\phi_2}{\phi} + \frac{\phi_3}{\phi}$$
(10.21a)

$$x_f = x_{f,1}\frac{\phi_1}{\phi} + x_{f,2}\frac{\phi_2}{\phi} + x_{f,3}\frac{\phi_3}{\phi}$$
(10.21b)

$$0 = \frac{1}{V_1}\frac{\phi_1}{\phi} + \frac{1}{V_2}\frac{\phi_2}{\phi} + \frac{1}{V_3}\frac{\phi_3}{\phi}.$$
(10.21c)

275

Table 10.4 Refractive and diffractive Abbe numbers for three infrared lens materials. The wavelength band of interest is 8–12 μm

Material	V_r	V_d
Ge	858.1	−2.5
ZnSe	58.6	−2.5
ZnS	22.8	−2.5

Solution of equations (10.21a)–(10.21c) for a 100 mm focal length ZnSe/ZnS/Ge athermal achromat that has $x_\mathrm{f} = 0$ ppm/°C in the 8–12 μm spectrum yields $\phi_\mathrm{ZnSe} = 0.019\,650$ mm^{-1}, $\phi_\mathrm{ZnS} = -0.007\,591$ mm^{-1} and $\phi_\mathrm{Ge} = -0.002\,059$ mm^{-1}. The performance of an $f/3$ lens constructed with these parameters is represented in Fig. 10.10. The spot diagrams and ray-intercept plots assume that the detector is placed at the $f(\lambda_\mathrm{min})$ and $f(\lambda_\mathrm{max})$ common focus, not at the minimum spot size. Thus, it is possible to see that athermalization and achromatization have been achieved at a cost in the secondary spectrum. Although in this example we set $x_\mathrm{f} = 0$ ppm/°C, any value of x_f can be achieved with a triplet.

Unfortunately, the same military and aerospace applications that require thermally stable, broadband performance also require lightweight, low-volume design. Because of the high costs of infrared lens materials, this makes the all-refractive triplet economically undesirable. However, because of the unique dispersive and thermal properties of diffractive surfaces, it is possible to replace one of the refractive lenses in the triplet with a diffractive lens (Hudyma and Kampe, 1992; Behrmann, 1993). Thus, equations (10.21a)–(10.21c) can be satisfied with two materials.

For a 100 mm focal length ZnSe/Ge/DOE athermal achromatic hybrid, the lens powers are $\phi_\mathrm{ZnSe} = 0.012\,009$ mm^{-1}, $\phi_\mathrm{Ge,r} = -0.002\,059$ mm^{-1} and $\phi_\mathrm{Ge,d} = -0.000\,505$ mm^{-1}. The performance of the hybrid at $f/3$ is represented in Fig. 10.11. Note that the secondary spectrum of the hybrid is less than that of the refractive triplet. However, it is necessary to point out that the spot sizes indicated were determined with geometrical optics. Because they are below the diffraction limit they do not represent the actual spot size that will be generated.

Although the hybrid element performs as a triplet, its effective mass and volume is that of a doublet. Thus, the hybrid solution offers savings in mass and volume of about 30% over the all-refractive triplet.

Note that the diffractive power in the two-material hybrid is weak. Thus, in contrast to athermalized single-material hybrids, the feature sizes are large and the diffractive surface is easy to fabricate. Because most high-performance infrared lenses already employ diamond-turned aspheric surfaces, it is relatively straightforward to add the fabrication of the diffractive surface to the overall fabrication process (Riedl and McCann, 1991). However, because the lens is subject to polychromatic illumination, consideration must be given to its diffraction efficiency across the input wavelength band.

In this section we have presented techniques for designing athermalized hybrid lenses that offer improved thermal performance with a reduced number of elements. For wide temperature range applications, these techniques offer an

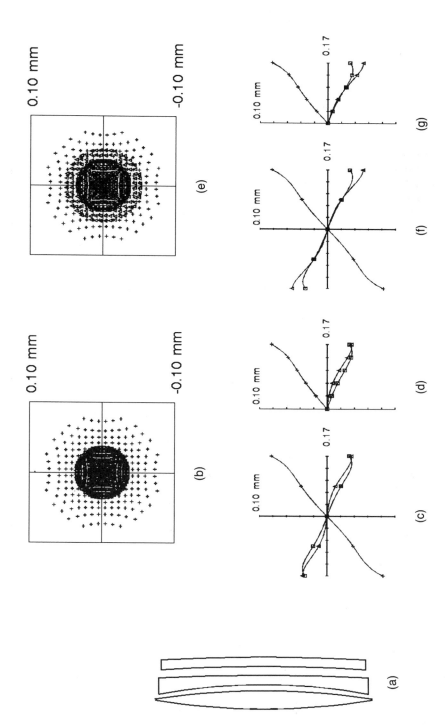

Figure 10.10 Performance of an *f*/3, 100 mm focal length athermal, achromatic triplet. (a) Cross-section of lens. Performance at 25°C: (b) on-axis spot diagram at best focus and (c) tangential plane and (d) sagittal plane ray-intercept plots when detector is placed at $f(\lambda_{min})$ and $f(\lambda_{max})$ common focus. (e)–(g) as in (b)–(d) except performance is at 50°C. $\lambda_{min} = 8.0 \ \mu m$ is indicated by □, $\lambda_0 = 10.0 \ \mu m$ by + and $\lambda_{max} = 12.0 \ \mu m$ by △.

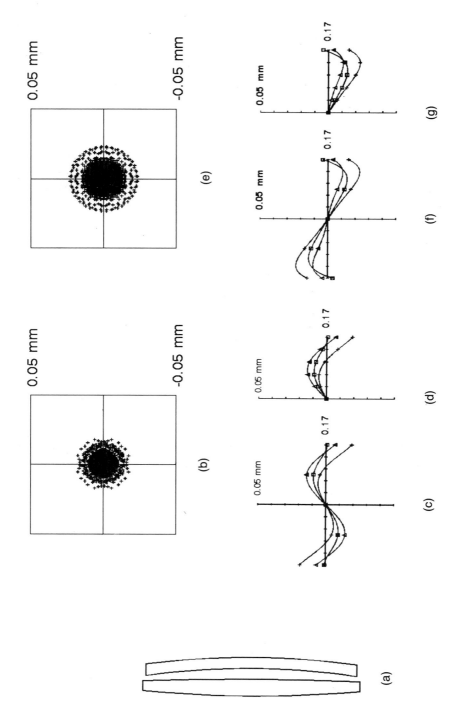

Figure 10.11 Performance of an *f*/3, 100 mm focal length athermal, achromatic hybrid triplet. (a)–(g) as in Fig. 10.10.

alternative to mechanical compensation. Hybrid athermalization has the potential to reduce the mass and volume of optical systems for infrared applications, which is an important consideration for military and aerospace systems; in commercial systems, hybrid athermalization can allow inexpensive materials, such as optical plastics, to be used over an extended temperature range.

10.4 Spherical Aberration

Many micro-optical systems require singlets that have short focal lengths and large apertures. Such lenses typically exhibit severe spherical aberration, the variation in focus with lens radius. For example, as shown in Fig. 10.12, the performance of an $f/2.5$ BK7 lens with a 50 mm focal length at $\lambda_0 = 632.8$ nm is degraded by a significant amount of third-order spherical aberration. At the edge of the pupil, the lens exhibits more than 33 waves OPD P–V.

Minimization of spherical aberration is one of the more significant challenges in lens design, particularly when one is limited to the design of spherical refractive surfaces. Traditional methods for reducing spherical aberration include solving for the optimal shape of the lens element, increasing the number of surfaces in the design, i.e. adding more lenses, and adding aspheric surfaces to the design (Kingslake, 1978). However, each of these techniques suffers from a number of disadvantages. The most significant are the increases in mass and volume that result from additional lens elements and the difficulties associated with the fabrication of aspheric surfaces. Due to their low fabrication costs, moulded plastic aspheres are used in some high-performance micro-optical systems (Muranaka *et al.*, 1988).

Alternatively, spherical aberration can be reduced through the use of diffractive surfaces. In fact, it was first demonstrated in 1965 that a hologram can reduce spherical aberration in a refractive lens system (Upatnieks *et al.*, 1966). In the previous sections of this chapter, we discussed hybrid lens design techniques that consider only the power of a diffractive lens. Our model of a diffractive lens has been based on the Fresnel zone plate, which is a simple model that assumes that the element has only focusing power. We can extend this model in more general terms by representing the diffractive lens in terms of a phase polynomial

$$\Phi(r) = \frac{2\pi}{\lambda_0} \sum_{n=1}^{N} a_{2n} r^{2n} \tag{10.22}$$

where a_2 specifies the lens focus and the remaining a_{2n} are aspheric coefficients at λ_0. If $a_2 = -1/2f$ and all other coefficients are zero, equation (10.22) simplifies to

$$\Phi(r) = -\frac{\pi r^2}{\lambda_0 f}$$

which is the phase function of a spherical lens (Goodman, 1968). However, it is the specification of the higher-order phase coefficients that allows DOEs to be designed with aspheric phase functions, such as those that can reduce spherical aberration introduced by a refractive surface (Swanson, 1989).

For example, evaluation of the Seidel sums from surface to surface in an optical

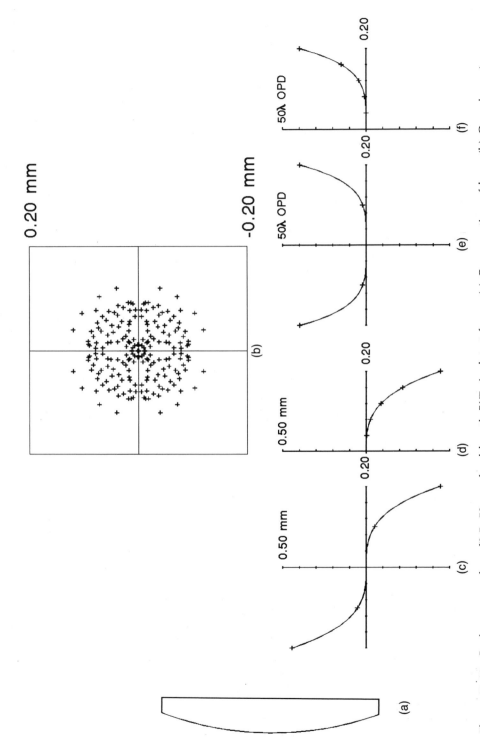

Figure 10.12 Performance of an f/2.5, 50 mm focal length BK7 singlet at focus. (a) Cross-section of lens. (b) On-axis spot diagram. Ray-intercept plot in (c) tangential and (d) sagittal planes. Optical path difference in (e) tangential and (f) sagittal planes.

system allows one to determine the value of the fourth-order phase coefficient a_4 that is necessary to eliminate third-order spherical aberration (Kleinhans, 1977; Sweatt, 1977). However, for complex optical systems this process can be tedious (Buralli and Morris, 1992a). Fortunately, most commercial lens design programs now allow DOEs to be modelled in the form of equation (10.22) and optimization routines can be used to determine the value of the coefficients that reduce the level of aberrations (Buralli, 1991).

The fourth-order phase coefficient necessary to reduce third-order spherical aberration of the BK7 singlet was determined by lens design software to be

$$\Phi(r) = (2.18 \times 10^{-6}) \frac{2\pi r^4}{\lambda_0}.$$

The radius r and λ_0 are in millimetres. Much like a refractive aspheric surface, the DOE provides the additional OPD necessary to eliminate the aberration. The performance of the hybrid is represented in Fig. 10.13. Note that, in comparison to Fig. 10.12, the scales in Figs 10.13(b) and (c) are reduced by a factor of 100. However, the scales for the spot diagrams are the same. It is important to remember that the spot size indicated in Fig. 10.13(d) was determined with geometrical optics; because it is almost three times smaller than the diffraction limit, it is not the actual spot size that would be achieved. In addition, if the performance of this lens were limited by the presence of higher-order spherical aberration, additional phase coefficients can be added to improve the performance.

Again, as for all hybrid design methods, consideration should be given to lens performance when the illumination is polychromatic (Swanson, 1989). None the less, this technique has been successfully applied to the design of focusing singlets for CO_2 laser machining applications where DOEs have been used to produce spot sizes smaller than those achieved by conventional convex-plano and meniscus lenses (Gruhlke *et al.*, 1992).

10.5 Hybrid Array Generators

The previous sections demonstrate that combining diffractive lenses with refractive lenses can realize a hybrid lens that is athermal and achromatic. However, a diffractive lens is only one specific diffractive element. A fundamental characteristic of diffractive optics is its ability to realize arbitrary functions and transformations. In this section we consider the design of hybrid elements that combine refractive lenses with diffractive array generators (Mait, 1997). In such a combination, the diffractive component does not improve the basic function of the lens: rather, the refractive and diffractive components perform different functions for which they are especially well-suited (Sauer *et al.*, 1992). The refractive lens is used to collimate and to focus, whereas the DOE is used to achieve beam splitting and deflection. Applications of array generation include multiple imaging (Dammann and Görtler, 1977), beam splitting for optical interconnects and optical switching (Kress and Lee, 1993; Morrison *et al.*, 1993; Jahns, 1994), and beam combining for summing semiconductor lasers (Leger *et al.*, 1987).

We first present the design of array generators that are athermal and then the

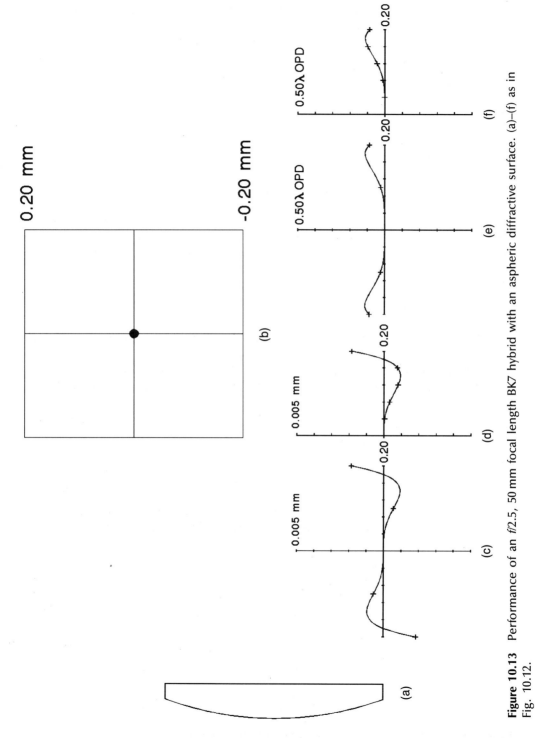

Figure 10.13 Performance of an f/2.5, 50 mm focal length BK7 hybrid with an aspheric diffractive surface. (a)–(f) as in Fig. 10.12.

design of an array generator system that is achromatic. We do not consider a specific array generator in our analysis because the deflection angle, which is related only to the grating period Λ of the array generator, is the more critical parameter. The deflection angle θ of the mth diffracted order is determined according to the grating equation

$$\sin \theta = \frac{m\lambda}{\Lambda}. \tag{10.23}$$

10.5.1 *Athermal Array Generators*

To analyze the thermal behaviour of an array generator, it is necessary to derive an expression for its opto-thermal expansion coefficient. Consider Fig. 10.14, which represents a plane wave normally incident onto the DOE $P(u, v)$. Under the assumption that the propagation of the transmitted wavefield is over a distance z_0 that satisfies the approximations of Fresnel diffraction, the wavefield $f(x, y)$ at z_0 is (Goodman, 1968)

$$f(x, y) = \exp\left[j\frac{\pi}{\lambda z_0}(x^2 + y^2)\right]$$
$$\times \iint P(u, v)\exp\left[j\frac{\pi}{\lambda z_0}(u^2 + v^2)\right]\exp\left[-j\frac{2\pi}{\lambda z_0}(ux + vy)\right]du\,dv. \tag{10.24}$$

A constant scale factor and linear phase term have been ignored.

Thermal expansion of the DOE in a direction transverse to the direction of light propagation can be modelled as

$$P_T(u, v) = P\left[\frac{u}{a(\Delta T)}, \frac{v}{a(\Delta T)}\right] \tag{10.25a}$$

where

$$a(\Delta T) = 1 + \alpha_g \Delta T. \tag{10.25b}$$

It has been shown that the transverse expansion of the DOE is the dominant thermal effect. The effects due to temperature-dependent refractive index changes

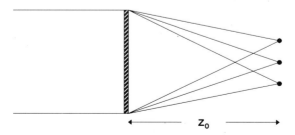

Figure 10.14 Application of a DOE in an optical processor architecture that satisfies conditions of Fresnel diffraction.

and thermal expansion of the substrate in the direction of light propagation are negligible (Behrmann *et al.*, 1993).

From section 10.3 we know that a diffractive lens experiences a change in focal length when it is subjected to a temperature change and we expect similar behaviour for more general Fresnel DOEs. Thus, under a change in temperature, we expect the DOE response $f(x, y)$ to shift from a plane located at z_0 to one located at z_t:

$$
f_T(x, y) = \exp\left[j \frac{\pi}{\lambda z_t} (x^2 + y^2) \right]
$$

$$
\times \iint P\left(\frac{u}{a}, \frac{v}{a} \right) \exp\left[j \frac{\pi}{\lambda z_t} (u^2 + v^2) \right] \exp\left[-j \frac{2\pi}{\lambda z_t} (ux + vy) \right] du\, dv
$$

$$
= a^2 \exp\left[j \frac{\pi}{\lambda z_t} (x^2 + y^2) \right]
$$

$$
\times \iint P(\zeta, \eta) \exp\left[j \frac{\pi a^2}{\lambda z_t} (\zeta^2 + \eta^2) \right] \exp\left[-j \frac{2\pi a}{\lambda z_t} (\zeta x + \eta y) \right] d\zeta\, d\eta. \quad (10.26)
$$

Equation (10.26) simplifies to

$$
f_T(x, y) = a^2 f\left(\frac{x}{a}, \frac{y}{a} \right) \tag{10.27a}
$$

when

$$
z_t = a^2 z_0 \approx (1 + 2\alpha_g \Delta T) z_0 \tag{10.27b}
$$

from which the opto-thermal expansion coefficient x_{opt} can be derived:

$$
x_{opt} = \frac{1}{z} \frac{dz}{dT} = 2\alpha_g. \tag{10.28}
$$

In addition to a displacement of the response along the optic axis, the response is scaled in amplitude and is scaled spatially in a direction transverse to the optic axis. The opto-thermal image expansion coefficient x_{img} is

$$
x_{img} = \frac{1}{x} \frac{dx}{dT} = \frac{1}{y} \frac{dy}{dT} = \alpha_g. \tag{10.29}
$$

We note that the expression for x_{opt} for a general Fresnel regime DOE is in agreement with that for a diffractive lens (see equation (10.16)). Thus, the design procedure for athermalization developed in section 10.3 for diffractive lenses is also valid for the design of more general hybrid elements. For example, a DOE that generates an array of spots at a distance of 25 mm from the element and is fabricated in fused silica is athermal to aluminium, so long as it contains negative refractive power, $\phi_r = -0.039$ mm^{-1}, and positive diffractive power, $\phi_d = 0.079$ mm^{-1}, in combination with the array generator. The performance of such an element at each diffracted order is similar to that represented in Fig. 10.8.

As a consequence of athermalization the image expansion coefficient equals the thermal expansion of the system, $x_{img} = \alpha_m$. However, because α_m is measured in parts per million, even for a 100°C change in temperature, the change in scale is only of the order of 10^{-3}, which is negligible.

10.5.2 *Achromatic Array Generators*

Unlike the thermal compensation of an array generator, the chromatic aberrations of an array generator cannot be corrected by any lens, refractive, diffractive or hybrid. To see this consider equation (10.23). If the illumination incident upon the array generator is polychromatic, each wavelength generates a different set of deflection angles for the diffracted orders. As represented in Fig. 10.15, if the array generator is placed in contact with a refractive lens, dispersion causes the arrays generated by the different wavelengths to shift axially and laterally from one another. If the lens is diffractive, dispersion causes the arrays to be shifted only axially. An achromat, either refractive or hybrid, provides a common focal distance for each wavelength, which results in a lateral shift between arrays generated by different wavelengths.

The achromatization of an array generator therefore requires the two-lens system represented in Fig. 10.16 (Schwab *et al.*, 1994). Its optical power is

$$\phi = \phi_2 - \phi_1 + \phi_1 \phi_2 d_1 \tag{10.30a}$$

and its dispersion in terms of Abbe number is

$$\frac{\phi}{V} = \frac{\phi_2}{V_2} - \frac{\phi_1}{V_1} + \phi_1 \phi_2 d_1 \left(\frac{1}{V_2} + \frac{1}{V_1} \right). \tag{10.30b}$$

In addition, to ensure that the output plane remains the focal plane of the system over the band of input wavelengths (Schwab *et al.*, 1994), the system must satisfy the conditions

$$0 = 1 - \phi_2 d_2 - \phi_1 [d_2 + d_1 (1 - \phi_2 d_2)] \tag{10.30c}$$

$$0 = \frac{\phi_2 d_2}{V_2} (1 - \phi_1 d_1) + \frac{\phi_1}{V_1} [d_2 + d_1 (1 - \phi_2 d_2)]. \tag{10.30d}$$

Solution of equations (10.30a)–(10.30d) for the lens powers ϕ_1 and ϕ_2 and distances d_1 and d_2 requires specification of the system power ϕ, system dispersion $t = 1/V$, and lens dispersions $t_1 = 1/V_1$ and $t_2 = 1/V_2$:

$$\phi_1 = \phi \frac{(1+t)t_2}{t + (1+t)t_2 - t_1} \tag{10.31a}$$

$$\phi_2 = -\phi \frac{(t - t_1)^2}{t_1 [t + (1+t)t_2 - t_1]} \tag{10.31b}$$

$$d_1 + d_2 = \frac{1}{\phi} \left[1 + \frac{t}{(1+t)t_2} \right] \tag{10.31c}$$

$$d_2 = \frac{1}{\phi} \frac{t_1}{t_1 - t}. \tag{10.31d}$$

The system power ϕ is fixed by system requirements and it is possible to express the lens powers and distances in terms of ϕ. Thus, only specification of the system and lens dispersions remains.

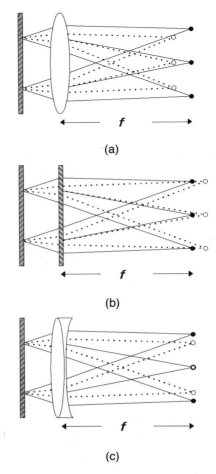

Figure 10.15 Effect of different optical elements placed in contact with an array generator. Dotted lines indicate a short wavelength and solid lines a long wavelength. (a) Refractive lens. (b) Diffractive lens. (c) Achromat. (After Schwab *et al.*, 1994.)

From equation (10.19), if the lenses are hybrids it is possible to design them to have arbitrary dispersion. However, selection of the dispersions t, t_1 and t_2 is dependent upon practical considerations in the manufacture of the optical elements and the construction of the optical system. In the example considered by Schwab *et al.* (1994), $\lambda_0 = 840$ nm and $\Delta\lambda = 10$ nm, which yields $|\Delta\lambda/\lambda_0| = 1/84$. Thus, the dispersion for a diffractive lens in this wavelength band is $t_d = -1/84$ and for two common optical materials $t_{BK7} = 1/2428$ and $t_{SF59} = 1/1285$. For moderate values of optical power, the dispersion of a hybrid doublet is limited in magnitude between 0 and $2t_d$ and that of a refractive doublet, between 0 and $t_d/10$ (Schwab *et al.*, 1994).

As shown in Fig. 10.17, the condition $d_1 + d_2 > d_2 > 0$ limits the choice of dispersions. To narrow our selection further, note that equation (10.31c) indicates that, to keep the overall system length small, the second lens should be dispersive,

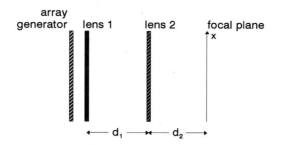

Figure 10.16 Optical system that consists of two lenses and two distances necessary to achromatize an array generator. (After Schwab *et al.*, 1994.)

which can be achieved if the second lens is purely diffractive, i.e. $t_2 = -1/84$.

In addition, to provide achromatization, the system dispersion must compensate for the dispersion of the array generator; thus, the system must also be highly dispersive:

$$|t| = \frac{\Delta\lambda}{\lambda_0} = \frac{1}{84}. \tag{10.32}$$

Such dispersion provides lateral coincidence in the focal plane and only axial shifts of the arrays generated by different wavelengths. The sign of the dispersion affects only the placement of the arrays: negative dispersion causes long wavelengths to experience a shorter focal length than short wavelengths. The opposite is true for positive dispersion.

From Fig. 10.17(a), if the system dispersion is positive, $t = 1/84$ and $t_2 = -1/84$, then t_1 is small, which implies that the power of the second lens is large. If system dispersion is negative, $t = -1/84$, then either $t_1 < -2/84$ or $t_1 > 0$. Because of the developing nature of DOE fabrication, fabrication of the diffractive components is a more critical issue than the fabrication of the refractive components. Therefore, the total diffractive power in the system should be kept to a minimum to ensure

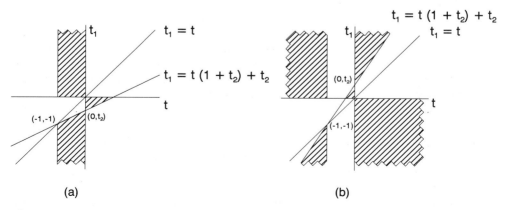

Figure 10.17 Representation of solution space for selection of system dispersion t and lens dispersions t_1 and t_2. (a) $t_2 < 0$. (b) $t_2 > 0$.

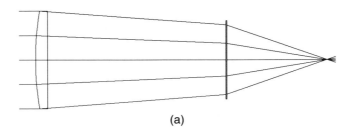

(a)

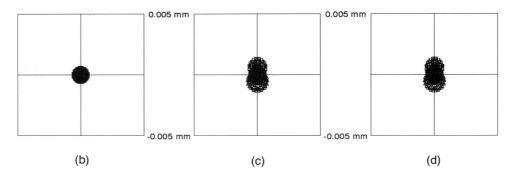

(b)   (c)   (d)

Figure 10.18 Representation of an achromatized array generator system. (a) Hybrid array generator system. Spot diagrams for (b) zeroth, (c) first and (d) second diffracted orders. $\lambda = 835$ nm is indicated by $\square$, $\lambda = 840$ nm by $+$, and $\lambda = 845$ nm by $\triangle$.

system manufacturability. If the first lens is constructed from SF59, the system diffractive optical power is minimum when $t_1 = 0.0339 \approx -3t$ (Schwab *et al.*, 1994). The resulting lens powers and distances are $\phi_1 = 0.205\phi$ ($\phi_{1,r} = 0.740\phi$ and $\phi_{1,d} = -0.535\phi$), $\phi_2 = 1.075\phi$, $d_1 = 1.272/\phi$, and $d_2 = 0.740/\phi$. The achromatized system and generated spots for an input illumination of 840 nm and a 10 nm bandwidth are represented in Fig. 10.18. The array generator has a 1 mm grating period and the system optical power is 0.04 mm^{-1}.

To compensate for spherical aberration, Schwab *et al.* (1994) used two refractive aspheres to realize the refractive optical power in the first lens. However, as indicated in section 10.4, one can correct spherical aberration by including high-order terms in the phase polynomials that describe the diffractive elements. The advantage of using diffractive aspheres over refractive is that, whereas fabrication costs of the diffractive elements are comparable, overall material costs are less. The phase polynomial for the first DOE, as determined by lens design software, is

$$\phi_{d,1}(r) = \frac{2\pi}{\lambda_0}[0.0107r^2 + (7.8891 \times 10^{-6})r^4 - (4.2173 \times 10^{-8})r^6]$$

and for the second

$$\phi_{d,2}(r) = \frac{2\pi}{\lambda_0}[-0.0215r^2 - (6.4599 \times 10^{-6})r^4 + (2.8433 \times 10^{-7})r^6]$$

where r and λ_0 are in millimetres.

As this example shows, correction of the chromatic aberrations of an array generator is not a problem in single hybrid element design, but is in hybrid system design, the proper treatment of which requires a chapter of its own. However, as we have shown, as long as the DOE can be characterized in a manner similar to that used for refractive optical elements, standard analysis and design techniques can be used.

10.6 Conclusions

Improvements in fabrication technology have made DOEs a realistic consideration for systems. As a consequence, applications of hybrid optical systems are being published and presented at an ever-increasing rate. The utility of such systems is highest when one is faced with the requirement for reduced weight or volume, or when the number of lens materials is limited.

In this chapter we have presented design techniques for the most common applications of hybrid optics. We hope that we have presented a methodology that is useful for determining if a hybrid technique is appropriate for a particular application. However, in general, we feel that traditional design techniques can be employed as long as the DOE can be described in terms that are similar to those used for refractive materials. In this way it is possible to see that the most unique feature of DOEs is their ability to synthesize optical components that have arbitrary dispersions and arbitrary thermal properties.

We have presented the basic techniques for designing achromatic lens systems and made comparisons between an achromatic doublet and a hybrid achromat. The dispersive nature of the diffractive surface reduces the fabrication requirements on both the refractive and diffractive lenses and thus allows for a reduction in mass and volume. However, when a system is exposed to broadband illumination, consideration should be given to the effects of stray light on system performance.

We showed that, even for lenses made from the same material, the thermal behaviour of refractive and diffractive lenses is unique. Therefore, single-material hybrid athermats can be designed to achieve previously unattainable thermal behaviour. In comparison to traditional design methods for broadband systems, hybrid athermalization techniques offer similar (in our hybrid triplet example, better) levels of performance but with the additional benefit of reduced mass and volume. In addition, as shown by the example of the array generator, this benefit is not limited strictly to imaging applications.

We further showed that when a diffractive element is described by a high-order polynomial it is possible to realize an aspheric surface and thus correct for monochromatic aberrations. As indicated by the hybrid system design to achromatize an array generator, the ability to model DOEs in this manner is critical to the development of future applications of DOEs, which is being facilitated by the development of commercial lens design software.

Acknowledgements

We would like to acknowledge the diffractive optics research community for its enthusiasm and creativity. Special thanks must be given to John P. Bowen of the

Rochester Photonics Corporation for valuable discussions and research collaboration.

References

BEHRMANN, G. P. (1993) Color correction in athermalized hybrid lenses. In *OSA Technical Digest Series: Optical Design for Photonics*, Vol. 9, pp. 67–70. Washington, D.C.: OSA.

BEHRMANN, G. P. and BOWEN, J. P. (1993) Influence of temperature on diffractive lens performance. *Appl. Opt.* **32**, 2483–2489.

BEHRMANN, G. P., BOWEN, J. P. and MAIT, J. N. (1993) Thermal properties of diffractive optical elements and design of hybrid athermalized lenses. In Lee, S. H. (ed.) *Diffractive and Miniaturized Optics*, Vol. CR49, pp. 212–233. Bellingham: SPIE.

BEHRMANN, G. P., TAYAG, T. J., PRATHER, D. W. and SARAMA, S. D. (1994) *Hybrid optics for athermalization and achromatization*. Presentation at the 1994 Annual Meeting of the Optical Society of America, Dallas, Texas.

BURALLI, D. A. (1991) Optical design with diffractive lenses. *Sinclair Optics: Design Notes* **2**, 1–4.

BURALLI, D. A. and MORRIS, G. M. (1992a) Design of two and three element diffractive Keplerian telescopes. *Appl. Opt.* **31**, 38–43.

(1992b) Effects of diffraction efficiency on the modulation transfer function of diffractive lenses. *Appl. Opt.* **31**, 4389–4396.

CHEN, C. W. and ANDERSON, J. S. (1993) Imaging by diffraction: grating design and hardware results. In Lee, S. H. (ed.) *Diffractive and Miniaturized Optics*, Vol. CR49, pp. 77–97. Bellingham: SPIE.

CLARK, P. P. and LONDOÑO, C. (1989) Production of kinoforms by single point diamond machining. *Opt. News* **15**, 39–40.

CZICHY, R. H. (1993) *Hybrid Optics for Space Applications*, European Space Agency Technical Report, ESA SP-1158.

DAMMANN, H. and GÖRTLER, K. (1977) High-efficiency in-line multiple imaging by means of multiple phase holograms. *Opt. Commun.* **3**, 312–315.

DAVIDSON, N., FRIESEM, A. A. and HASMAN, E. (1993) Analytic design of hybrid diffractive–refractive achromats. *Appl. Opt.* **32**, 4770–4774.

FRITZ, T. A. and COX, J. A. (1989) Diffractive optics for broadband infrared imagers: Design examples. In Cindrich, I. N. and Lee, S. H. (eds) *Holographic Optics: Computer and Optically Generated*, Vol. 1052, pp. 25–31. Bellingham: SPIE.

GAN, M. (1990) Kinoforms long focal objectives for astronomy. In Schulte in den Baeumen, J. and Tyson, R. K. (eds) *Adaptive Optics and Optical Structures*, Vol. 1271, pp. 330–338. Bellingham: SPIE.

GOODMAN, J. W. (1968) *Introduction to Fourier Optics*. New York: McGraw-Hill.

GOTO, K., HIGUCHI, Y. and MORI, K. (1989) Plastic grating collimating lens. In Wilson, T. (ed.) *Optical Storage and Scanning Technology*, Vol. 1139, pp. 169–176. Bellingham: SPIE.

GRUHLKE, R., GIAMONA, L. and SULLIVAN, W. (1992) Diffractive optics for industrial lasers. In *Diffractive Optics: Design, Fabrication, and Applications*, Vol. 9, pp. 61–63. Washington, D.C.: OSA.

HECHT, E. (1987) *Optics*, 2nd edn, Reading, Massachusetts: Addison-Wesley.

HUDYMA, R. M. and KAMPE, T. U. (1992) Hybrid refractive/diffractive elements in lenses for staring focal plane arrays. In Krumwiede, G. C., Genberg, V. L., Aikens, D. M. and Thomas, M. J. (eds) *Design of Optical Instruments*, Vol. 1690, pp. 90–91. Bellingham: SPIE.

JAHNS, J. (1994) Diffractive optical elements for optical computers. In Jahns, J. and Lee,

S. H. (eds) *Optical Computing Hardware*, pp. 137–167. San Diego: Academic Press.

JAHNS, J., LEE, Y. H., BURRUS JR., C. A. and JEWELL, J. (1992) Optical interconnects using top-surface-emitting microlasers and planar optics. *Appl. Opt.* **31**, 592–597.

JAMIESON, T. H. (1981) Thermal effects in optical systems. *Opt. Eng.* **20**, 156–160.

KINGSLAKE, R. (1978) *Lens Design Fundamentals*. San Diego: Academic Press, Inc.

KLEINHANS, W. A. (1977) Aberrations of curved zone plates and Fresnel lenses. *Appl. Opt.* **16**, 1701–1704.

KRESS, B. and LEE, S. H. (1993) Iterative design of computer generated Fresnel holograms for free space optical interconnections. In *Optical Computing Technical Digest 1993*, Vol. 7, pp. 22–25. Washington, D.C.: OSA.

KUBALAK, D. and MORRIS, G. M. (1992) A hybrid diffractive/refractive lens for use in optical data storage. In *Diffractive Optics: Design, Fabrication, and Applications*, Vol. 9, pp. 93–95. Washington, D.C.: OSA.

LEE, S. H. (ed.) (1993) *Diffractive and Miniaturized Optics*. Vol. CR49. Bellingham: SPIE.

LEGER, J. R., SWANSON, G. J. and VELDKAMP, W. B. (1987) Coherent laser addition using binary phase gratings. *Appl. Opt.* **26**, 4391–4399.

LONDOÑO, C. and CLARK, P. P. (1992) Modeling diffraction efficiency effects when designing hybrid diffractive lens systems. *Appl. Opt.* **31**, 2248–2252.

LONDOÑO, C., PLUMMER, W. T. and CLARK, P. P. (1993) Athermalization of a single-component lens with diffractive optics. *Appl. Opt.* **32**, 2295–2302.

MAIT, J. N. (1997) Fourier Array Generators. In Herzig, H. P. (ed.) *Micro-optics*, pp. 293–323. London: Taylor & Francis.

MORRISON, R. L., WALKER, S. L., McCORMICK, F. B. and CLOONAN, T. J. (1993) Practical applications of diffractive optics in free-space photonic switching. In Lee, S. H. (ed.) *Diffractive and Miniaturized Optics*, Vol. CR49, pp. 265–289. Bellingham: SPIE.

MURANAKA, M., TAKAGI, M. and MARUYAMA, T. (1988) Precision molding of aspherical plastic lens for cam-corder and projection TV. In Riedl, M. L. (ed.) *Replication and Molding of Optical Components*, Vol. 896, pp. 123–131. Bellingham: SPIE.

RIEDL, M. J. and McCANN, J. T. (1991) Analysis and performance limits of diamond turned diffractive lenses for the 3–5 and 8–12 micrometer regions. In Hartmann, R. and Smith, W. J. (eds) *Infrared Optical Design and Fabrication*, Vol. CR38, pp. 153–163. Bellingham: SPIE.

ROGERS, P. J. (1991) Athermalization of IR optical systems. In Hartmann, R. and Smith, W. J. (eds) *Infrared Optical Design and Fabrication*, Vol. CR38, pp. 69–94. Bellingham: SPIE.

SASIAN, J. M. and CHIPMAN, R. A. (1993) Staircase lens: a binary and diffractive field curvature corrector. *Appl. Opt.* **32**, 60–66.

SAUER, F., JAHNS, J., FELDBLUM, A. Y., NIJANDER, C. and TOWNSEND, W. (1992) Diffractive–refractive microlens arrays for beam permutation. In *Diffractive Optics: Design, Fabrication, and Applications*, Vol. 9, pp. 33–35. Washington, D.C.: OSA.

SCHWAB, M., LINDLEIN, N., SCHWIDER, J., AMITAI, Y., FRIESEM, A. A. and REINHORN, S. (1994) Achromatic diffractive fan-out systems. In Cindrich, I. and Lee, S. H. (eds) *Diffractive and Holographic Optics Technology*, Vol. 2152, pp. 14–20. Bellingham: SPIE.

SHAFER, D. R. (1993) *Unusual binary-optics telescope design*. Presentation at the 1993 Annual Meeting of the Optical Society of America, Toronto, October.

SMITH, W. J. (1990) *Modern Optical Engineering*, 2nd edn. New York: McGraw-Hill.

STONE, T. and GEORGE, N. (1988) Hybrid diffractive–refractive lenses and achromats. *Appl. Opt.* **27**, 2960–2971.

SWANSON, G. J. (1989) *Binary Optics Technology: The Theory and Design of Multi-Level Diffractive Optical Elements*, MIT Lincoln Laboratory, Tech. Rep. 854.

SWEATT, W. C. (1977) Describing holographic optical elements as lenses. *J. Opt. Soc. Am.* **67**, 803–808.

UPATNIEKS, J., VANDER LUGT, A. and LEITH, E. (1966) Correction of lens aberrations by means of holograms. *Appl. Opt.* **5**, 589–593.

WOOD, A. P. (1992) Design of infrared hybrid refractive–diffractive lenses. *Appl. Opt.* **31**, 2253–2258.

11

Fourier Array Generators

J. N. MAIT

11.1 Introduction

As technologies for the fabrication of micron and submicron structures have matured (Zaleta *et al.*, 1993; Gale, 1997; Stern, 1997), interest in phase-only diffractive optical elements (DOEs) has increased. Many discussions of DOEs concentrate primarily on their application (Morrison *et al.*, 1993; Jahns, 1994) and others on the analysis of DOE performance using diffraction theory (Glytsis *et al.*, 1993) Although DOE analysis is critical to an understanding of the technology, equally important for widespread application of DOEs is an understanding of their design, or, in contrast to DOE analysis, DOE synthesis.

In this chapter we consider the design of an element that is capable of producing an array of point sources with illumination from a single point source (Fig. 11.1). The element can function as a multiple imager (Dammann and Görtler, 1971; Dammann and Klotz, 1977), beam splitter (Killat *et al.*, 1982), or, when operated in reverse, as a beam combiner (Veldkamp *et al.*, 1986; Leger *et al.*, 1987). Although the varied applications complicate nomenclature, the function of the element is referred to here as array generation and the diffractive element itself as an array generator.

Various techniques for achieving array generation have been addressed by Streibl (1994), one of which is the placement of a phase-only diffraction grating in the Fourier plane of a coherent optical processor such that its point spread function is the desired array (Fig. 11.2). Of the array generation techniques described by Streibl, only the Fourier processor is capable of performing a convolution between an input object and the generated spot array and, thus, must be used for multiple imaging and signal processing applications.

If the source illumination in Fig. 11.2 is spatially coherent and quasi-monochromatic with wavelength λ, a Fourier transform relationship exists between the diffractive element $P(u,v)$ and its response $p(x,y)$ in the image plane:

$$p(x,y) = \iint P(u,v) \exp[j2\pi(ux+vy)]\, du\, dv. \tag{11.1}$$

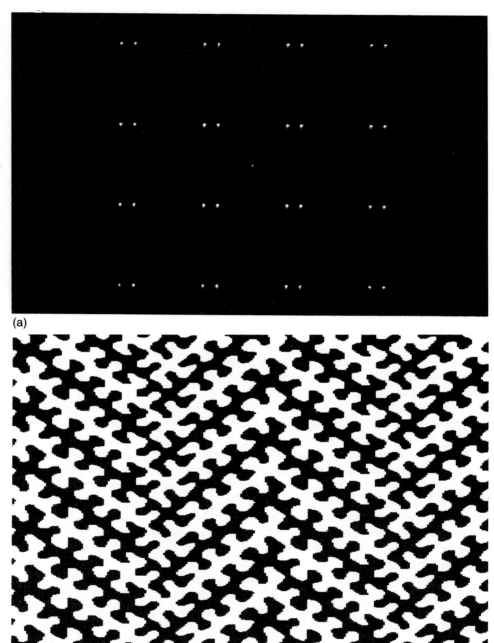

(a)

(b)

Figure 11.1 Array generation for addressing an 8×4 semiregular array of symmetric self-electrooptical effect devices (SEEDs). (a) Generated array. Spacing between spots in a single pair is $35\,\mu$m and spacing between spot pairs, $210\,\mu$m. Illumination wavelength is 850 nm. (b) Portion of the corresponding binary-phase Fourier array generator showing 2×6 periods. Entire array generator is 14×80 periods. Single period is $758.2\,\mu\text{m} \times 126.4\,\mu\text{m}$ and consists of 256×256 cells. Each cell is $3.0\,\mu\text{m} \times 0.5\,\mu\text{m}$. Effective focal length of Fourier transform lens is 15.6 mm. (Morrison *et al.*, 1995)

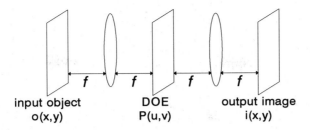

input object   DOE   output image
o(x,y)   P(u,v)   i(x,y)

Figure 11.2 Coherent optical processor.

For simplicity, unity magnification, $f_1 = f_2 = f$, and reduced coordinates, $f\lambda = 1$, have been assumed.

It is further assumed that $P(u, v)$ is phase-only, i.e.

$$P(u, v) = \exp[j\theta(u, v)].$$
(11.2)

Synthesis of the array generator entails the determination of the phase $\theta(u, v)$ such that $p(x, y)$ contains the desired array $q(x, y)$:

$$q(x, y) = \sum_{n=1}^{N} q_n \delta(x - x_n, y - y_n)$$
(11.3a)

where N is the total number of spots in the array and the coordinates (x_n, y_n) indicates their location. The magnitude and phase of each spot are given by the phasor q_n. The Fourier transform of $q(x, y)$ is

$$Q(u, v) = \sum_{n=1}^{N} q_n \exp[-j2\pi(ux_n + vy_n)].$$
(11.3b)

If the spot locations (x_n, y_n) are rational, then $Q(u, v)$ is periodic and, therefore, the array generator $P(u, v)$ must also be periodic:

$$P(u, v) = \tilde{P}(u, v) ** \text{repl}(u, v)$$
(11.4a)

where $\tilde{P}(u, v)$ is a single period representation of the array generator and $**$ represents two-dimensional convolution. A single period has unity extent along both axes. As shown in Fig. 11.3, the convolution of $\tilde{P}(u, v)$ with $\text{repl}(u, v)$ tiles the Fourier plane. As a consequence, the points in image space at which the generated array has value are determined by the lattice array $\text{samp}(x, y)$:

$$p(x, y) = \tilde{p}(x, y) \, \text{samp}(x, y).$$
(11.4b)

The functions $\text{repl}(u, v)$ and $\text{samp}(x, y)$ form a Fourier transform pair.

We consider in this chapter the design of a diffractive element that achieves array generation in the Fourier architecture of Fig. 11.2 subject to the constraints of phase-only fabrication. For example, the binary-phase diffractive element that generated the array of Fig. 11.1(a) is represented in Fig. 11.1(b). In lens design the functions that achieve focusing and correct for aberrations are themselves phase functions; thus, phase-only diffractive lens design can be achieved by setting the phase of the DOE equal to the desired phase function. However, few, if any, spot arrays of interest have transforms that are phase-only. Thus, array generator design must encode, either explicitly or implicitly, the Fourier magnitude $|Q(u, v)|$ into

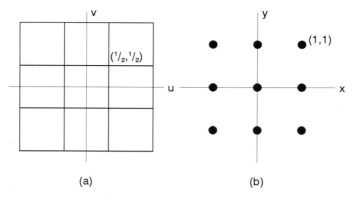

(a)   (b)

Figure 11.3 Transform relationship between Fourier replication function and latttice of image sampling points it generates. (a) Tiling of Fourier plane with $P(u, v)$ by repl(u, v). (b) Lattice of image sample points samp(x, y) generated by repl(u, v).

$\theta(u, v)$. The design is further complicated by the constraints imposed by fabrication. All fabrication techniques, even those that are capable of realizing a continuum of phase values, impose some limitations on the design. Diamond turning, for example, is incapable of realizing an instantaneous phase change of 2π.

We begin in section 11.2 with a brief consideration of the applications of array generators. We continue in section 11.3 with one-dimensional designs and, in particular, one-dimensional binary-phase gratings and consider thereafter extensions to continuous-phase gratings. As long as the design is separable, extensions of one-dimensional concepts to two dimensions are straightforward. In section 11.4 we discuss the special concerns of two-dimensional non-separable designs. Fabrication issues are considered briefly in section 11.5 and concluding remarks are presented in section 11.6.

11.2 Applications

Applications of Fourier array generators can be classified into three basic categories: power splitting, power combining and multiple imaging. In the first two applications no spatial information is encoded on the beams. The objective of power splitting is to redistribute power from a single input to multiple outputs; in power combining multiple inputs are summed coherently to produce a single output. In multiple imaging the input beam is encoded with spatial information and a superposition of this information is performed for purposes of information processing.

Power splitting is an application that is not unique to Fourier array generators. The techniques reviewed by Streibl (1994) are all capable of distributing power from a single source to multiple receivers. The first reported use of a Fourier array generator as a power splitter was as a fibre optic star coupler by Killat *et al.* (1982). More recent applications of power splitting include clock distribution for computing systems (Prongué *et al.*, 1990; Walker and Jahns, 1992) and addressing optoelectronic devices for optical switching (Vasara *et al.*, 1992; Morrison *et al.*, 1993, 1995).

The unique feature of a Fourier array generator is its multiple imaging ability. In fact, the first application of an array generator was the multiple imaging of a single pattern to expose integrated circuit masks (Dammann and Görtler, 1971). An array of images was produced by convolving the input pattern with the spot array. The separation between spots was such that there was no overlap between adjacent images. However, more recent applications make use of the convolution with a spot array to perform image processing, e.g. symbolic substitution (Mait, 1989a; Thalmann *et al.*, 1989) and morphology (Athale *et al.*, 1992). In these applications the spot array represents a pattern or shape for which a search is made in the input image.

The Fourier system's ability to perform coherent summations also allows Fourier array generators to be used to sum the outputs of independent lasers (Veldkamp *et al.*, 1986; Leger *et al.*, 1987); i.e. a Fourier array generator can be used to perform power combining, the inverse operation of power splitting. Under the assumption that N lasers of equal amplitude are input to a Fourier array generator, the outputs from the lasers can be summed coherently to produce a transmitted beam that is approximately N times the intensity of the individual lasers. In essence, the system performs a correlation between the lasers and the generated spot array to produce a correlation peak on-axis. The desire is that the intensity of the peak be as large as possible.

11.3 One-dimensional Array Generators

11.3.1 Binary-phase Array Generators

One of the first array generators reported in the literature is a binary-phase multiple imager demonstrated by Dammann and Görtler (1971) and Dammann and Klotz (1977). The success of his design has been acknowledged by the common use of the term Dammann grating to refer to a $\{0, \pi\}$-binary-phase array generator. For historical and pedagogical reasons, the first design technique discussed is Dammann's, from which more general techniques are derived.

The need for sophisticated array generator design is highlighted by Fig. 11.4. The transform of the desired three-spot array in Fig. 11.4(a) is represented in Fig. 11.4(b). Zero phase is assumed for each spot. Although a binary-phase grating can be generated from the transform via quantization, as indicated in Figs 11.4(c) and (d), the imposition of a nonlinearity in the Fourier domain introduces errors in the generated array.

To achieve a uniform intensity spot array, Dammann recognized that a one-dimensional $\{0, \pi\}$-binary-phase array generator can be represented completely by the locations z_k at which its phase transitions occur. Consider Fig. 11.5 and the representation of $\tilde{P}(u)$ in terms of a binary amplitude function $\Sigma(u)$:

$$\tilde{P}(u) = 2\Sigma(u) - \text{rect}(u) \tag{11.5a}$$

where

$$\Sigma(u) = \sum_{\substack{k=1 \\ k\,\text{odd}}}^{K-1} \text{rect}\left[\frac{u - (z_{k+1} + z_k)/2}{z_{k+1} - z_k}\right]. \tag{11.5b}$$

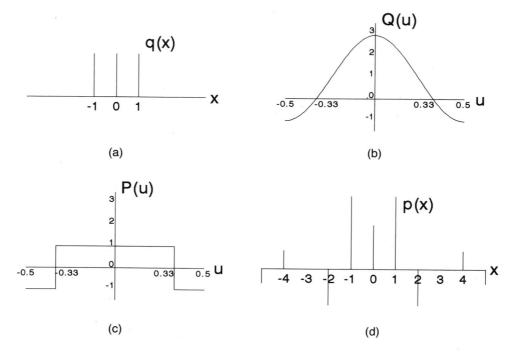

Figure 11.4 Performance of an array generator designed using quantization. (a) Desired three-spot array. (b) Fourier transform of (a). (c) Binary-phase array generator resulting from quantization of (b). (d) Resulting generated array.

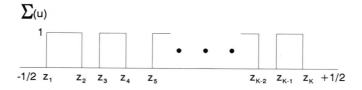

Figure 11.5 Binary amplitude distribution $\Sigma(u)$ used to describe a one-dimensional binary-phase array generator.

If we assume, as Dammann did, that $\mathrm{repl}(u) = \mathrm{comb}(u)$, where

$$\mathrm{comb}(u) = \sum_{n=-\infty}^{\infty} \delta(u - n)$$

then, by virtue of the self-transforming properties of the comb function, the generated array is

$$p(x) = \sum_{n=-\infty}^{\infty} [2\sigma(n) - \mathrm{sinc}(n)] \, \delta(x - n) \tag{11.6a}$$

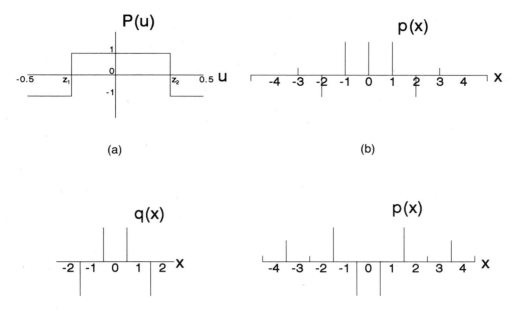

Figure 11.6 Dammann design of binary-phase array generators. (a) Form of the binary-phase generator. (b) Generated array when $-z_1 = z_2 = 0.3676$. (c) Desired four-spot array. (d) Generated array when $-z_1 = z_2 = 0.0544$ and π-phase is applied to alternate periods. The array is inverted because α is negative.

where

$$\sigma(n) = \sum_{\substack{k=1 \\ k \, \text{odd}}}^{K-1} (z_{k+1} - z_k) \, \text{sinc}[(z_{k+1} - z_k)n] \exp[j\pi(z_{k+1} + z_k)n]. \qquad (11.6b)$$

Inspection of equation (11.6b) reveals that $\sigma(n)$ is Hermitian, i.e. $\sigma(n) = \sigma^*(-n)$; thus $p(x)$ is also Hermitian. Therefore, for a $\{0, \pi\}$-binary-phase grating $q(x)$ cannot be completely arbitrary; rather it must satisfy the symmetry conditions imposed by the grating.

The phase transition locations are determined by setting the generated array $p(x)$ proportional to the desired array $q(x)$. The method is illustrated in Fig. 11.6 by the design of the three-spot array in Fig. 11.4(a). The $\{0, \pi\}$-binary-phase grating in Fig. 11.6(a) has only two phase transitions and, due to symmetry, the spot array $q(x)$ can be described by just two equations, thus:

$$p(0) = 2(z_2 - z_1) - 1 = \alpha$$
$$p(1) = 2(z_2 - z_1) \, \text{sinc}[(z_2 - z_1)] \exp[j\pi(z_2 + z_1)] = \alpha$$

where α is a constant of proportionality between $p(x)$ and $q(x)$. The scale α indicates the inability to predict the absolute amplitude of the generated output. Due to the Hermitian symmetry imposed by a $\{0, \pi\}$-binary-phase array generator,

$p(0)$ must have zero phase, which implies that α is real. Thus, the two complex-valued equations correspond to three real-valued equations that are used to solve for the three unknowns: $\alpha = 0.4706$, $z_1 = -0.3676$ and $z_2 = 0.3676$. The generated array $p(x)$ is represented in Fig. 11.6(b).

The comb-function replication used by Dammann generates a spot array that assumes values at integer locations in the image plane and, from an application of symmetry arguments, generates an array that has an odd number of spots. To generate an even number of spots it is necessary to use an alternative replication (Morrison, 1992):

$$P(u) = \bar{P}(u) * \exp(j\pi u)\,\mathrm{comb}(u) \tag{11.7a}$$

which generates

$$p(x) = \bar{p}(x)\,\mathrm{comb}(x - 1/2). \tag{11.7b}$$

The π-phase change between replicas acts as a spatial carrier to translate all spots off-axis without changing the spacing between sample points. In conjunction with the symmetry imposed by the grating, this implies that the array has an even number of spots.

As an example, consider the generation of the four-spot array represented in Fig. 11.6(c), which also requires an array generator that has only two phase transitions. The equations for its design are

$$p(1/2) = 2(z_2 - z_1)\,\mathrm{sinc}[(z_2 - z_1)/2]\exp[j\pi(z_2 + z_1)/2] = \alpha$$

$$p(3/2) = 2(z_2 - z_1)\,\mathrm{sinc}[3(z_2 - z_1)/2]\exp[j3\pi(z_2 + z_1)/2] = -\alpha.$$

Because an even-numbered spot array has no on-axis spot, all of its generated spots are complex valued, which implies that α is complex. Thus, α accounts for two of the four real-valued unknowns to be determined: $|\alpha| = 0.4202$, $\arg\{\alpha\} = \pi$, $z_1 = -0.0544$ and $z_2 = 0.0544$. The generated array is represented in Fig. 11.6(d).

The general procedure for Dammann's design of a $\{0, \pi\}$-binary-phase array generator is summarized below.

For a $(2N + 1)$-spot array, set $p(x) = \alpha q(x)$, for $x = [0, N]$, and solve the $2N + 1$ real-valued equations for real α and z_k, $k = [1, 2N]$. For a $2N$-spot array, set $p(x) = \alpha q(x)$, for $x = [1/2, 3/2, \ldots, (2N + 1)/2]$, and solve the $2N$ real-valued equations for complex α and z_k, $k = [1, 2N - 2]$.

We note that, unlike the solution of a set of linear equations, the solution of a set of nonlinear equations does not require that the number of equations and number of unknowns be equal. In Dammann's design maintaining an equal number of unknowns and equations must be seen as a rule of thumb rather than a necessity. The system of nonlinear equations can be solved using standard numerical analysis techniques, such as Newton–Raphson (Jahns *et al.*, 1989) and steepest-descent (Krackhardt and Streibl, 1989).

To quantify the performance of an array generator it is helpful to examine Fig. 11.4. We note first that the source array $q(x)$ is specified only over a finite region X, whereas the grating reponse $p(x)$ has infinite extent. As mentioned, a few spot arrays of interest have Fourier transforms that are phase only, thus it is not possible for $p(x)$ to be non-zero everywhere outside of X. Since the signal window X is

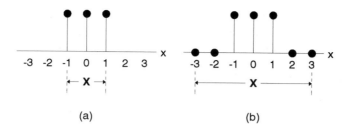

Figure 11.7 Significance of signal window X. Spot arrays with (a) $X = [-1, 1]$ and (b) $X = [-3, 3]$.

not the entire image plane, the diffraction efficiency η with which $q(x)$ is generated is less than 100%, where

$$\eta = \frac{\int_{x \in X} |p(x)|^2 \, dx}{\int_{-\infty}^{\infty} |p(x)|^2 \, dx} = \frac{\int_{x \in X} |p(x)|^2 \, dx}{\int_{-1/2}^{1/2} |\tilde{P}(u)|^2 \, du} = \int_{x \in X} |p(x)|^2 \, dx. \tag{11.8}$$

Parseval's theorem and the assumption that $\tilde{P}(u)$ is phase only have been used to obtain the final expression for η.

The importance of the signal window X in design is indicated in Fig. 11.7. In contrast to Fig. 11.7(a), where the magnitude and phase of the spot array outside $x = [-1, 1]$ are unspecified, Fig. 11.7(b) indicates that the value of the spot array at the locations $x = \{-3, -2, 2, 3\}$ must be zero. Figures 11.7(a) and (b) therefore represent two different design problems (Mait, 1990).

Diffraction efficiency measures only the amount of energy within X and, as a consequence, indicates that unimpeded free space can achieve 100% diffraction efficiency. Thus, in addition to diffraction efficiency, it is necessary to measure the error with which the energy is distributed within X as compared to the desired distribution. The signal-to-noise ratio (SNR) in X is

$$\text{SNR}_{\text{coh}} = \frac{\eta - e_{\text{coh}}}{e_{\text{coh}}} \tag{11.9a}$$

where

$$e_{\text{coh}} = \int_{x \in X} |p(x) - \beta_{\text{coh}} q(x)|^2 \, dx \tag{11.9b}$$

and

$$\beta_{\text{coh}} = \frac{\int_{x \in X} \text{Re}[p(x) q^*(x)] \, dx}{\int |q(x)|^2 \, dx}. \tag{11.9c}$$

The subscript coh indicates a coherent design where both magnitude and phase of $q(x)$ are specified and are to be faithfully reconstructed. Magnitude and intensity error measures can also be used. The presence of error makes it necessary to distinguish the generated scale factor β from the design scale factor α. The value of β is determined to minimize the error between $p(x)$ and $q(x)$.

For the quantization design represented in Fig. 11.4(d), $\eta = 71.9\%$, $\beta_{\text{coh}} = 0.4787$, $e_{\text{coh}} = 0.0317$, and $\text{SNR}_{\text{coh}} = 21.68$ and for the Dammann design in Fig. 11.6(b), $\eta = 66.4\%$, $\beta_{\text{coh}} = 0.4706$, $e_{\text{coh}} = 0.0$, and $\text{SNR}_{\text{coh}} = \infty$. In the remainder we assume that the objective of array generator design is the generation of $q(x)$ with maximum diffraction efficiency and minimum error. But, as seen in the three-spot design example, trade-offs exist between the two that must be balanced for a particular application.

At first glance the efficiencies of 66.4 and 71.9% for the two three-spot array generators appear low. However, rather than a comparison to the absolute maximum efficiency of 100%, the efficiencies need to be compared to a more realistic upper bound that is specific to $q(x)$ (Wyrowski, 1990)

$$\eta_{ub} = \frac{[\int_{-1/2}^{1/2}|Q(u)|\,du]^2}{\int_{-1/2}^{1/2}|Q(u)|^2\,du} \geq \eta. \tag{11.10}$$

Note that the diffraction efficiency upper bound measures the degree to which the Fourier magnitude is a constant, i.e. a nearly flat $|Q(u)|$ has an upper bound close to 100%.

For the three-spot example the upper bound is 68.7%. The quantization design technique achieves an efficiency greater than this bound because the array it generates contains error, and determination of the diffraction efficiency upper bound assumes zero error within the signal window. When compared to the upper bound, the diffraction efficiency for both designs is acceptable.

It is known from signal processing that $|Q(u)|$ is sensitive to the phase of $q(x)$ (Oppenheim and Lim, 1981); thus, if the phase of $q(x)$ is unimportant, it can be used as a design freedom to flatten the Fourier magnitude and thereby make it easier to design for high diffraction efficiency. For example, if the phase of the on-axis spot in the three-spot array is set to $\pi/2$, the diffraction efficiency upper bound increases to 93.8%. To realize a non-zero on-axis phase for odd-numbered spot arrays, the π-phase is changed to some arbitary phase value ϕ (Killat *et al.*, 1982):

$$\bar{P}(u) = [1 - \exp(j\phi)]\,\Sigma(u) + \exp(j\phi)\,\text{rect}(u) \tag{11.11a}$$

which produces the array

$$p(n) = [1 - \exp(j\phi)]\,\sigma(n) + \exp(j\phi)\,\text{sinc}(n). \tag{11.11b}$$

Design of a $\{0, \phi\}$-binary-phase grating is similar to that of a $\{0, \pi\}$-grating except that α is now complex and ϕ is unknown. Solution of Dammann's equations for a three-spot array that has arbitrary on-axis phase yields $-z_1 = z_2 = 0.25$, $\alpha = 0.5370 \exp(-j0.5669)$, and $\phi = 2.0078$. The diffraction efficiency is 86.5%.

The entire array phase can be used to further increase diffraction efficiency. For example, the diffraction efficiency upper bound of a nine-spot fan-out, i.e. an array that has equal magnitude spots, increases from 39.3% when the array has zero phase to 99.3% when non-zero array phase is allowed. Dammann's technique for designing $\{0, \pi\}$-binary-phase gratings when array phase is a free parameter is summarized below.

> For a $(2N+1)$-spot array, set $|p(x)| = \alpha|q(x)|$, for $x = [0, N]$, and solve the $N+1$ real-valued equations for z_k, $k = [1, N+1]$. For a $2N$-spot array, set $|p(x)| = \alpha|q(x)|$, for $x = [1/2, 3/2, \ldots, (2N+1)/2]$, and solve the N real-valued equations for z_k, $k = [1, N]$.

Because it is possible to find multiple solutions to this problem, rather than solve

Table 11.1 Dammann fan-out gratings

N	ϕ	η_{ub}	η	z_1	z_2	z_3	z_4	z_5	z_6
2	π	81.06	81.06	$-0.500\,00$	$0.500\,00$				
3	π	68.74	66.42	$-0.367\,63$	$0.367\,63$				
	2.008	93.82	86.52	$-0.250\,00$	$0.250\,00$				
4	π	72.05	70.64	$-0.054\,40$	$0.054\,40$				
5	π	83.80	77.39	$-0.367\,66$	$-0.019\,30$	$0.019\,30$	$0.367\,66$		
	2.993	87.20	77.38	$-0.471\,41$	$-0.133\,42$	$0.133\,42$	$0.489\,04$		
6	π	85.28	82.45	$-0.302\,31$	$-0.121\,77$	$0.116\,23$	$0.496\,22$		
7	π	83.07	78.63	$-0.337\,65$	$-0.237\,14$	$0.237\,14$	$0.469\,06$		
	2.473	89.62	84.48	$-0.430\,30$	$-0.215\,13$	$0.215\,13$	$0.439\,14$		
8	π	83.06	74.55	$-0.427\,74$	$-0.181\,67$	$0.179\,35$	$0.294\,42$		
9	π	80.57	70.26	$-0.281\,21$	$-0.158\,39$	$-0.077\,74$	$0.123\,82$	$0.189\,19$	$0.500\,00$
	2.535	87.74	80.78	$-0.351\,84$	$-0.173\,83$	$-0.134\,13$	$-0.058\,80$	$0.358\,54$	$0.500\,00$
10	π	83.31	74.40	$-0.475\,76$	$-0.249\,20$	$-0.002\,05$	$0.118\,71$	$0.268\,80$	$0.342\,11$
11	π	82.11	78.40	$-0.363\,94$	$-0.295\,63$	$-0.153\,21$	$0.083\,51$	$0.166\,79$	$0.500\,00$
	2.589	89.03	84.44	$-0.412\,88$	$-0.282\,04$	$-0.154\,91$	$-0.045\,84$	$0.217\,16$	$0.500\,00$
12	π	86.16	77.96	$-0.381\,19$	$-0.335\,09$	$-0.050\,12$	$0.173\,29$	$0.274\,66$	$0.417\,79$

for α, the solution is subject to the constraint of finding the solution that has maximum α. The problem can be posed equivalently for $\{0, \phi\}$-binary-phase generators, in which case ϕ is also an unknown.

Dammann-designed binary-phase generators for fan-out arrays that number from 2 to 12 are listed in Table 11.1. An exhaustive search to find a solution that has maximum diffraction efficiency was not made. However, each solution is within 10% of the diffraction efficiency upper bound. Some of the $\{0, \phi\}$-binary designs are from Killat *et al.* (1982). Also, any error in magnitude is at least three orders of magnitude lower than the signal, which, we feel, effectively constitutes zero error.

When multiple (>2) phase levels are available, Dammann's technique requires that the phases be ordered in a fixed sequence so that only the phase transitions need be determined. For example, Walker and Jahns (1990) considered a blazed array generator that has M levels of phase distributed evenly between 0 and 2π:

$$\tilde{P}(u) \sum_{k=0}^{K-1} \exp(j2\pi k/M) \operatorname{rect}\left[\frac{u - (z_{k+1} + z_k)/2}{z_{k+1} - z_k}\right].$$

One disadvantage of an off-axis array generator is the need to fabricate a high-frequency spatial carrier. To overcome this, more general techniques for array generator design, discussed in section 11.4, can be applied.

11.3.2 *Continuous-phase Array Generators*

Although many DOE fabrication techniques produce multilevel phase quantized structures (Stern, 1997), systems that are capable of fabricating continuous-phase

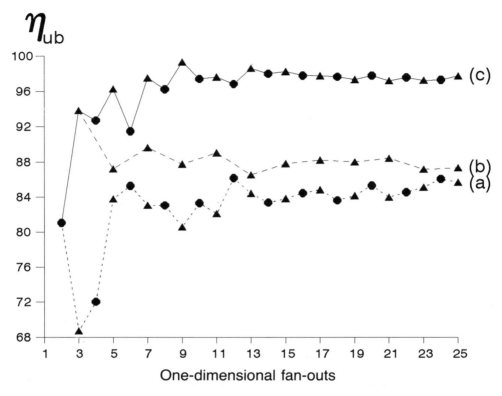

Figure 11.8 Upper bound on diffraction efficiency of phase-only fan-out generators. Triangles indicate odd-numbered fan-outs and circles indicate even-numbered fan-outs. (a) $\{0, \pi\}$-binary-phase array generators. (b) $\{0, \phi\}$-binary-phase array generators. (c) Continuous-phase array generators.

array generators have been demonstrated (Faklis *et al.*, 1993; Streibl *et al.*, 1993; Gale, 1997). Continuous-phase DOEs lend themselves conveniently to parametrization, which reduces computation costs and also reduces the size of intermediate storage necessary before fabrication. Continuous-phase array generators also ensure higher diffraction efficiency than quantized phase array generators, as indicated by the graph of diffraction efficiency upper bounds in Fig. 11.8 (Krackhardt *et al.*, 1992).

In lens design, phase functions can be parametrized as polynomials in x and y. Unfortunately, this representation is not amenable for array generator design and we consider instead three alternate parametric representations: a virtual source array, poles and zeroes in a complex plane, and a weighted sum of orthogonal basis functions.

Virtual Source Array Representation

Dammann's technique for binary-phase array generator design was developed in response to the poor designs that result from binary quantization. The inefficiencies of quantization design are no less severe for continuous-phase designs.

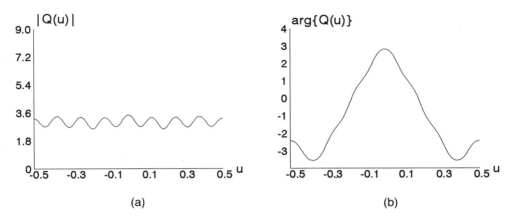

Figure 11.9 Fourier transform of upper bound solution for nine-spot fan-out. (a) Magnitude. (b) Phase.

For example, consider that the nine-spot fan-out $q(x)$ that has phase values

$$\theta_0 = 3.142, \ \theta_1 = 2.455, \ \theta_2 = 3.887, \ \theta_3 = 0.135, \ \theta_4 = 1.772,$$

$$\theta_{-1} = 2.455, \ \theta_{-2} = 3.887, \ \theta_{-3} = 0.135, \ \theta_{-4} = 1.772$$

has a diffraction efficiency upper bound of 99.3%. That the transform $Q(u)$ is nearly phase only is indicated in Fig. 11.9. However, if $|Q(u)|$ is quantized to unity and an array generator is constructed only from the phase of $Q(u)$, i.e. $P(u) = \exp[j\theta(u)]$, where

$$\theta(u) = \arg\{Q(u)\} \tag{11.12}$$

the magnitude of the generated spot array, represented in Fig. 11.10(a), contains error. The diffraction efficiency is 99.38%.

To compensate for this error it is possible to modify equation (11.12) by replacing the desired array $q(x)$ with a virtual source array $v(x)$ (Herzig *et al.*, 1990; Prongué *et al.*, 1992):

$$v(x) = \sum_{n=N_1}^{N_2} v(n) \, \delta(x - n) \tag{11.13a}$$

$$V(u) = \sum_{n=N_1}^{N_2} v(n) \exp(-j2\pi un). \tag{11.13b}$$

The array generator phase is therefore

$$\theta(u) = \arg\{V(u)\}. \tag{11.14}$$

The virtual source array $v(x)$ is used only to specify the phase function that describes the array generator $P(u)$, but is not a 'real' array; i.e. it is itself never generated. Note that the phase $\theta(u)$ is specified completely by the $N_2 - N_1$ magnitudes and $N_2 - N_1$ phases of the virtual array $v(x)$.

The phase of $v(x)$ is determined to achieve high diffraction efficiency; however, its magnitude is determined so as to reduce any residual errors in the response (Farn, 1991; Prongué *et al.*, 1992). For the nine-spot fan-out, a virtual source array

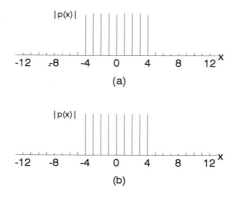

Figure 11.10 Continuous-phase array generator design for nine-spot fan-out. (a) Magnitude of spot array generated by quantization of $Q(u)$. (b) Magnitude of spot array generated by quantization of $V(u)$.

$v(x)$ was determined that has the same phase as the upper bound solution; however, its magnitude is

$$|v(0)| = 1.022, |v(1)| = 0.998, |v(2)| = 0.987, |v(3)| = 0.957, |v(4)| = 1.059,$$

$$|v(-1)| = 0.998, |v(-2)| = 0.987, |v(-3)| = 0.957, |v(-4)| = 1.059.$$

The spot array magnitude generated by the array generator of equation (11.14) is represented in Fig. 11.10(b). There is no magnitude error and the diffraction efficiency is reduced only minimally to 99.28%. It should be noted that modification of the array magnitude $|q(x)|$ to $|v(x)|$ does not make the Fourier transform phase only, rather it precompensates the array for Fourier magnitude quantization.

Pole–Zero Representation

Examination of equation (11.13b) reveals that the generator function $V(u)$ can be expressed as a polynomial in $\exp(j2\pi u)$ of order $N_2 - N_1$. Thus, instead of using the magnitudes and phases of $v(x)$, it is possible to represent the array generator in terms of the zeros of this polynomial. The complex nature of the polynomial coefficients, which are the spot array complex-wave amplitudes $v(n)$, implies that the polynomial roots are also complex and, in general, contain no symmetry. The complex roots can be determined by using the z-transform to continue $V(u)$ analytically into the complex plane:

$$V(z) = \sum_{n=N_1}^{N_2} v(n)z^{-n}$$

$$= v(N_1)z^{-N_1} \prod_{k=1}^{N_2-N_1} (1 - w_k z^{-1}) \tag{11.15}$$

where $z = r\exp(j2\pi u)$ and $w_k = r_k\exp(j\Omega_k)$. The multiplicative factors indicate that the array generator zeros are unaffected by the absolute scale or spatial location of the array. Thus, for purposes of design, $v(N_1)$ can be set to unity and N_1 can be set to zero. The complex-plane zeros of the nine-spot virtual source

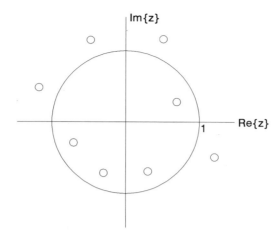

Figure 11.11 Representation of complex-plane zeros of virtual source array for nine-spot fan-out.

array from above are represented in Fig. 11.11. The fact that its zeros exhibit a high degree of symmetry is perhaps one explanation for the array generator's high diffraction efficiency; however, this needs to be investigated quantitatively.

The representation of the generator function in terms of zeros allows us to generalize to a representation in terms of both poles and zeros (Mait, 1991c):

$$V(z) = \frac{\prod\limits_{k=1}^{K} (1 - w_k z^{-1})}{\prod\limits_{l=1}^{L} (1 - w_l z^{-1})}. \tag{11.16}$$

The array generator phase is determined by evaluating the phase of $V(z)$ on the unit circle, $r = 1$:

$$\theta(u) = \arg\{V[\exp(j2\pi u)]\}. \tag{11.17}$$

In a linear shift-invariant system, the presence of a pole in its system function indicates that the system response is infinite in extent, which is the case for a phase-only array generator. Because the presence of a single pole affects the entire generator response, design of an array generator that has poles may help reduce the number of parameters necessary for design.

One advantage to pole–zero design is that the phase-only constraint can be satisfied explicitly. If w_k is a pole of $V(z)$ and $1/w_k^*$ is a zero, then $V(z)$ has constant magnitude:

$$P(z) = V(z) = \prod_{k=1}^{K} |w_k| \frac{[1 - (1/w_k^*) z^{-1}]}{(1 - w_k z^{-1})}. \tag{11.18a}$$

The array generator phase is

$$\theta(u) = \sum_{k=1}^{K} \tan^{-1}\left\{ \frac{\sinh[\ln(r_k)] \sin(2\pi u - \Omega_k)}{1 - \cosh[\ln(r_k)] \cos(2\pi u - \Omega_k)} \right\}. \tag{11.18b}$$

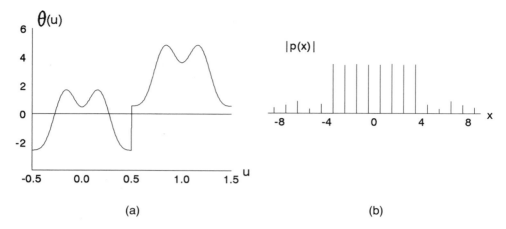

(a)   (b)

Figure 11.12 Continuous-phase array generator design using Legendre polynomials for eight-spot fan-out. (a) Array generator phase. (b) Magnitude of generated array.

In contrast to the virtual source representation where the number of zeros is one less than the number of sources, the number of complex parameters K for the pole–zero representation is independent of the number of spots in the generated array.

Orthonormal Decomposition

A weighted sum of orthogonal basis functions is an alternate representation of $\theta(u)$ that allows the number of parameters to be independent of the number of spots:

$$\theta(u) = \sum_{m=0}^{M} c_m t_m(u). \tag{11.19}$$

Candidates for the functions $t_m(u)$ include Legendre polynomials and Chebyshev polynomials, both of which are orthogonal over a finite region. The order of the expansion M is unknown and can either be included in the optimization or fixed arbitrarily.

The design of array generators using Legendre polynomials has been addressed by Morrison (1991). The phase $\theta(u)$ represented in Fig. 11.12(a) results from summing Legendre polynomials up to order $M = 3$ with coefficients $c_0 = 0$, $c_1 = 2.3262$, $c_2 = -1.0490$ and $c_3 = -0.8396$. The magnitude of the eight-spot fan-out that $\theta(u)$ generates is represented in Fig. 11.12(b). The diffraction efficiency is 94.2% and the magnitude error is low. Because the number of spots is even, the phase is replicated with a π-phase shift between periods.

Design for Large-numbered Spot Arrays

As spot array size increases so too does the number of parameters required for a design. However, if two small-sized spot arrays are appropriately scaled spatially, their convolution results in a single large-sized array (McCormick, 1989):

$$q_3(x) = q_1(x) * q_2(x). \tag{11.20}$$

Table 11.2 Diffraction efficiency upper bounds for separable and non-separable two-dimensional fan-outs

$N \times N$	$(0, \pi)$-binary-phase design		Continuous-phase design	
	separable	non-separable	separable	non-separable
3×3	47.25	75.71	88.00	94.06
5×5	70.22	75.00	92.70	93.99
7×7	69.00	78.56	95.10	95.22
9×9	64.92	77.89	98.66	98.68

If phase elements $P_1(u) = \exp[j\theta_1(u)]$ and $P_2(u) = \exp[j\theta_2(u)]$ have been designed to generate the arrays $q_1(x)$ and $q_2(x)$, the array generator for $q_3(x)$ is

$$P_3(u) = \exp\{j[\theta_1(u) + \theta_2(u)]\}. \tag{11.21}$$

However, errors in the individual designs propagate into the final design and are multiplicative.

11.4 Two-dimensional Array Generators

In this section we consider the design of two-dimensional array generators. We note, however, that for spot arrays whose magnitude is separable and whose phase is unimportant both separable and non-separable designs are possible. The trade-off in design between the two is the complexity of design versus generated diffraction efficiency. Based on a comparison of the diffraction efficiencies in Table 11.2, we can conclude that non-separable designs are recommended for binary-phase array generators and separable designs for continuous phase (Mait, 1991b). For this reason we consider only the design of non-separable binary-phase array generators for two-dimensional spot arrays. However, the design techniques presented are sufficiently general that they are also applicable to the design of one- and two-dimensional multilevel array generators.

A critical limitation of Dammann's technique is that it does not extend easily into two dimensions. Even for binary-phase designs it is necessary to assume a two-dimensional pattern for the locations of phase transitions (Mait, 1989b); i.e. there is no way to allow for arbitrary patterns. This problem is compounded by multilevel designs which, even for one-dimensional designs, require that the available phase levels be placed into a fixed sequence. Finally, Dammann's technique is analytic, i.e. it does not explicitly call for optimization, and, as the number of parameters increases, coupling between equations makes root-finding difficult (Jahns *et al.*, 1989).

The application of Dammann's technique to a magnitude design highlights the fact that array generator design is more appropriately cast as a problem in optimization; i.e. a performance measure that characterizes the array generator is optimized subject to the constraints imposed by its fabrication. For example, design of a multilevel phase element that generates a spot array with high

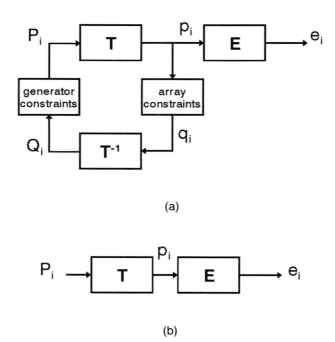

Figure 11.13 Optimization algorithms used for design. (a) Bidirectional. (b) Unidirectional.

diffraction efficiency and minimum error in intensity can be formulated as

$$\text{minimize } e = \iint_{(x,y)\in X} \left| \, |p(x,y)|^2 - \alpha_{ub}^2 |q(x,y)|^2 \, \right|^2 dx\,dy \qquad (11.22)$$

$$\text{subject to } \tilde{P}(u,v) = \sum_{k=0}^{N_p-1} \sum_{l=0}^{N_p-1} \exp(j\theta_{k,l}) \, \text{rect}\left(\frac{u-k\Delta}{\Delta}, \frac{v-l\Delta}{\Delta} \right)$$

where $\theta = 2\pi m/M$, $m = [0, M-1]$, N_p is the number of cells into which the unit period is divided, $\Delta = 1/N_p$, and α_{ub} is the scale factor that corresponds to the diffraction efficiency upper bound. The metric e is the absolute distance between the array generator intensity response $|p(x,y)|^2$ and the spot array intensity $\alpha_{ub}^2 |q(x,y)|^2$, which is a fixed, upper limit in the space of solutions. In contrast to Dammann's technique, here we have assumed that the locations of phase transitions are fixed and the phase values are unknown.

11.4.1 *Design Algorithms*

Two general optimization algorithms for the solution of this problem are represented in Fig. 11.13 (Fienup, 1980). Whereas Fienup referred to the algorithms in Figs 11.13(a) and (b) as error-reduction and input–output, to indicate the nature of data flow, we prefer the nomenclature bidirectional and unidirectional.

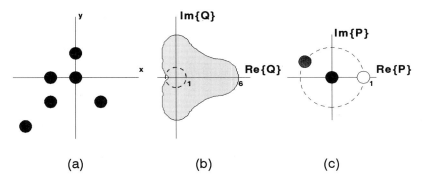

Figure 11.14 Two-dimensional ternary-valued array generator design. (a) Desired asymmetric fan-out $q(x, y)$. (b) Complex plane representation of $Q(u, v)$ assuming zero array phase. (c) Complex plane representation of a ternary-valued array generator constructed from a $\{0, 1\}$-binary-amplitude mask and $\{0, 5\pi/6\}$-binary-phase mask.

Use of a bidirectional algorithm implies not only an understanding of the effects that variations in the array generator have on its response (the operator T), but also an understanding of how variations in the response affect the array generator (the inverse operator T^{-1}). For the system shown in Fig. 11.2, under the assumption of scalar diffraction, T is the inverse Fourier transform (equation (11.1)) and the inverse operator T^{-1} is the forward Fourier transform. Given these models, Fig. 11.13(a) is the iterative Fourier transform algorithm (IFTA) (Gallagher and Liu, 1973; Fienup, 1980; Wyrowski, 1991). The operator E is the metric used to assess system performance, e.g. the integral in equation (11.22).

As an example of IFTA design, we consider the generation of the two-dimensional fan-out $q(x, y)$ represented in Fig. 11.14(a). To emphasize the general applicability of optimization techniques to the design of array generators with arbitrary constraints, we consider an array generator constructed from the combination of a $\{0, 1\}$-binary-amplitude mask and a $\{0, 5\pi/6\}$-binary-phase mask. The array generator is represented in the complex plane in Fig. 11.14(c). The complex plane representation of $Q(u, v)$ is represented in Fig. 11.14(b). Zero spot phase has been assumed and only the locations of Fourier values are indicated. A more complete representation would indicate the distribution of values within this region.

If we assume an intensity design, Fig. 11.14(a) is a pictorial representation of the response constraints and Fig. 11.14(c) a pictorial representation of the array generator constraints. Although the IFTA is simple in its implementation, the naive application of these constraints will most certainly cause the algorithm to stagnate in a local minimum.

To avoid this, it is helpful to recall three important points from the discussion of Dammann's technique: there exists a variability in the scale factor α, array phase can be used to increase diffraction efficiency, and, when characterizing performance, it is necessary only to consider the response generated within a defined signal window. These points were first realized during the developmental years of computer-generated holography (Lee, 1978) and are reflected in the error measure of equation (11.22). The theory for their general application to design

311

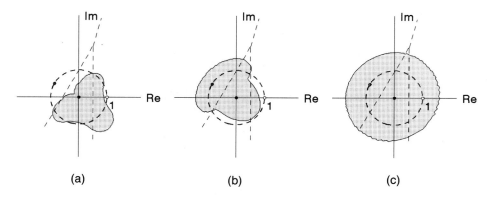

Figure 11.15 Representation in the complex plane of the effect of using freedoms in design. (a) Scale freedom. (b) Scale and phase freedom. (c) Scale, phase and amplitude freedom.

was later codified by Wyrowski (1991), who referred to them as scale freedom, phase freedom, and complex-wave amplitude freedom, that is, outside of the signal window there exists a freedom in the specification of the response complex-wave amplitude.

The application of each of these freedoms to the design example is indicated in Fig. 11.15. The boundaries indicate the threshold for quantization design. In Fig. 11.15(a) a complex scale factor is used to rotate and scale $Q(u, v)$ to reduce quantization error and in Fig. 11.15(b) phase freedom is used to redistribute $Q(u, v)$ so that it is more phase-like. Scale freedom has also been used to reduce quantization error. The distribution represented in Fig. 11.15(c) is achieved using scale, phase and amplitude freedoms. Amplitude freedom was exercised by shifting the location of the fan-out array off-axis, which multiplies $Q(u, v)$ by a linear-phase term. The result is a more even distribution of values in the complex plane.

If the array generator were phase-only, the distribution in Fig. 11.15(c) would be an acceptable input for quantization. Our design, however, requires a distribution that is matched to the ternary values of the array generator, which, as shown in Fig. 11.16, can be achieved using the IFTA. Scale, phase and amplitude freedoms were exercised to obtain the distribution shown in Fig. 11.16(a), which, upon quantization, yields the array generator represented in Fig. 11.16(b). The corresponding spot array is represented in Fig. 11.16(c). The application of amplitude freedom to shift the array off-axis is apparent in the response.

Instead of redistributing Fourier values, the philosophy behind unidirectional algorithms is that if the array generator can be characterized by a finite set of quantized parameters, e.g. array generator phase levels, then there exists a finite (albeit large) number of permutations of these parameters. The objective is to find, in an efficient manner, the permutation that achieves the optimization. Examples of unidirectional algorithms include gradient descent methods and annealing algorithms (Feldman and Guest, 1989; Jennison *et al.*, 1989; Turunen *et al.*, 1989). In these algorithms the present array generator $P_i(u, v)$ is a function of the previous array generator and the error generated by it:

$$P_i(u, v) = P_{i-1}(u, v) + G[e_{i-1}(u, v)]. \tag{11.23}$$

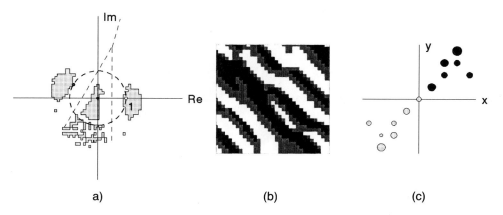

Figure 11.16 Design of ternary-valued array generator using IFTA with scale, phase and amplitude freedoms. (a) Complex-plane representation of Fourier values before final quantization. (b) Array generator $\tilde{P}(u, v)$. (c) Spot array generated by (b). Desired spot array is represented by black circles. Relative intensity is indicated by size.

For example, simulated annealing algorithms typically begin with an array generator that has a random distribution of allowed values. Each value in turn is altered to another allowed value and the effect of this change on the response is evaluated. If the performance improves, the change is accepted unconditionally. However, if the performance decreases, the change is accepted conditionally, based upon a probability that is a function of the number of iterations. An array generator designed in such a manner is shown in Fig. 11.17. The design freedoms of scale, phase and amplitude were still exercised.

The performance of the IFTA-designed and the annealed-designed array generators is compared in Table 11.3. Although the annealed array generator performs slightly better than the iterated array generator, a general statement about the performance of the two algorithms relative to one another cannot be made. Consider the $\{0, \pi\}$-binary-phase array generator design of a 3×3 fan-out array indicated in Fig. 11.18, the performance of which is also summarized in Table 11.3. Design freedoms were fully exercised in both algorithms, which produced comparable results. The question as to which algorithm is preferred is one of the computational costs involved. Figure 11.18 also indicates that the exercise of amplitude freedom does not necessarily imply the use of a spatial carrier.

If care is not exercised, both bidirectional and unidirectional algorithms are prone to finding local minima and not global ones. To overcome this, it is possible at each iteration to modify several array generators in a manner such as equation (11.23), which is the philosophy behind simplex (Press *et al.*, 1986) and genetic algorithms (Johnson *et al.*, 1993).

We note in conclusion that the principles presented here for Fourier array generator design are equally applicable to the design of Fresnel array generators; i.e. generators that produce the spot array at some plane behind the generator without the use of a transform lens (Hatakoshi and Nakamura, 1993; Kress and Lee, 1993; Wood *et al.*, 1993).

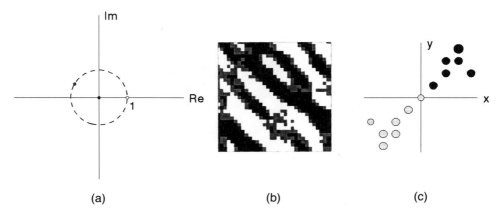

(a)   (b)   (c)

Figure 11.17 Design of ternary-valued array generator using simulated annealing with scale, phase and amplitude freedoms. Figures correspond to those in Fig. 11.16.

Table 11.3 Summary of two-dimensional designs

	η (%)	e
Asymmetric fan-out $\eta_{ub} = 43.35\%$		
Iterative design	29.16	0.0003
Annealed design	31.26	0.0002
3×3 fan-out $\eta_{ub} = 75.71\%$		
Iterative design	67.51	0.0002
Annealed design	67.62	0.0007

Further, if rigorous diffraction theory is used to model the array generator, because it is difficult to model inverse diffraction rigorously, a unidirectional algorithm must be used for design. This is the approach taken by Noponen *et al.* (1992) to design binary-phase array generators with features that are comparable in size to the illumination wavelength. Under these conditions, the response of the array generator as described by scalar Fourier theory (equation (11.1)) is no longer valid.

The representation of the binary-phase array generator is similar to that used by Dammann; i.e. the parameters used in the vector design algorithm are the locations of phase transitions. The initial set of phase transitions are determined by quantizing to binary phase a high-frequency spatial carrier that is modulated by a continuous-phase scalar-designed array generator. The phase transition locations are modified based on the vector response of the array generator.

Designs using this technique indicate that it is possible to exceed the scalar diffraction efficiency upper bounds indicated in Fig. 11.8. For example, Noponen *et al.* (1992) designed a binary-phase array generator that generates nine spots with 93.3% diffraction efficiency. (The scalar upper bound is 80.6%.) The periodicity of the array generator is 4.86λ, its depth 1.26λ, and the locations of phase transition

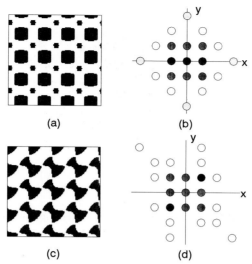

Figure 11.18 Design of $\{0, \pi\}$-binary-phase element to generate 3×3 fan-out array. (a) Array generator designed using IFTA. (b) Spot array generated by (a). Relative intensity is indicated by grey level. (c) Array generator designed using simulated annealing. (d) Spot array generated by (c).

$\{0.00, 0.69\lambda, 1.41\lambda, 1.87\lambda, 2.59\lambda, 3.28\lambda\}$. In addition to providing increased diffraction efficiency, resonance domain elements can also be designed as polarization-sensitive devices.

Unfortunately, the vector design of array generators is more intensive than that of scalar. Further, the fabrication of elements with features of the order of a wavelength can also be more difficult. Consider that the nine-spot resonance domain array generator has a minimum feature of 0.69λ. For long visible and near infrared wavelengths ($0.8-1.2\,\mu\text{m}$), this implies a minimum feature of the order of $0.6-0.8\,\mu\text{m}$. Even with present technology this still represents a fabrication challenge. Nonetheless, Turunen *et al.* (1993) have fabricated one-dimensional fan-outs with $0.5\,\mu\text{m}$ features that operate at 850 nm. The minimum feature for far infrared wavelengths ($8-10\,\mu\text{m}$), however, is well within the specifications of most fabrication techniques.

11.4.2 *Alternate Sampling Lattices*

Up to this point we have assumed a rectangular base period and replication functions that produce a rectilinear lattice of sample points. However, it is possible to design spot arrays for alternate sampling lattice geometries and we consider here techniques for generating non-rectilinear sampling lattices.

The simplest means for creating a spot array that is non-rectilinear is the geometric transformation of a rectilinear array generator (Streibl, 1989). The transformed array generator and corresponding sampling lattice are represented in Fig. 11.19. However, if the facility for fabricating DOEs is rectilinear, the phase

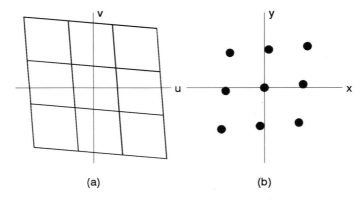

(a)   (b)

Figure 11.19 Generation of non-rectilinear sample lattice by scaling and skewing a rectilinear array generator. (a) Scaled and skewed array generator and (b) corresponding lattice.

transitions of the scaled and skewed array generator will have to be interpolated onto a rectangular grid.

Alternatively it is possible to produce a tilted array of spots by introducing a translational shift between neighbouring columns (or rows) in a rectilinear replicating function (Mait, 1991a). As represented in Fig. 11.20(a), v_0 is the amount of shift between columns. The tilt angle γ in the corresponding lattice shown in Fig. 11.20(b) is given by $\gamma = \tan^{-1}(1/v_0)$.

Hexagonal replication and sampling, represented in Fig. 11.21, can be achieved via a translational shift along one axis and a change in scale along the orthogonal axis. If the desired spot array contains either circular or hexagonal symmetry, hexagonal replication with a hexagonal base period reduces the dimensionality of the design from two to one (Mait, 1991a).

The previous replication functions all assume that the unit period is replicated infinitely in Fourier space, which is, of course, an unrealistic assumption. However, it is possible to account for the finite number of replicas that are actually fabricated in the definition of the replication function. For a rectilinear sample lattice, the replication function is

$$\text{repl}(u, v) = \sum_{k=0}^{K-1} \sum_{l=0}^{L-1} \delta(u - k, v - l) \tag{11.24a}$$

which generates the sampling function

$$\text{samp}(x, y) = \frac{\sin(K\pi x)}{\sin(\pi x)} \frac{\sin(L\pi y)}{\sin(\pi y)}. \tag{11.24b}$$

A constant-phase term has been ignored. The function $\text{samp}(x, y)$ is comb-like, i.e. it assumes its peak value of KL at integer values of x and y; however, it is non-zero in-between the peaks. The width of a single peak is $1/K \times 1/L$.

As long as the number of replicas K and L is large, the effects of truncation can be ignored, which is certainly true of array generators that are fabricated as fixed elements. However, for applications in real-time image processing, array generators have also been produced dynamically using spatial light modulators

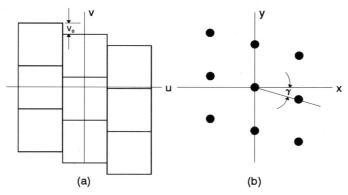

Figure 11.20 Generation of non-rectilinear sample lattice by shifting columns of a rectilinear array generator. (a) Array generator with shifted columns and (b) corresponding lattice.

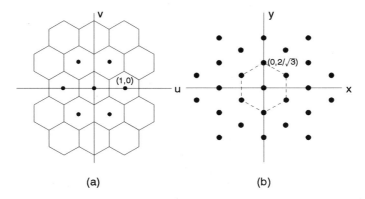

Figure 11.21 Generation of hexagonal sample lattice using hexagonal replication. (a) Array generator and (b) corresponding lattice.

(SLMs) and the space-bandwidth product of currently available SLMs (of the order of 512×512) limits the number of replicas that can be displayed.

11.5 Fabrication Considerations

Up to this point, we have considered fabrication only from the perspective of the transmission values that a particular fabrication technique can achieve. However, the fidelity of the optically generated spot array is affected by the resolution of the fabrication facility and the accuracy with which it can realize the designed array generator.

The array generator of Fig. 11.1, for example, was fabricated by photolithographic etch of fused silica (Morrison *et al.*, 1995). An e-beam generated mask that had $0.5\,\mu$m minimum feature was used to create the etched binary pattern. Although the spot array was generated with a diffraction efficiency within

2% of the theoretically predicted 74.7%, the intensity uniformity varied between ±15% from the average. Further, the intensity of the central order was 50% of the average intensity and approximately 1% of the total diffracted energy.

Because DOE fabrication is the subject of other chapters within this volume (Gale, 1997; Stern, 1997), we present here only some general considerations on fabrication and its relationship to Fourier array generators. Although the analysis is presented for one-dimensional elements, the results are also valid in two dimensions.

The Fourier nature of the array generator system allows us to observe that, in general, the array response is more sensitive to local errors, i.e. space variant errors, than to global errors. For example, the relative change in intensity of a single spot $\Delta I(n)$ for a continuous-phase fan-out array generator fabricated with a small ($<\pi$) space-varying phase error $\theta_\varepsilon(u)$ is (Jahns *et al.*, 1989):

$$\frac{\Delta I(n)}{I(n)} = \sqrt{2}\frac{N}{\eta}\Delta\theta \tag{11.25a}$$

where η is the diffraction efficiency of the array generator, N is the number of spots in the fan-out, and $\Delta\theta$ is the variance of the phase error:

$$\Delta\theta = \int\theta_\varepsilon^2(u)\,\mathrm{d}u - \{\int\theta_\varepsilon(u)\,\mathrm{d}u\}^2. \tag{11.25b}$$

For a 10% deviation in spot intensity, even a one-dimensional nine-spot array generator, which has a diffraction efficiency upper bound of 99.3%, requires a phase error variance of less than 0.007.

In contrast, a constant error in phase depth scales the spot array in a global fashion. From equation (11.11b), a $\{0, \phi\}$-binary-phase array generator designed for a one-dimensional odd-numbered spot array and fabricated with a constant phase error ε generates the intensity spot array

$$I_\varepsilon(n) = I(n) + 2[\cos(\phi + \varepsilon) - \cos\phi]\sigma(n)[\text{sinc}(n) - \sigma(n)]. \tag{11.26}$$

The relative difference in intensity between the on-axis spot and other spots in a fan-out array is

$$\frac{\Delta I}{I} = \frac{I_\varepsilon(0) - I_\varepsilon(n)}{I} = \frac{N}{\eta}\left[\frac{\cos(\phi + \varepsilon) - \cos\phi}{1 - \cos\phi}\right]. \tag{11.27}$$

For $\phi = \pi$, $\Delta I/I = 10\%$ and $\eta = 70\%$, $\varepsilon < 0.05\pi$ for $N > 11$. In comparison to space variant errors, the global nature of the phase error reduces the tolerances on phase fabrication. Nonetheless, the tolerances are still strict.

For a similar phase error ε in the fabrication of a $\{0, \pi\}$-binary-phase element for even-numbered spot array generation, no spots are generated at integer locations except for the on-axis spot, the value of which is

$$I_\varepsilon(0) = \sin^2(\varepsilon/2). \tag{11.28a}$$

The spots generated at half integers are again simply scaled:

$$I_\varepsilon(n + 1/2) = \cos^2(\varepsilon/2)I(n + 1/2). \tag{11.28b}$$

If it is possible not to have a detector on-axis, the effects of the phase error can be completely ameliorated, which is one significant advantage of even-numbered spot arrays over odd-numbered arrays.

Although the response of a binary-phase array generator is relatively tolerant to a constant phase error, it is less tolerant to errors in the position of phase transition locations, which are space-variant. From equation (11.6b), a binary-phase array generator fabricated with a position error dz in the phase transition location z_k yields

$$p_{dz}(n) = p(n) + \exp(j2\pi z_k n)\, dz. \tag{11.29}$$

For a fan-out, the error in position generates a relative intensity error (Jahns *et al.*, 1989)

$$\frac{\Delta I(n)}{I(n)} = 4\sqrt{\frac{N}{\eta}}\, N\, dz. \tag{11.30}$$

Because a single period of the array generator is assumed to be unity, the position error dz is a normalized error. If the mask is fabricated by an e-beam writer that has a position error of $0.25\,\mu$m and a single period of the grating is $1\,$mm, $dz = 0.25 \times 10^{-3}$. If, in addition, $\Delta I(n)/I(n) = 10\%$ and $\eta = 70\%$, the maximum number of spots that can be generated with fidelity is 19.

Even when the masks used to expose photoresist contain no errors in the locations of phase transitions, it is possible to introduce errors in the generated array due to improper exposure (O'Shea, 1995). Over- or underexposure of the masks into photoresist either dilates or erodes all features by a constant amount. Although this can be modelled using the previous analysis, the error in the array generator is global and, therefore, at least to first order, manifests itself primarily in the on-axis spot.

For an odd-numbered fan-out array the change in area that generates a fractional change γ_o in the intensity of the on-axis spot is

$$\Delta A_o = \frac{1 - \sqrt{1 - \gamma_o}}{2}\sqrt{\frac{\eta}{N}}. \tag{11.31a}$$

Because an even-numbered array has zero intensity on-axis, a change in area increases the intensity on-axis by a fraction γ_e of the intensities of the other spots; thus

$$\Delta A_e = \frac{\sqrt{\gamma_e}}{2}\sqrt{\frac{\eta}{N}}. \tag{11.31b}$$

The areas ΔA_o and ΔA_e are defined over a single period. Note that to maintain the same level of variation in intensity as the number of fan-outs increases, the change in area that the array generator can tolerate decreases.

For a change in on-axis spot intensity that is 10% of the nominal fan-out intensity, an array generator for even-numbered fan-outs can tolerate an area change that is approximately six times that of an odd-numbered fan-out. For example, if we assume a nominal diffraction efficiency of 75% for a binary-phase element, a 16-spot array generator can tolerate a 3.42% change in area, whereas a 17-spot array generator can tolerate only a 0.54% change. The even-numbered array generator is less sensitive to changes in area due to the inversion between replicas which moderates any increase in area associated with a single phase value.

From Table 11.1 the number of phase transitions required to generate N spots is of the order of $N/2$, therefore $\Delta A = N \Delta z/2$, where Δz is the scale of the erosion or dilation normalized to unit period. If the period of the array generator is 1 mm, the 16-spot array generator can tolerate an erosion or dilation by 3.80 μm, but the 17-spot array generator by only 0.54 μm. Except for the fact that $\Delta A = N \Delta z$, the above analysis extends without modification to two-dimensional array generators, for which the effect on the on-axis spot is more noticeable (O'Shea, 1995).

For multilevel phase array generators, perhaps more critical than errors introduced in mask generation, are errors generated by misalignment between masks during multiple exposure and etch steps. Such misalignments introduce both translation and phase errors into the array generator that must be confronted each time an element is fabricated. Although an analytic expression is not available to assess the affect of misalignments on array generator response, a stochastic analysis indicates that the array uniformity and diffraction efficiency fall off linearly with misalignment (Miller *et al.*, 1993).

The total number of features used to construct the array generator, which is referred to as its space-bandwidth product (SBWP), also indicates the fidelity of a generated spot array. The SBWP is limited by the size of the minimum feature and the size of the field over which the features can be accurately placed. An array generator that has an SBWP of $M \times M$ can have $K \times K$ number of replicas; each replica consists of $M/K \times M/K$ features. Thus, the smallest normalized position error dz is K/M, which indicates that to reduce the sensitivity to position errors, the number of replicas should be as small as possible. However, as seen in the previous section, keeping the number of replicas small when the SBWP is fixed increases the size of each generated spot. This trade-off between spot size and sensitivity to position errors is not as pronounced for an array generator that has a SBWP of $10^4 \times 10^4$, e.g. a 1×1 cm^2 element fabricated with 1 μm features, as it is for a SLM that has an SBWP of only 512×512.

11.6 Conclusions

In this chapter we have discussed techniques for Fourier array generator design. Although Dammann's design technique is a pedagogically useful tool, it is not useful for applications that require two-dimensional spot array generation nor is it useful for multilevel designs. In such applications, optimization techniques are more appropriate, but, to achieve the best performance possible, it is necessary to make full use of all design freedoms available. To achieve high diffraction efficiency, phase freedom is the most critical. For two-dimensional designs in which phase freedom is available, non-separable designs are recommended for binary-phase array generators; and separable designs for continuous and multilevel phase array generators. In some instances, the geometry of the spot array can allow a two-dimensional non-separable design to be reduced to one dimension. Research issues in array generator design that require more attention include the development of efficient and effective algorithms that incorporate rigorous diffraction theory and the development of hybrid refractive–diffractive designs to account for aberrations and broadband applications.

Acknowledgements

The author is grateful to his colleagues who have influenced his understanding and presentation of this material and have provided their support: Frank Wyrowski, Antti Vasara, Jari Turunen, Norbert Streibl, Willi Stork, Don O'Shea, Johannes Schwider, Dennis Prather, John Pellegrino, Rick Morrison, Jim Leger, Ulrich Krackhardt, Jürgen Jahns, Hans Peter Herzig, Michael Feldman, Michael Farn, and Karl Heinz Brenner.

References

ATHALE, R. A., MAIT, J. N. and PRATHER, D. W. (1992) Optical morphological image processing with acoustooptic devices. *Opt. Commun.* **87**, 99–104.

DAMMANN, H. and GÖRTLER, K. (1971) High-efficiency in-line multiple imaging by means of multiple phae holograms. *Opt. Commun.* **3**, 312–315.

DAMMANN, H. and KLOTZ, E. (1977) Coherent optical generation and inspection of two-dimensional periodic structures. *Opt. Acta* **24**, 505–515.

FAKLIS, D., BOWEN, J. P. and MORRIS, G. M. (1993) Continuous phase diffractive optics using laser pattern generation. *Holography: SPIE's International Technical Working Group Newsletter* **3**, 1.

FARN, M. W. (1991) New iterative algorithm for the design of phase-only gratings. In Cindrich, I. N. and Lee, S. H. (eds) *Holographic Optics: Computer and Optically Generated*, Vol. 1555, pp. 34–42. Bellingham: SPIE.

FELDMAN, M. R. and GUEST, C. C. (1989) High efficiency hologram encoding for generation of spot arrays. *Opt. Lett.* **14**, 479–481.

FIENUP, J. R. (1980) Iterative method applied to image reconstruction and to computer-generated holograms. *Opt. Eng.* **19**, 297–306.

GALE, M. T. (1997) Direct writing of continuous-relief micro-optics. In Herzig, H. P. (ed.) *Microoptics*, pp. 87–126. London: Taylor & Francis.

GALLAGHER, N. C. and LIU, B. (1973) Method for computing kinoforms that reduces image reconstruction error. *Appl. Opt.* **12**, 2328–2335.

GLYTSIS, E. N., GAYLORD, T. K. and BRUNDRETT, D. L. (1993) Rigorous coupled-wave analysis and applications of grating diffraction. In Lee, S. H. (ed.) *Diffractive and Miniaturized Optics*, Vol. CR49, pp. 3–31. Bellingham: SPIE.

HATAKOSHI, G. and NAKAMURA, M. (1993) Grating lenses for optical branching. *Appl. Opt.* **32**, 3661–3669.

HERZIG, H. P., PRONGUÉ, D. and DÄNDLIKER, R. (1990) Design and fabrication of highly efficient fan-out elements. *Japan. J. Appl. Phys.* **29**, L 1307–1309.

JAHNS, J. (1994) Diffractive optical elements for optical computers. In Jahns, J. and Lee, S. H. (eds) *Optical Computing Hardware*, pp. 137–167. San Diego: Academic Press.

JAHNS, J., DOWNS, M. M., PRISE, M. E., STREIBL, N. and WALKER, S. J. (1989) Dammann gratings for laser beam shaping. *Opt. Eng.* **28**, 1267–1275.

JENNISON, B. K., ALLEBACH, J. P. and SWEENEY, D. W. (1989) Iterative approaches to computer-generated holography. *Opt. Eng.* **28**, 629–637.

JOHNSON, E. G., KATHMAN, A. D., HOCHMUTH, D. H., COOK, A. L., BROWN, D. R. and DELANY, B. (1993) Advantages of genetic algorithm optimization methods in diffractive optic design. In Lee, S. H. (ed.) *Diffractive and Miniaturized Optics*, Vol. CR49, pp. 54–74. Bellingham: SPIE.

KILLAT, U., RABE, G. and RAVE, W. (1982) Binary phase gratings for star couplers with high splitting ratio. *Fiber Integ. Opt.* **4**, 159–167.

KRACKHARDT, U. and STREIBL, N. (1989) Design of Dammann-gratings for array generation. *Opt. Commun.* **74**, 31–34.

KRACKHARDT, U., MAIT, J. N. and STREIBL, N. (1992) Upper bound on the diffraction efficiency of phase-only fan-out elements. *Appl. Opt.* **31**, 27–37.

KRESS, B. and LEE, S. H. (1993) Iterative design of computer generated Fresnel holograms for free space optical interconnections. In *Optical Computing Technical Digest 1993*, Vol. 7, pp. 22–25. Washington, D.C.: OSA.

LEE, W.-H. (1978) Computer-generated holograms: Techniques and applications. In Wolf, E. (ed.), *Progress in Optics*, Vol. 16, pp. 119–223. Amsterdam: North-Holland.

LEGER, J. R., SWANSON, G. J. and VELDKAMP, W. B. (1987) Coherent laser addition using binary phase gratings. *Appl. Opt.* **26**, 4391–4399.

MAIT, J. N. (1989a) Design of Dammann gratings for optical symbolic substitution. In Goodman, J. W., Chavel, P. and Roblin, G. (eds) *Optical Computing 88*, Vol. 963, pp. 646–652. Bellingham: SPIE.

(1989b) Design of Dammann gratings for two-dimensional, nonseparable, noncentro-symmetric responses. *Opt. Lett.* **14**, 196–198.

(1990) Design of binary-phase and multiphase Fourier gratings for array generation. *J. Opt. Soc. Am. A* **7**, 1514–1528.

(1991a) Design for two-dimensional nonseparable array generators. In Cindrich, I. N. and Lee, S. H. (eds) *Computer and Optically Generated Holographic Optics*, Vol. 1555, pp. 43–52. Bellingham: SPIE.

(1991b) Upper bound on the diffraction efficiency of phase-only array generators. In Cindrich, I. N. and Lee, S. H. (eds) *Computer and Optically Generated Holographic Optics*, Vol. 1555, pp. 53–62. Bellingham: SPIE.

(1991c) *Complex plane representation and design of array generators*. Presentation at the 1991 Annual Meeting of the Optical Society of America, San Jose, November.

MCCORMICK, F. B. (1989) Generation of large spot arrays from a single laser beam by multiple imaging with binary phase gratings. *Opt. Eng.* **28**, 299–304.

MILLER, J. M., TAGHIZADEH, M. R., TURUNEN, J. and ROSS, N. (1993) Multi-level-grating array generators: fabrication analysis and experiments. *Appl. Opt.* **32**, 2519–2525.

MORRISON, R. L. (1991) private communication.

(1992) Symmetries that simplify design of spot array phase gratings. *J. Opt. Soc. Am. A* **9**, 464–471.

MORRISON, R. L. WOJCIK, M. J. and BUCHHOLZ, D. B. (1995) Nonseparable, surface-relief gratings that generate large, arbitrary intensity beam arrays. In Lee, S. H. and Hinton, H. S. (eds) *Optoelectronic Interconnects III*, Vol. 2400, 32–41. Bellingham: SPIE.

MORRISON, R. L., WALKER, S. L., MCCORMICK, F. B. and CLOONAN, T. J. (1993) Practical applications of diffractive optics in free-space photonic switching. In Lee, S. H. (ed.) *Diffractive and Miniaturized Optics*, Vol. CR49, pp. 265–289. Bellingham: SPIE.

NOPONEN, E., VASARA, A., TURUNEN, J., MILLER, J. M. and TAGHIZADEH, M. R. (1992) Synthetic diffractive optics in the resonance domain. *J. Opt. Soc. Am. A* **9**, 1206–1213.

OPPENHEIM, A. V. and LIM, J. S. (1981) The importance of phase in signals. *Proc. IEEE* **69**, 529–541.

O'SHEA, D. C. (1995) Reduction of the zero order intensity in binary Dammann gratings. *Appl. Opt.* **34**, 6533–6537.

PRESS, W. H., FLANNERY, B. P., TEUKOLSKY, S. A. and VETTERLING, W. T. (1986) *Numerical Recipes: The Art of Scientific Computing.* Cambridge: Cambridge University Press.

PRONGUÉ, D., HERZIG, H. P., DÄNDLIKER, R. and GALE, M. T. (1992) Optimized kinoform structures for highly efficient fan-out elements. *Appl. Opt.* **31**, 5706–5711.

STERN, M. (1997) *Binary optics fabrication.* In Herzig, H.-P. (ed.) *Micro-optics*, pp. 53–85. London: Taylor & Francis.

STREIBL, N. (1989) Beam shaping with optical array generators. *J. Mod. Opt.* **36**, 1559–1573.

— (1994) Multiple beamsplitters. In Jahns, J. and Lee, S. H. (eds) *Optical Computing Hardware*, pp. 227–248. San Diego: Academic Press.

STREIBL, N., SCHWIDER, J., SCHRADER, M. and KRACKHARDT, U. (1993) Synthetic phase holograms written by laser lithography. *Opt. Eng.* **32**, 781–785.

THALMANN, R., PEDRINI, G., ACKLIN, B. and DÄNDLIKER, R. (1989) Optical symbolic substitution using diffraction gratings. In Goodman, J. W., Chavel, P. and Roblin, G. (eds) *Optical Computing 88*, Vol. 963, pp. 635–641. Bellingham: SPIE.

TURUNEN, J., VASARA, A. and WESTERHOLM, J. (1989) Kinoform phase relief synthesis: a stochastic method. *Opt. Eng.* **28**, 1162–1167.

TURUNEN, J., BLAIR, P. B., MILLER, J. M., TAGHIZADEH, M. T. and NOPONEN, E. (1993) Bragg holograms with binary synthetic surface-relief profile. *Opt. Lett.* **18**, 1022–1024.

VASARA, A., TAGHIZADEH, M. R., TURUNEN, J., WESTERHOLM, J., NOPONEN, E., ICHIKAWA, H., MILLER, J. M., JAAKKOLA, T. and KUISMA, S. (1992) Binary surface relief gratings for array illumination in digital optics. *Appl. Opt.* **31**, 3320–3336.

VELDKAMP, W. B., LEGER, J. R. and SWANSON, G. J. (1986) Coherent summation of laser beams using binary phase gratings. *Opt. Lett.* **11**, 303–305.

WALKER, S. J. and JAHNS, J. (1990) Array generation with multilevel phase gratings. *J. Opt. Soc. Am. A* **7**, 1509–1513.

— (1992) Optical clock distribution using integrated free-space optics. *Opt. Commun.* **90**, 359–371.

WOOD, D., MCKEE, P. and DAMES, M. (1993) Multiple-imaging and multiple-focusing Fresnel lenses with high numerical aperture. In Denisyuk, Y. N. and Wyrowski, F. (eds) *Holographics International '92*, Vol. 1732, pp. 307–316. Bellingham: SPIE.

WYROWSKI, F. (1990) Characteristics of diffractive optical elements/digital holograms. In Cindrich, I. N. and Lee, S. H. (eds) *Computer and Optically Formed Holographic Optics*, Vol. 1211, pp. 2–10. Bellingham: SPIE.

— (1991) Digital holography as part of diffractive optics. *Rep. Prog. Phys.* **54**(12), 1481–1571.

ZALETA, D. E., DASCHNER, W., LARSSON, M., KRESS, B. C., URQUHART, K. S. and LEE, S. H. (1993) Diffractive optics fabricated by electron-beam direct write methods. In Lee, S. H. (ed.) *Diffractive and Miniaturized Optics*, Vol. CR49, pp. 117–137. Bellingham: SPIE.

Polarization Transformation Properties of High Spatial Frequency Surface-relief Gratings and their Applications

C. W. HAGGANS AND R. K. KOSTUK

12.1 Introduction

The polarization transformation properties of gratings have many potential applications in optical design. In principle, a full range of functions such as polarization beam splitters, retarders and rotators can be realized (Fig. 12.1). In addition, the grating surface can modify the isotropic nature of most dielectric materials producing an artificial birefringent layer. With the advent of high-resolution microlithography and holographic techniques it should be possible to integrate many of these functions using the planar optic format described previously in Chapter 7.

Gratings fall into three basic categories with respect to their effect on the polarization of light. These may be specified as gratings with: (1) a period (Λ) several times larger than the wavelength (λ) of the illuminating field ($\Lambda \gg \lambda$); (2) a period much smaller than the wavelength ($\Lambda \ll \lambda$); and (3) a period of the same order as the wavelength ($\Lambda \approx \lambda$). In the first category the grating has relatively little effect on the polarization properties of light, and is often referred to as the scalar regime. In the second category the wavelength to grating period ratio (λ/Λ) is large enough that only the zero transmitted and reflected orders propagate from the grating substrate, and hence are called zero-order gratings. In this case a rigorous analysis of the vector components of the electromagnetic fields must be considered in order to obtain a full understanding of their polarization properties. However, in some instances effective medium theory can be applied to approximate their birefringent properties. In the third category several diffraction orders may propagate and the efficiency and polarization state of these orders will change with the polarization of the reconstruction beam. This will typically occur when $0.2 < \lambda/\Lambda < 2$, and is appropriate for applications where high diffraction efficiency is required for a limited number of non-zero diffraction orders.

A theoretical description of electromagnetic field interactions with gratings is required to understand their polarization and energy distribution properties. The history of the electromagnetic theory of surface-relief gratings up to 1984 is

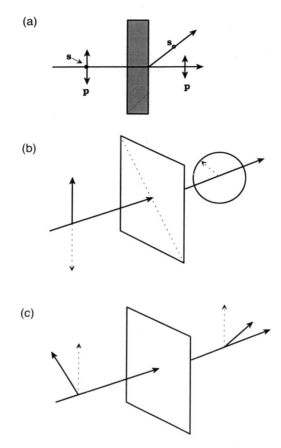

Figure 12.1 Examples of polarization control elements formed with gratings. (a) Polarization beam splitter. (b) Polarization retardation element. (c) Polarization rotator.

documented by Petit (1980) and Maystre (1984). During the past 20 years, numerous approaches for describing the vector nature of the electromagnetic field in the grating region have been completed. When these approaches are combined with numerical techniques which can be implemented on high-speed computers, they provide a powerful tool for analyzing the polarization properties of gratings.

In general, when a beam incident on a grating contains both orthogonal polarization components (Fig. 12.2), the diffracted orders are elliptically polarized. This occurs because the efficiencies of the two polarization components for a given order normally differ, and a phase difference between the components is usually introduced. Most researchers studying the polarization properties of surface-relief gratings have analyzed the dependence of the diffracted order efficiencies on the incident polarization state. However, the polarization state of diffracted beams can also be controlled using gratings with different structures and incidence conditions. Interest in the polarization properties of surface-relief diffractive structures has

326

increased dramatically with the improved lithographic techniques developed during the 1980s. This interest is evident from the number of publications related to this topic (see for example the recent reviews on diffractive optics theories and applications given in *J. Opt. Soc. Am.* A, **12**(5) (1995), and *Appl. Opt.* **34**(14) (1995)).

In this chapter we investigate the polarization properties of high spatial-frequency surface-relief gratings. Emphasis is placed on the design and application of lamellar gratings in the zero-order diffraction regime. Higher-order diffractive elements have already been discussed in Chapter 7, and the methods presented in this chapter can readily be extended to these components. Background definitions and parameters for the theoretical analysis are provided in section 12.2. In section 12.3 the polarization properties of zero-order surface-relief type gratings are evaluated using exact theories and compared to the results from approximate theories. Applications of these gratings as polarization control elements are also illustrated.

12.2 Definition of Polarization Parameters

The electromagnetic theory of diffraction gratings has been discussed in detail in Chapter 2. In this section we define terms useful for describing the polarization properties of these elements.

Figure 12.2 shows the general diffraction geometry for a surface-relief type grating. Twelve independent parameters are required to describe fully the grating diffraction problem for lamellar profile gratings: (1) grating depth (d), (2) period (Λ), (3) duty cycle (D), (4) input refractive index (n_1), (5) refractive index of area 1 in the grating region (n_{g1}), (6) refractive index of area 2 in the grating region (n_{g2}), (7) substrate (external) refractive index (n_2), (8) polar angle of incidence (Θ_i), (9) azimuthal angle of incidence (Φ_i), (10) vacuum wavelength of the incident wave (λ), (11) incident polarization amplitude ratio angle (α_i), and (12) incident polarization phase difference angle δ_i.

The plane of incidence is defined by the incident propagation vector

$$\boldsymbol{k}_i = k_g n_1 \hat{k}_i = 2\pi n_1 \hat{k}_i/\lambda \tag{12.1}$$

and the grating normal

$$\hat{n} = -\hat{z}.$$

The incident and diffracted field components are related through the K-vector closure condition

$$\boldsymbol{k}_n = \boldsymbol{k}_i - n\boldsymbol{K} \tag{12.2}$$

where $\boldsymbol{k}_n$ is the wave vector of the nth diffracted order and $\boldsymbol{K}$ is the grating vector with magnitude equal to $2\pi/\Lambda$, and a direction perpendicular to the grating planes (the x-direction in Fig. 12.2).

The s polarization component of the incident plane wave is perpendicular to the plane of incidence and has a unit vector

$$\hat{s} = \frac{\boldsymbol{k}_i \times \hat{n}}{|\boldsymbol{k}_i \times \hat{n}|}. \tag{12.3}$$

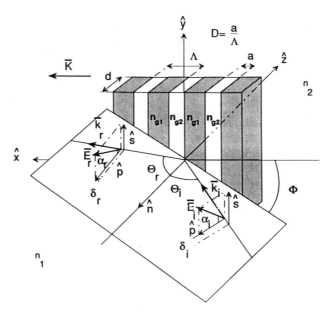

Figure 12.2 Grating and incidence geometry.

The p polarized component of the incident plane wave lies in the plane of incidence and has unit vector

$$\hat{p} = \frac{\hat{s} \times k_i}{|\hat{s} \times k_i|}. \tag{12.4}$$

The electric field vector of the incident plane wave has the form

$$E_i = sA_s \exp[i(k_i \cdot r - \omega t - \phi_s)] + pA_p \exp[i(k_i \cdot r - \omega t - \phi_p)]. \tag{12.5}$$

In this equation $A_s(A_p)$ and $\phi_s(\phi_p)$ are respectively the magnitude and phase of the complex amplitudes

$$\begin{aligned} \bar{A}_s &= A_s \exp(-i\phi_s) \\ \bar{A}_p &= A_p \exp(-i\phi_p) \end{aligned}. \tag{12.6}$$

The state of polarization of the incident plane wave and the diffracted orders is described using the polarization angles α and δ where

$$\alpha = \tan^{-1}\left(\frac{A_s}{A_p}\right) \tag{12.7}$$

with $0° \leq \alpha \leq 90°$, and

$$\delta = \phi_s - \phi_p \tag{12.8}$$

with $-180° \leq \delta \leq 180°$. We will also use the convention that α_i and δ_i refer to the incident beam, $\alpha_{1,n}$ and $\delta_{1,n}$ refer to the nth reflected diffracted order, and $\alpha_{2,n}$ and $\delta_{2,n}$ refer to the nth transmitted order.

In the following discussion of zero-order surface-relief gratings and the resulting form birefringence we will be concerned with constraints on the grating parameters to realize this condition. This is accomplished using the grating equation and the existing material parameters to determine the grating period (in reflection)

$$\Lambda = \frac{m\lambda/n_1}{\sin\Theta_i\cos\Phi_i - \sin\Theta_m\cos\Phi_m} \tag{12.9}$$

where m corresponds to the diffraction order and i to the incident beam angle in medium n_1. The zero-order condition in reflection and transmission occurs when $|\sin\Theta_m| \geq 1$, therefore

$$\Lambda_{zo} < \min\left[\frac{\lambda}{\cos\Phi_i}\left(\frac{\lambda}{n_2\ n_1\sin\Theta_i}\right), \frac{1}{\cos\Phi_i}\left(\frac{\lambda}{n_1 + n_1\sin\Theta_i}\right)\right] \tag{12.10}$$

In the sections to follow we will consider the effects of varying d, D, and to a lesser extent n_2 and n_{2g}, on the field properties η, α and δ for either transmitted or reflected beams. For these evaluations we will use both coupled wave methods (CWM) (Moharam and Gaylord, 1982, 1983) and modal methods (MM) (Li, 1993).

12.3 Polarization Properties of Zero-order Lamellar Surface-relief Gratings and Applications

12.3.1 *Polarization Properties*

In this section we compare the diffraction properties of three different types of zero-order surface-relief gratings using rigorous vector diffraction analysis (Moharam and Gaylord, 1983). The three gratings shown in Figs 12.3(a)–(c) are: (a) a dielectric grating with $n_1 = n_{g1} < n_2 = n_{g2}$; (b) a dielectric–metallic grating; and (c) a dielectric grating with $n_1 = n_{g1} > n_2 = n_{g2}$, and $\theta > \theta_c$ where θ_c is the critical angle of the interface. Each grating satisfies the zero-order diffraction condition described by equation (12.10). For the first type of grating we consider the phase δ, efficiency η, and polarization angle α of the transmitted zero-order beam ($\delta_{2,0}$, $\eta_{2,0}$, $\alpha_{2,0}$), and for the second two types we consider these parameters for the zero-order beam reflected into the incident medium ($\delta_{1,0}$, $\eta_{1,0}$, $\alpha_{1,0}$). For this comparison we fix the following parameters: $\lambda = 0.78\ \mu m$, $\Lambda = 0.3\ \mu m$, $\alpha_i = 45°$, $\delta_i = 0°$, and $\Phi_i = 0°$ for all three grating types. For the dielectric grating with $n_1 < n_2$, we assume that $n_1 = 1.0$, $n_2 = 1.5$ and $\theta_i = 0°$. For the dielectric–metallic grating we assume that $\theta_i = 45°$ and $n_1 = 1.5$, $n_2 = 0.175 + i4.91$ which are typical values for a glass–gold interface. For the dielectric grating with $n_1 > n_2$ it is assumed that $n_1 = 1.51$ and $n_2 = 1.0$ and $\theta_i = 45°$. In order to provide some intuition for the trends in the polarization properties of high-frequency gratings we will concentrate on interpreting graphical data computed from rigorous diffraction theory. Elements of this theory can be found in Chapter 2 and in the cited references (Botten *et al.*, 1981a and 1981b; Moharam and Gaylord, 1982, 1983). In section 12.3.2 we will consider the design tolerances for different types of polarization elements using these parameters.

Grating Type 1

Dielectric zero-order gratings with $n_1 < n_2$ (Type 1) can be formed in several ways, including holographic exposure of photoresist materials deposited on glass

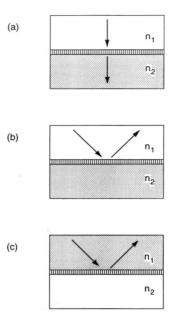

Figure 12.3 (a) Dielectric–dielectric grating ($n_1 < n_2$). (b) Dielectric–metallic grating. (c) Dielectric–dielectric grating ($n_1 > n_2$ and $\theta > \theta_c$).

substrates or by using lithographic patterning of photoresist followed by etching a dielectric substrate. Graphical data of $\delta_{2,0}$, $\eta_{2,0}$ and $\alpha_{2,0}$ as a function of grating depth (d) with a duty cycle of 0.5 for the Type 1 dielectric grating are shown in Figs 12.4–12.6. Similar plots for these beam parameters as a function of duty cycle (D) with a fixed depth of 0.5 μm are shown in Figs 12.7–12.9. Examination of these figures illustrates several characteristics. First the plot of efficiency vs. depth (Fig. 12.5) shows that the transmittance can be increased to 0.9975 (at $d = 0.175\ \mu$m) which is significantly greater than the value for an unmodulated surface (0.96). A grating of this type can be used as an antireflection coating (Gaylord *et al.*, 1987), and can be very useful for optical surfaces which are difficult to match with conventional coatings. Another characteristic is that the phase difference (Fig. 12.4) is approximately linear with grating depth, which is analogous to a uniaxial anisotropic film (Cescato *et al.*, 1990). The results also show that the polarization ratio angle $\alpha_{2,0}$ changes in an oscillatory manner with depth (Fig. 12.6). This results from the different rate of change in the efficiency for the two orthogonal polarizations due to the periodicity of the field structure in the grating region as a function of grating depth (Popov *et al.*, 1990). Varying the duty cycle (D) introduces some asymmetry to the phase with respect to $D = 0.5$ (equal areas of glass and air). For this grating the minimum $\delta_{2,0}$ occurs when $D = 0.45$ (Fig. 12.7). $\eta_{2,0}$ also shows an asymmetric behaviour with a maximum value occurring at $D = 0.4$. The plot of $\alpha_{2,0}$ versus D shows that s polarized light is preferentially transmitted for $D < 0.4$ since $\alpha_{2,0} > 45°$, and p light transmission is greater for $D > 0.4$. These preferential transmittance characteristics are an indication of resonance behaviour in the grating region.

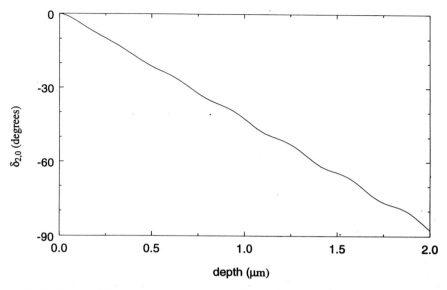

Figure 12.4 Phase difference versus grating depth for a grating at a glass–air surface with $\lambda = 0.78 \,\mu m$, $\Lambda = 0.3 \,\mu m$, $\alpha_i = 45°$, $\delta_i = 0°$, $\Phi_i = 0°$, $n_{g1} = n_1 = 1.0$, $n_{g2} = n_2 = 1.5$, $\theta_i = 0°$, $D = 0.5$.

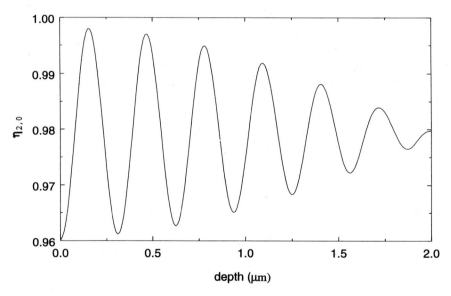

Figure 12.5 Same as Fig. 12.4 except for zeroth transmitted order diffraction efficiency.

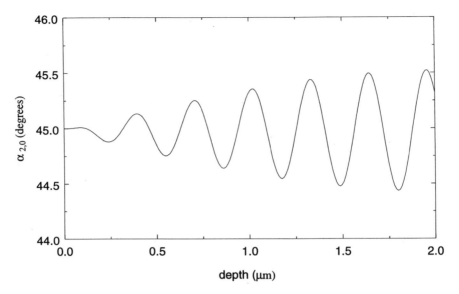

Figure 12.6 Same as Fig. 12.5 except for zeroth transmitted order amplitude ratio angle.

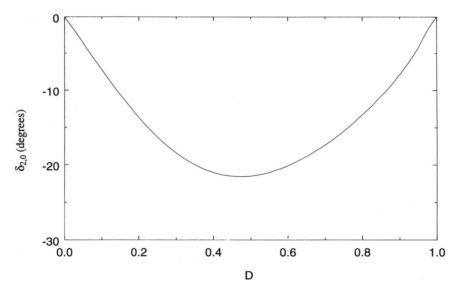

Figure 12.7 Phase difference of the zeroth transmitted order as a function of duty cycle for $d = 0.5\ \mu m$ and $N = 45$ (CWM). All other parameters are as in Fig. 12.4.

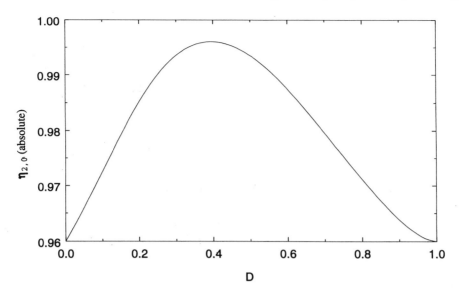

Figure 12.8 Same as Fig. 12.7 for zeroth transmitted order diffraction efficiency.

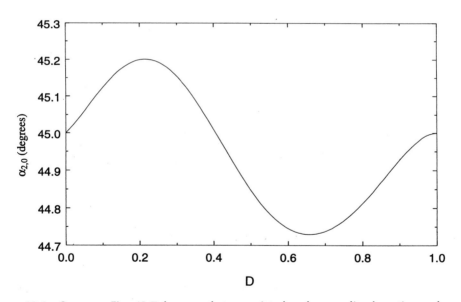

Figure 12.9 Same as Fig. 12.7 for zeroth transmitted order amplitude ratio angle.

It should be noted that for a Type 1 grating, δ, η and α can be modified to some extent by changing the refractive index in the grating region (n_{g2}) from that of the substrate (n_2) as would exist for a photoresist grating formed on a glass substrate. However, the maximum value of $\delta_{2,0}$ is limited by the practical fabrication limits of grating depth (gratings with a depth several times the length of the period are difficult to fabricate). For this example the maximum phase difference that can practically be obtained is $<90°$.

Grating Type 2

Metallic zero-order gratings have been studied for use as totally absorbing components (Gaylord *et al.*, 1987), as retardation elements (Kok and Gallagher, 1988; Naqvi and Gallagher, 1990; Haggans *et al.*, 1993b), and as polarization conversion elements (Haggans *et al.*, 1993b). Calculated plots of $\delta_{1,0}$, $\eta_{1,0}$ and $\alpha_{1,0}$ as a function of grating depth are illustrated in Figs 12.10–12.12 with $\Phi_i = 0°$ and the parameters section 12.3.1 for a metallic grating. Similar plots with $\Phi_i = 90°$ are shown in Figs 12.13–12.15, and a plot of $\delta_{1,0}$ as a function of duty cycle is shown in Fig. 12.16. Figure 12.10 shows that when $\Phi_i = 0$, $\delta_{1,0}$ is weakly dependent on d. Figures 12.10–12.12 show a resonance at a grating depth of $0.22\,\mu$m, and suggests a surface plasmon type effect. For plasmon excitation, the polarization component with a non-zero projection along the grating vector is absorbed (Inagaki *et al.*, 1983). Hence we would expect the p polarization component to be absorbed if this resonance is due to a surface plasmon. Figure 12.12 confirms that the p component is absorbed at resonance because the reflected beam is strongly s polarized ($\alpha_{1,0} \gg 45°$). Figure 12.13 shows that when $\Phi_i = 90°$, $\delta_{1,0}$ is strongly dependent on d. The efficiency η_i (Fig. 12.14) shows strong absorption resonances near $d = 0.08\,\mu$m and $0.35\,\mu$m. Resonances are also visible in Fig. 12.15 for $\alpha_{1,0}$ and indicate that the s polarization component is preferentially absorbed in agreement with the idea that the projection along the grating vector is being absorbed. In Fig. 12.16 the phase is plotted as a function of duty cycle with $d = 0.21\,\mu$m, and varies by $360°$ when the duty cycle changes from $D = 0$ to $D = 1$.

Up to this point, we have only considered situations with the reconstruction beam incident with $\Phi_i = 0$ or $90°$. If Φ_i is allowed to vary between $0°$ and $90°$, additional control of the diffraction properties of zero-order gratings can be achieved. In particular, we will discuss the mechanisms responsible for the modification of the amplitude ratio angle $\alpha_{1,0}$, i.e. selective absorption due to the excitation of surface plasmons (Inagaki *et al.*, 1983, 1986; Gupta, 1987; Bryan-Brown *et al.*, 1990), and polarization conversion with minimal absorption (Haggans *et al.*, 1993b) due to the amisotropic nature of gratings.

Figure 12.17 shows a contour plot of the polarization component amplitude ratio angle $\alpha_{1,0}$ versus Φ and d for the glass–gold grating of Fig. 12.10 with $D = 0.5$. The plane wave incident on the grating is linearly polarized ($\alpha_i = 45°$, $\delta_i = 0°$). In this figure, $\alpha_{1,0}$ differs significantly from α_i in two regions: a relatively symmetrical minimum of $\alpha_{1,0}$ centred near $\Phi = 72°$ and $d = 0.1\,\mu$m, and an asymmetrical maximum of $\alpha_{1,0}$ centred near $\Phi = 27°$ and $d = 0.05\,\mu$m. In the following paragraphs, we will demonstrate that the minimum of $\alpha_{1,0}$ is a result of polarization conversion ($\alpha_{1,0} \neq \alpha_i$ with minimal absorption). We show that the second of these regions ($\alpha_{1,0} \to 90°$) is due to two phenomena: a combination of polarization conversion (with a maximum centred near $\Phi = 30°$ and $d = 0.05\,\mu$m) and preferential absorption due to the excitation of a surface plasmon (centred at $\Phi = 18°$ and $d = 0.02\,\mu$m). First, we consider the behaviour of $\alpha_{1,0}$ and $\eta_{1,0}$ for a grating geometry where polarization conversion does not occur ($\Phi_i = 90°$). Figure 12.15 is a plot of $\alpha_{1,0}$ versus d for the grating of Fig. 12.10 (with $\alpha_i = 45°$, $\delta_i = 0°$). Figure 12.14 is a plot of zeroth-order efficiency for the same grating. These figures show that the s polarization component is selectively absorbed at the dips in the $\alpha_{1,0}$ curve of Fig. 12.15. Thus, rather than polarization conversion, selective

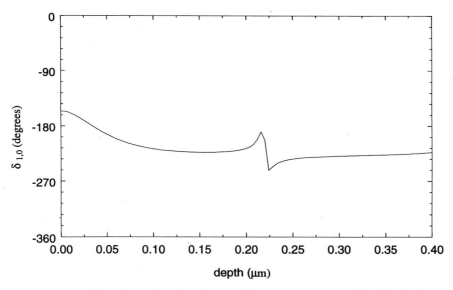

Figure 12.10 Zeroth reflected order phase difference versus grating depth for a glass–gold grating with $\lambda = 0.78\ \mu m$, $\Lambda = 0.3\ \mu m$, $\alpha_i = 45°$, $\delta_i = 0°$, $\Phi_i = 0°$, $n_{g1} = n_1 = 1.51$, $n_{g2} = n_2 = 0.175 + i4.91$, $\theta_i = 45°$, $D = 0.5$.

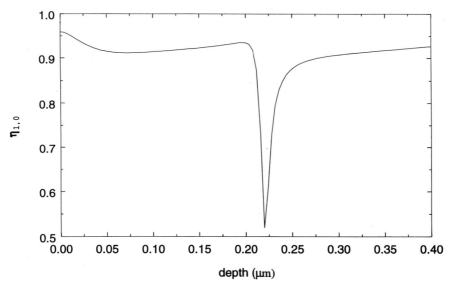

Figure 12.11 Same as Fig. 12.10 for zeroth reflected order diffraction efficiency.

absorption is the origin of the change of $\alpha_{1,0}$ with respect to α_i. We now investigate the change of $\alpha_{1,0}$ which occurs for non-zero azimuthal angles of incidence. As discussed above, Fig. 12.17 is a contour plot of $\alpha_{1,0}$ versus d and Φ (determined from modal analysis) which exhibits one maximum and one minimum of $\alpha_{1,0}$. Figure 12.18 is the corresponding contour plot for total reflectivity ($\eta_{1,0}$). Note

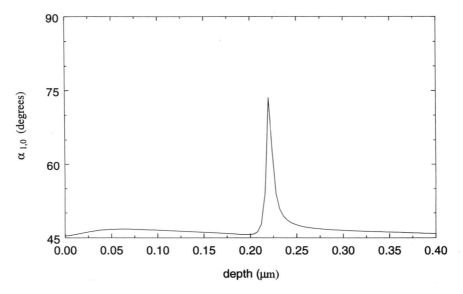

Figure 12.12 Same as Fig. 12.10 for the *s–p* amplitude ratio angle of the zeroth reflected order.

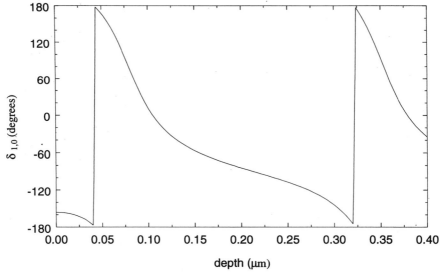

Figure 12.13 Phase difference versus grating depth for the zeroth order reflected from the grating of Fig. 12.10 with $\Phi_i = 90°$. Data generated using MM.

that the first *s* polarization absorption dip shown in Fig. 12.14 is visible across the top of this plot. Also, note that minimal absorption occurs at the point of minimum $\alpha_{1,0}$ in Fig. 12.17 ($\Phi_i = 72°$ and $d = 0.1\ \mu$m). This indicates that $s \rightarrow p$ polarization conversion (as opposed to selective absorption) occurs at this point.

The second extremum in Fig. 12.17 is more complex due to the absorption maximum (at $\Phi_i = 18°$ and $d = 0.02\ \mu$m in Fig. 12.18) which occurs near the point of maximum $\alpha_{1,0}$. The maximum value of $\alpha_{1,0}$ ($\approx 90°$) occurs near $\Phi_i = 27°$ and

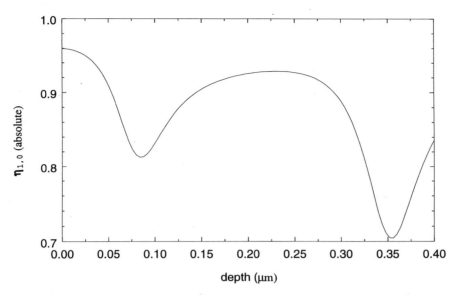

Figure 12.14 Same as Fig. 12.13 for the efficiency of the zeroth reflected order.

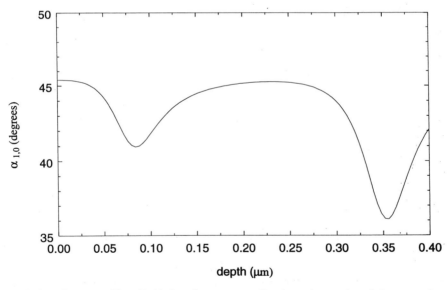

Figure 12.15 Same as Fig. 12.13 for the *s–p* amplitude ratio angle of the zeroth reflected order.

$d = 0.05\ \mu$m (see Fig. 12.17). In order for this maximum to be caused by selective absorption, the reflectivity ($\eta_{1,0}$) at this point would have to drop significantly. Figure 12.18 shows no dip in reflectivity at these values of d and Φ_i. Thus, this point of maximum $\alpha_{1,0}$ is due to $s \to p$ polarization conversion, not surface-wave excitation.

We now analyze the asymmetry of this peak in $\alpha_{1,0}$. Figure 12.19 is a plot of

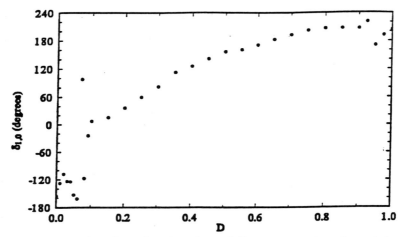

Figure 12.16 Zeroth reflected order phase difference as a function of duty cycle for the grating of Fig. 12.10. Data generated with CWM for $N = 79$.

$\alpha_{1,0}$ as a function of d at $\Phi_i = 18°$ (note that the scale for d in this plot has much higher resolution than that used in Fig. 12.17 which allows the peak for $\alpha_{1,0}$ to be resolved more accurately). The point of maximum $\alpha_{1,0}$ occurs at $d = 0.02\,\mu\text{m}$. Figure 12.20 is a similar plot for total reflectivity. Note that the total reflectivity drops significantly at $d = 0.02$. The excitation of a surface plasmon at this point can be anticipated due to phase matching relations. As shown in Yariv and Yeh (1984), the magnitude of the propagation vector for a surface plasmon (sp) at the interface between a dielectric (n_1) and a metal $(n_2 = n - i\kappa)$ (exp$(i\omega t)$ time convention) is

$$|k_{\text{sp}}| = \left[\frac{n_1^2(n^2 - \kappa^2)}{n_1^2 + (n^2 - \kappa^2)} \right]^{1/2} k_0. \tag{12.11}$$

For a wavelength of $0.780\,\mu\text{m}$ and the refractive indices of the glass–gold grating under consideration, this relation gives $|k_{\text{sp}}| = 1.595k_0 = 1.0514k_1$. The condition for grating-assisted plasmon excitation by the free-space beam is

$$k_{\text{sp}} = (k_i \sin \Theta_i \cos \Phi_i - mK)\,\hat{x} + k_i \sin \Theta_i \sin \Phi_i\,\hat{y} \tag{12.12}$$

(with all parameters as defined in section 12.2). The square modulus of this relation is

$$|k_{\text{sp}}|^2 = [(k_1 \sin \Theta_i \cos \Phi_i - mK)^2 + k_1^2 \sin^2 \Theta_i \sin^2 \Phi_i]^{1/2}. \tag{12.13}$$

Substituting (12.11), $m = 1$ (first-order coupling), and the grating parameters into (12.13), gives $\Phi_i = 15.05°$ as the azimuthal angle of incidence for maximum plasmon excitation. This agrees with the Φ_i value for the absorption maximum in Fig. 12.18, and suggests that the origin of this absorption peak is the excitation of a surface plasmon. (This analysis is approximate because it assumes that the propagation constant of the plasmon for an unmodulated surface remains the same for non-zero grating depths. Although this is not rigorously true, this assumption

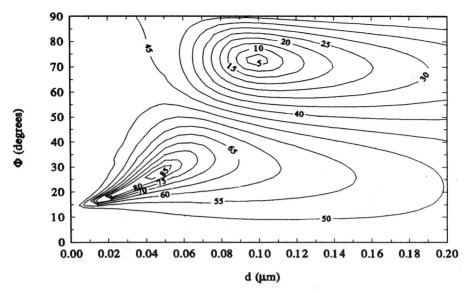

Figure 12.17 Contour plot of $\alpha_{1,0}$ as a function of Φ and d for the grating of Fig. 12.10. Data generated using MM.

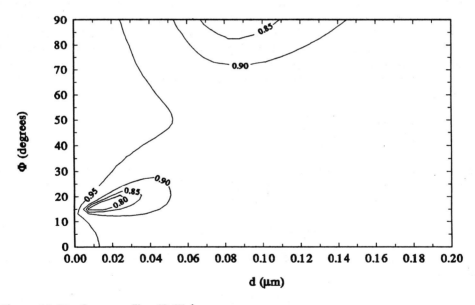

Figure 12.18 Same as Fig. 12.17 for $\eta_{1,0}$.

has been shown to be valid for small grating depths (Inagaki *et al.*, 1983). Neviere (1980) has rigorously verified that absorptions of this type are due to surface plasmon excitation.

Figure 12.21 is a contour plot of $\delta_{1,0}$ for the grating of Fig. 12.17. There are two regions in this figure where the phase difference varies rapidly. These regions are at the points of maximum polarization conversion in Fig. 12.17. When passing

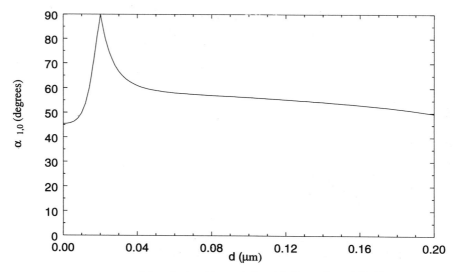

Figure 12.19 $\alpha_{1,0}$ versus d for the grating of Fig. 12.10 at $\Phi_1 = 18°$. Data generated using MM.

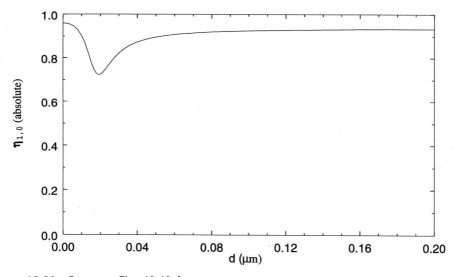

Figure 12.20 Same as Fig. 12.19 for $\eta_{1,0}$.

through the upper phase singularity ($\Phi_i = 72°$ and $d = 0.1 \, \mu$m) as a function of d or Φ_i, a phase shift of 180° occurs. Similar behaviour has been experimentally observed and theoretically analyzed by Simon and Simon (1984), Depine *et al.* (1987), and Andrewartha *et al.* (1979a, 1979b). At the point of minimum $\alpha_{1,0}$ in Fig. 12.17, the *s* polarization component vanishes. Intuitively, a singularity in $\delta_{1,0}$ at $\alpha_{1,0} = 0°$ or 90° is to be expected, because $\delta_{1,0}$ is not unique for purely *p* or *s* polarized light.

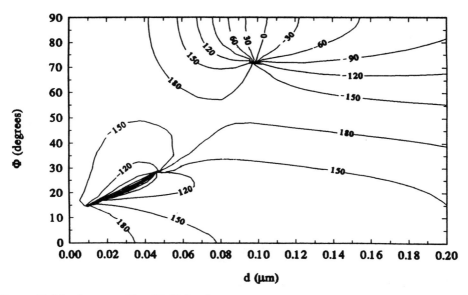

Figure 12.21 Same as Fig. 12.17 for $\delta_{1,0}$.

The behaviour is more complicated at the minimum of $\alpha_{1,0}$ ($\Phi_i = 27°$ and $d = 0.05\ \mu$m) in Fig. 12.21 due to the existence of the surface plasmon resonance centred at $\Phi_i = 18°$ and $d = 0.02\ \mu$m (see Fig. 12.18) and the polarization conversion maximum at $\Phi_i = 27°$ and $d = 0.0475\ \mu$m. At both of these points, a 180° phase shift is observed.

In summary, zeroth-order metallic gratings (grating Type 2) have been shown to exhibit strong polarization properties. Specifically, large amounts of retardation and polarization conversion occur for very shallow metallic gratings. Absorption resonances are prominent in the efficiency and amplitude ratio angle curves for these gratings. In addition two of these resonances coincide with the excitation of surface plasmons.

Grating Type 3

The last type of zero-order grating considered has dielectric incident and substrate media with $n_1 > n_2$, and Θ_i greater than the critical angle of medium n_1 (Fig. 12.3(c)). Total internal reflection occurs at the grating interface and is specularly reflected with 100% efficiency. Recently, gratings of this type were proposed for use as retarders and polarization rotators operating on the specularly reflected order (Haggans *et al.*, 1993b).

To demonstrate the properties of this type of zero-order grating, coupled wave method calculations were made using the parameters described earlier for a grating at a glass–air interface. For this grating geometry, $\eta_{1,0} = 1.0$ because of the total internal reflection condition. Additionally, $\alpha_{1,0}$ is 45° when $\Phi = 0°$ or 90° because the reflection coefficients for s and p polarization are both 1.0.

Figure 12.22 is a plot of $\delta_{1,0}$ as a function of d. Note that the phase difference

341

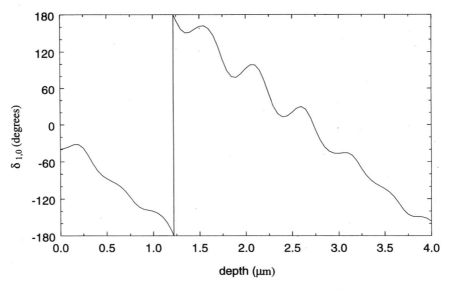

Figure 12.22 Plot of $\delta_{1,0}$ versus d for a grating at a glass–air interface for the condition $n_1 > n_2$ and $\Theta_i > \Theta_c$ with $\lambda = 0.78 \, \mu m$, $\Lambda = 0.3 \, \mu m$, $\alpha_i = 45°$, $\delta_i = 0°$, $\Phi_i = 0°$, $n_{g1} = n_1 = 1.51$, $n_{g2} = n_2 = 1.0$, $\theta_i = 45°$, $D = 0.5$.

introduced to the zeroth reflected order as a function of grating depth for this geometry greatly exceeds that introduced to the transmitted order by the grating of Fig. 12.4 (where n_1 was less than n_2). Figure 12.23 is a similar plot for $\Phi = 90°$. Note that the rate of change of phase difference with depth exceeds that of Fig. 12.22. However, the dependence of $\delta_{1,0}$ on d for both of these figures is very weak compared to the metallic grating of Fig. 12.13. Figure 12.24 is a plot of $\delta_{1,0}$ versus D for the grating of Fig. 12.22 with $d = 0.5 \, \mu m$. In this case, the maximum value of $|\delta_{1,0}|$ occurs near $D = 0.5$. In this figure, no absolute phase difference occurs between $D = 0$ and $D = 1$. This differs from the metallic grating case in which a 360° phase change occurs. The peak in $\delta_{1,0}$ occurring at $D = 0.82$ is due to the resonance nature of this grating arrangement, and thus cannot be explained from intuitive arguments.

Figure 12.25 is a contour plot of the polarization angle $\alpha_{1,0}$ versus d and Φ_i for the grating of Fig. 12.22 with $D = 0.7$. Linear polarization is incident on the grating ($\alpha_i = 45°$ and $\delta_i = 0°$). In this conical mount, at a depth of approximately $1.72 \, \mu m$ and $\Phi_i = 57°$, the reflected wave is almost completely p polarized ($\alpha_{1,0} \approx 0°$). Since the reflectivity is 1.0 for this grating (there are no propagating transmitted orders), this means that polarization conversion occurs upon reflection from the grating. This conversion is analogous to that observed for reflection from a uniaxial film (see Section 12.3.3)..

Figure 12.26 is a similar plot for the s–p phase difference $\delta_{1,0}$. In this plot, a sudden phase change of approximately 180° occurs near the depth and azimuthal angle of incidence corresponding to maximum $s \rightarrow p$ polarization conversion. This behaviour is typical of that observed upon reflection from any medium for which one of the two orthogonal polarization components vanishes. A detailed discussion

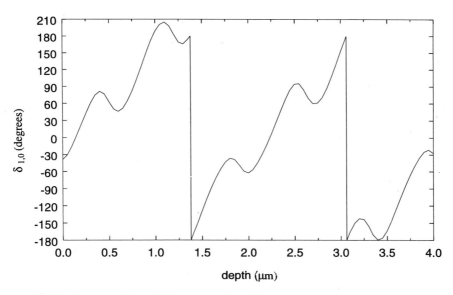

Figure 12.23 Same as Fig. 12.22 except $\Phi_i = 90°$.

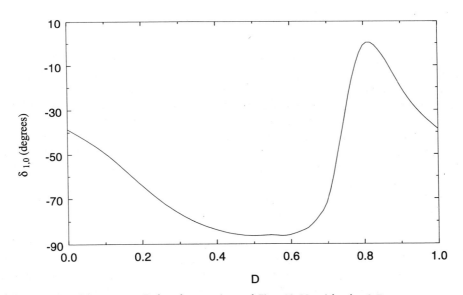

Figure 12.24 $\delta_{1,0}$ versus D for the grating of Fig. 12.22 with $d = 0.5\,\mu$m.

of the origin of this phase behaviour can be found in papers by Andrewartha *et al.* (1979a, 1979b) and Depine *et al.* (1987).

In summary for dielectric zero-order gratings, retardation and polarization conversion occurs upon reflection. However, very deep gratings are required to equal the performance of much shallower metallic gratings. The absence of absorption resonances and the total internal reflection condition for this grating type yields 100% efficiency in reflection.

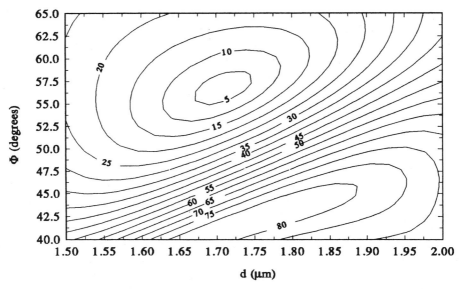

Figure 12.25 Contour plot of $\alpha_{1,0}$ versus d for the grating of Fig. 12.22 with $D = 0.7$.

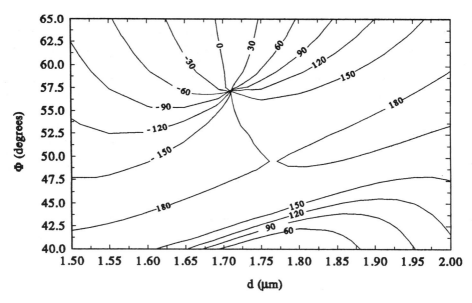

Figure 12.26 Same as Fig. 12.25 except for $\delta_{1,0}$.

12.3.2 *Application of Lamellar Zero-order Gratings as Polarization Components*

In this section we will use rigorous diffraction analysis to evaluate lamellar glass–air and dielectric–metal gratings for use as polarization components. The geometrical parameters of these gratings can be varied to control the s–p phase difference ($\delta_{1,0}$)

and s–p amplitude ratio ($\alpha_{1,0}$) of specularly reflected beams. Designs for optical wavelength phase compensators, quarter-wave and half-wave retarders, and polarization conversion elements are presented. These components can be integrated into polarization-sensitive planar optic devices (Jahns and Huang, 1989; Jahns and Brumback, 1990; Kostuk *et al.*, 1990). They have also been used for antireflection surfaces (Raguin and Morris, 1993) and for amplitude polarization components (Azzam *et al.*, 1991).

As discussed earlier, when light is reflected from a high spatial-frequency lamellar grating at a glass–air ($\Theta > \Theta_c$) or dielectric–metal interface, a phase shift between the orthogonal polarization components is introduced. No polarization conversion between the components occurs when the grating vector is contained in or is perpendicular to the plane of incidence. By utilizing this grating depth dependent phase shift (with no associated polarization conversion), gratings can function as phase compensation and retardation elements operating on a specularly reflected (zero-order) beam. In addition, if light is conically incident on a grating of this type (grating vector not in the plane of incidence), polarization conversion between the orthogonal polarization components occurs upon reflection (the ratio of the orthogonal polarization component amplitudes is modified with negligible loss of efficiency). For certain geometries, polarization conversion occurs without the introduction of a phase difference. For these cases, the gratings function as linear polarization rotators. For other geometries, if a phase compensation element is cascaded with the polarization conversion element, a linear polarization rotator can be obtained.

The performance of these gratings is limited by three factors. First, variation of geometrical parameters (period, depth, duty cycle and grating profile) from design values typically results from fabrication inaccuracies. Next, if a laser diode is used as the illuminating source, the wavelength will vary due to manufacturing tolerances and thermal effects. Finally, the angle of incidence on the grating will vary due to alignment tolerances. The effects of these three degrading factors on grating performance will also be considered in this section. Specific examples of a quarter-wave retardation grating at a glass–air interface and a polarization conversion grating at a glass–metal interface are used to determine which geometrical and operational parameters have the greatest effect on grating performance.

The diffraction geometry used in this section is the same as Fig. 12.2. The polar angle of incidence is large enough so that the zeroth transmitted order is evanescent (dielectric gratings with $n_1 > n_2$ are considered, as discussed in the previous section).

Design Considerations

Both dielectric and metallic gratings can function as phase compensators, quarter-wave and half-wave retarders, and polarization conversion gratings. Absorption upon reflection by metallic gratings makes dielectric gratings more desirable for retardation elements. Metallic gratings are more attractive for polarization conversion gratings because of the large depth required for dielectric conversion gratings. Therefore, dielectric designs for phase compensation and retardation gratings, and dielectric–metallic gratings for polarization conversion elements are considered.

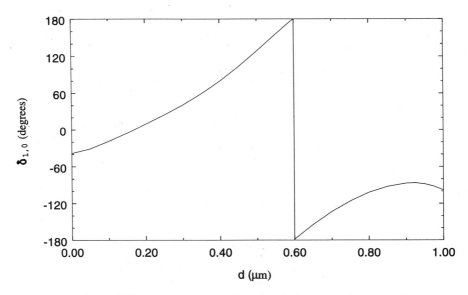

Figure 12.27 Phase difference versus grating depth for a grating at a glass–air interface. Data generated using CWM with $N = 57$.

A 45° polar angle of incidence is used for the following designs, resulting in total internal reflection for the dielectric gratings. To ensure that only the zeroth order propagates, a grating period of $0.3\,\mu m$ is used with $\lambda = 0.780\,\mu m$. The grating depth, duty cycle and azimuthal angle of incidence are varied to obtain the desired polarization properties. For the dielectric grating designs, the input and external (substrate) media are glass and air, respectively ($n_1 = 1.51$, $n_2 = 1.0$). Gold is used as the external medium for the metallic polarization conversion grating design ($n_2 = 0.175 + i4.91$ at $\lambda = 0.780\,\mu m$ (Palik, 1985)). The incident medium for the metallic grating is also glass ($n_1 = 1.51$), although retardation and polarization conversion can be demonstrated for incidence from any medium (e.g. air).

Phase Compensators

Figure 12.27 is a plot of $\delta_{1,0}$ as a function of grating depth for a grating at a glass–air interface. This is the same grating shown previously in Fig. 12.22, except that $D = 0.7$ ($\Theta = 45°$, $\Phi = 90°$, $\Lambda = 0.3\,\mu m$, $D = 0.7$, $n_1 = 1.51$, $n_2 = 1.0$, $\lambda = 0.780\,\mu m$, $\alpha_i = 45°$ and $\delta_i = 0°$). Note that output phase differences from $-180°$ to $-85°$ and $-40°$ to $180°$ can be obtained for grating depths between 0 and $1.0\,\mu m$. For this special conical mount ($\Phi = 90°$), no polarization conversion occurs. Thus, $\alpha_{1,0} = 45°$ and the zeroth reflected order diffraction efficiency is 1.0 because there is no transmission or absorption. All remaining phase difference values ($-85°$ to $-40°$) can be obtained with a slightly different dielectric grating design with $\Phi = 0°$ and $D = 0.5$ (Fig. 12.22). Since any phase difference between $\pm 180°$ can be imparted onto an incident beam after reflection from a dielectric grating, an elliptically polarized incident beam can be converted to a linearly polarized beam after reflection from a grating with the appropriate depth. This makes these gratings useful as phase compensators in polarization-sensitive substrate-mode devices (Haggans *et al.*, 1993b).

Total internal reflection at a glass–air interface introduces a phase difference between the orthogonal polarization components of the incident beam. Reflection with 100% efficiency, without introducing a phase difference, is also possible when a grating is present at the interface. An example of a grating that preserves linear polarization upon reflection ($\alpha_{1,0} = \alpha_i$ and $\delta_{1,0} = \delta_i = 0°$) is shown in Fig. 12.27 at a depth of approximately $0.18\ \mu m$. In addition to preserving the incident state of polarization, this grating has an efficiency of 1.0 because there is no absorption or transmission at the glass–air interface. Gratings with $\delta_{1,0} = \delta_i$ are useful in polarization-sensitive substrate-mode devices where the signal must propagate over large distances, requiring many internal reflections.

Quarter-wave and Half-wave Retarders

A lamellar dielectric grating at a glass–air interface operates as a quarter-wave retarder when $\delta_{1,0} = \pm 90°$. From Fig. 12.27, $\delta_{1,0} = -90°$ at a grating depth of approximately $0.85\ \mu m$. Reflection from a grating with this depth converts a linearly polarized incident beam ($\alpha_i = 45°$, $\delta_i = 0°$) into a left circularly polarized beam. Similarly, $\delta_{1,0} = 90°$ at a depth of approximately $0.42\ \mu m$. Reflection from this grating converts a linearly polarized beam ($\alpha_i = 45°$, $\delta_i = 0°$) into a right circularly polarized beam.

A lamellar dielectric grating at a glass–air interface operates as a half-wave retarder when $\delta_{1,0} = \pm 180°$. This phase difference is obtained at a depth of approximately $0.6\ \mu m$ for the grating of Fig. 12.27. Reflection from this grating rotates the field vector from $\alpha_{1,0}$ above the p axis to $\alpha_{1,0}$ below the p axis.

Polarization Conversion Gratings

Reflection of conically incident light ($\alpha_{1,0} \neq \alpha_i$) from a lamellar grating at a glass–gold interface introduces polarization conversion. Figure 12.17 is a contour plot of $\alpha_{1,0}$ versus d and Φ for a linearly polarized incident beam ($\alpha_i = 45°$, $\delta_i = 0°$) reflected from a grating at a glass–gold interface. Note that values of $\alpha_{1,0}$ from <5° to >85° can be obtained for $d < 0.2\ \mu m$. Near $\Phi = 75°$ and $d = 0.1\ \mu m$, the reflected beam is almost purely p polarized ($\alpha_{1,0} \approx 0°$). Similarly, near $\Phi = 27°$ and $d = 0.05\ \mu m$, the reflected beam is nearly s polarized ($\alpha_{1,0} \approx 90°$). By varying the azimuthal angle of incidence and the depth, gratings which give intermediate values of $\alpha_{1,0}$ can be selected. Figure 12.18 is a similar plot for total diffraction efficiency. This figure shows that high efficiency regions exist for the selection of most rotation angles. Figure 12.21 is a contour plot of $\delta_{1,0}$ versus d and Φ for the same grating geometry. Points lying on the 0° or 180° phase difference contours indicate designs for linear polarization rotators. Contours with phase differences other than 0° or 180° require that a phase compensator be cascaded with the conversion grating to obtain a linearly polarized output beam ($\delta_{1,0,\text{total}} = 0°$ or $\pm 180°$). When $\delta_{1,0,\text{total}} = 0°$, orientations of the field vector from 0° to 90° above the p axis (see Fig. 12.2) can be obtained using these cascaded elements. Similarly, when $\delta_{1,0,\text{total}} = \pm 180°$, the field vector of the reflected beam can be changed from 0° to 90° below the p axis.

There are two regions in Fig. 12.19 where the phase difference varies rapidly. These regions are near the points of maximum polarization conversion. Regions of rapid phase variation must be avoided in the design of polarization conversion elements because it is difficult to control the phase in these regions during fabrication. The exception to this occurs when $\alpha_{1,0} \approx 0°$ or $90°$. At these points, the beam is almost linearly polarized. Thus, the rapid variation of $\delta_{1,0}$ at these points is not as significant.

Parameter Tolerances for Dielectric Quarter-wave Retardation Gratings

In this section we consider the effects of varying the grating parameters of a dielectric lamellar grating quarter-wave plate with nominal values: $\Theta_i = 45°$, $d = 0.55 \, \mu m$, $\Phi_i = 0°$, $\Lambda = 0.3 \, \mu m$, $D = 0.5$, $n_1 = 1.51$, $n_2 = 1.0$, $\lambda = 0.780 \, \mu m$, $\alpha_i = 45°$ and $\delta_i = 0°$. The beam reflected from this grating would nominally be linearly polarized with $\alpha_{1,0} = 45°$ and $\delta_{1,0} = -90°$. For this tolerance analysis we consider the sensitivity of the polarization state of the reflected beam to: (1) d, (2) λ, (3) D, (4) Θ_i and (5) Φ_i relative to the retardation performance of a conventional high quality quarter-wave retarder ($\Delta\delta = 1.2°$ or $\lambda/300$). Results from coupled wave calculations for these grating parameters are summarized in Table 12.1.

Because of the large depth, fabricating a lamellar profile grating to this tolerance would be exceedingly difficult. However, studies have shown that similar retardation properties can be obtained with arbitrary profile (e.g. trapezoidal) gratings (Haggans *et al.*, 1993b). Given a repeatably manufacturable profile and good photoresist thickness process control, this depth tolerance is realistic.

If a laser diode is used as the illumination source, the quarter-wave retarder must be designed to function over a wide range of wavelengths. The wavelength of a typical high-power laser diode (15 mW Sharp LT021 (Sharp Corporation, 1988)) with a specified wavelength of $0.780 \, \mu m$, can be expected to vary from $0.768 \, \mu m$ to $0.796 \, \mu m$. For this retardation grating, only a $\lambda/100$ ($\pm 3.6°$) retardation error is maintained over this wavelength range. To obtain a $\lambda/300$ ($\pm 1.2°$) retardation tolerance the wavelength must be controlled within $\pm 10 \, nm$.

The polar angle of incidence (Θ_i) of light on the polarization elements varies in practice due to alignment tolerances. Analysis of this grating indicates that a $\pm 0.4°$ range of the polar angle of incidence reduces the retardation performance

Table 12.1 Design and tolerances of dielectric quarter-wave retardation grating

Variable	Nominal value	Tolerance for $\Delta\delta_{1,0} < 1.2°$
d	$0.55 \, \mu m$	$\pm 0.02 \, \mu m$
λ	$0.780 \, \mu m$	$-0.004 \, \mu m$
		$+0.006 \, \mu m$
D	0.5	-0.1
		$+0.16$
Θ	$45°$	$\pm 0.4°$
Φ	$0°$	$\pm 4°$

Table 12.2 Design and tolerances for a polarization conversion grating

Variable	Nominal value	Tolerance for $\Delta\alpha_{1,0} = \pm 2°$ and $\Delta\delta_{1,0} = \pm 5°$
d	0.103 μm	± 0.002 μm
λ	0.780 μm	$>\pm 0.015$ μm
D	0.7	± 0.1
Θ	45°	$\pm 5°$
Φ	56.5°	$\pm 3°$

to approximately $\lambda/300$ ($\pm 1.2°$). Thus, tight alignment tolerances must be maintained for system applications of this element.

Tolerances for a Metallic Polarization Conversion Grating

A grating which converts linearly polarized light ($\alpha_i = 0°$, $\delta_i = 0°$) into circularly polarized light ($\alpha_{1,0} = 45°$, $\delta_{1,0} = -90°$) would be useful in a write-once optical storage head (Marchant, 1990). One possible design for this element has incidence and grating parameters of $\Theta_i = 45°$, $d = 0.103$ μm, $\Phi_i = 56.5°$, $\Lambda = 0.3$ μm, $D = 0.5$, $n_1 = 1.51$, $n_2 = 0.175 + 4.91$, $\lambda = 0.780$ μm, $\alpha_i = 0°$ and $\delta_i = 0°$.

As before, a tolerance analysis of this element must consider the sensitivity of the polarization state to grating depth, wavelength, duty cycle, polar angle of incidence, and azimuthal angle of incidence. The allowable range over which each of these parameters can vary while equalling or exceeding the performance of a conventional quarter-wave retarder (QWR) (with a typical retardation tolerance of $\Delta\delta = \pm\lambda/100$ or 3.6°) is shown in Table 12.2. The conventional QWR is oriented with its fast axis at 45° to the p axis. In this orientation for the lamellar zero-order grating with $\alpha_i = 45°$, $\delta_i = 0°$, $\alpha_{2,0} = 45°$ and $\delta_{2,0} = -90°$ the tolerances are feasible for λ, Θ_i and Φ_i. However, the tolerance on the duty cycle is very demanding, and the required depth control is also impractical at visible wavelengths.

This particular design would be exceedingly difficult to fabricate using conventional techniques. However, by developing optimization methods for coupled wave and modal theory, it may be possible to obtain more realistic values for the depth and duty cycle by reducing larger tolerances on other design parameters. In the next section we examine an approximate technique for analyzing the performance of lamellar gratings which can help in the design optimization.

12.3.3 Approximate Design Methods for Zero-order Lamellar Gratings

In general, exact computation of the complex reflection and transmission coefficients for zero-order gratings requires the use of rigorous electromagnetic grating theories (Petit, 1980). However, in the quasistatic limit (the limit at which the grating period to wavelength ratio approaches zero ($\Lambda/\lambda \to 0$)), the optical properties of a grating are equivalent to those of a uniaxial film with the optic axis parallel to the grating vector (Bell *et al.*, 1982; McPhedran *et al.*, 1982;

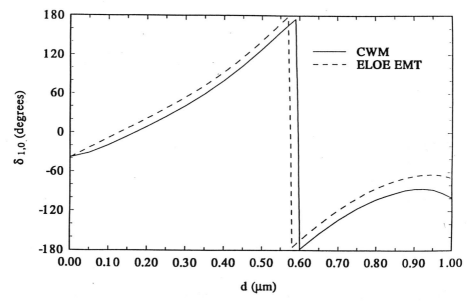

Figure 12.28 *s–p* phase difference versus grating depth for a dielectric grating with $D = 0.7$, $n_1 = 1.51$, $n_2 = 1.0$, $\alpha_i = 45°$, $\delta_i = 0°$, $\lambda = 780$ nm, $\lambda = 0.3\,\mu\text{m}$, $\Theta = 45°$ and $\Phi = 90°$.

Bouchitte and Petit, 1985). In this limit, the ordinary and extraordinary refractive indices (n_o and n_e) of an equivalent uniaxial film can be calculated, and the reflection and transmission coefficients of the equivalent film–substrate system can be determined using conventional stratified anisotropic medium analysis methods (Yeh, 1979; Mansuripur, 1990). This approach is known as an effective medium theory for the zeroth-order grating problem. The effective medium approach is attractive because it is intuitive and requires relatively little computation when compared to rigorous diffraction techniques.

At optical frequencies, the quasistatic limit is difficult to implement with conventional grating fabrication processes where the minimum Λ/λ that can be achieved is ~0.2. Recent work (Haggans *et al.*, 1993; Raguin and Morris, 1993) indicates that reflectances calculated using effective medium theory deviate significantly from those calculated with rigorous grating theories, especially as the grating period to wavelength ratio is increased. A more rigorous method for determining the reflection and transmission coefficients for lamellar gratings when the quasistatic limit is not satisfied is to solve Maxwell's equations in the grating region to obtain the propagation vector of the lowest-order eigenmode of the stratified medium. From this propagation vector magnitude, it is possible to calculate the indices for the effective uniaxial film. Rytov (1956) and Babin *et al.* (1993) discussed this approach for the case of normal incidence. Haggans *et al.* (1993) extended this approach to conical incidence geometries.

Figure 12.28 is a plot of the *s–p* phase difference $\delta_{1,0}$ computed using coupled wave theory and exact lowest-order eigenmode effective medium theory (ELOE EMT) (Haggans *et al.*, 1993) for this grating with $\Phi = 90°$, $D = 0.7$, $\lambda = 0.780\,\mu\text{m}$,

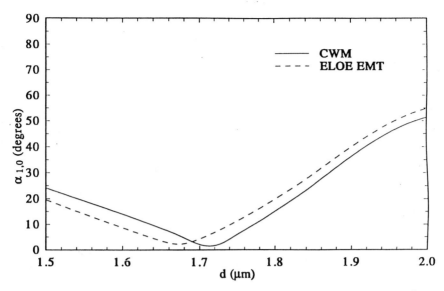

Figure 12.29 Amplitude ratio angle versus grating depth for the grating of Fig. 12.28 except that $\Phi = 57°$.

$\Lambda = 0.3\,\mu\text{m}$, $\Theta = 45°$, $n_1 = 1.51$, $n_2 = 1.0$, and linear polarization incident on the grating ($\alpha_i = 45°$ and $\delta_i = 0°$). For this grating, the effective indices of the ELOE EMT ($n_o = 1.1919$, $n_e = 1.1045$) give retardation values that agree with rigorous theory to within $\pm 20°$ for depths less than $0.8\,\mu\text{m}$.

Figure 12.29 is a plot of the polarization angle $\alpha_{1,0}$ for a plane wave reflected from a grating at a glass–air interface. The grating and beam parameters are $\lambda = 0.780\,\mu\text{m}$, $\Lambda = 0.3\,\mu\text{m}$, $\Theta = 45°$, $\Phi = 57°$, $D = 0.7$, $n_1 = 1.51$ and $n_2 = 1.0$, with linear polarization incident on the grating ($\alpha_i = 45°$ and $\delta_i = 0°$). The two curves in this figure are the predictions of coupled wave theory and the ELOE EMT ($n_o = 1.1954$, $n_e = 1.0933$). Although the minimum obtained with ELOE EMT effective indices occurs at a depth approximately $0.03\,\mu\text{m}$ less than the rigorous prediction, the quantitative agreement is within $\pm 5°$, even in this non-quasistatic regime ($\Lambda/\lambda = 0.385$).

This accuracy and the reduced complexity with respect to rigorous treatments suggests that ELOE EMT may be useful as a pre-design tool in the design of retardation elements at glass–air interfaces. To highlight this reduced complexity, we compare the use of ELOE EMT to CWM for the grating of Fig. 12.28. Using CWM, 45 spatial harmonics (N) were retained to give reasonable convergence of $\delta_{1,0}$. In contrast, the values of n_o and n_e obtained using ELOE EMT can be inserted into CWM with $N = 1$ (D is set equal to 1, and $\phi_{s,1,0}$ and $\phi_{p,1,0}$ are computed in separate runs). Since the run-time of CWM goes as N^3, use of ELOE EMT provides a significant saving in computational time.

12.4 Concluding Remarks

The purpose of this chapter was to provide an overview of the phase and polarization properties of grating structures and approaches for the design of optical components. Zero-order lamellar gratings were evaluated to demonstrate these properties, but the same approach can be applied to higher-order gratings with different surface profiles. Exact evaluation of these components requires rigorous vector diffraction analysis tools such as modal or coupled wave methods. Several designs for polarization components were presented using zero-order gratings along with their required fabrication tolerances. However, if specific design characteristics are required, optimization techniques must be incorporated with these models. The use of approximate effective medium theories is one way of guiding the evaluation and design of polarization-dependent zero-order gratings. With additional theoretical and fabrication process development, zero-order gratings can provide an important class of polarization control elements. The authors gratefully acknowledge Lifeng Li for the modal method (MM) and for his contributions to the manuscripts upon which this work is based.

References

ANDREWARTHA, J. R., FOX, J. R. and WILSON, I. J. (1979a) Resonance anomalies in the lamellar grating. *Optica Acta* **26**, 69–89.
(1979b) Further properties of lamellar grating resonance anomalies. *Optica Acta* **26**, 197–209.
AZZAM, R. M. A., GIARDINA, K. A. and LOPEZ, A. G. (1991) Conventional and generalized Mueller-matrix ellipsometry using the four-detector photopolarimeter. *Opt. Eng.* **30**, 1583–1588.
BABIN, S., HAIDNER, H., KIPFER, P., LANG, A., SHERIDAN, J. T., STORK, W. and STREIBL, N. (1993) Artificial index surface relief diffraction optical elements. *Proc. Soc. Photo-Opt. Instrum. Eng.* **1751**, 202–213.
BELL, J. M., DERRICK, G. H. and McPHEDRAN, R. C. (1982) Diffraction gratings in the quasistatic limit. *Optica Acta* **29**, 1475–1489.
BOTTEN, L. C., CRAIG, M. S., McPHEDRAN, R. C., ADAMS, J. L. and ANDREWARTHA, J. R. (1981a) The dielectric lamellar diffraction grating. *Optica Acta* **28**, 413–428.
BOTTEN, L. C., CRAIG, M. S. and McPHEDRAN, R. C. (1981b) Highly conducting lamellar diffraction gratings. *Optica Acta* **28**, 1103–1106.
BOUCHITTE, G. and PETIT, R. (1985) Homogenization techniques as applied in the electromagnetic theory of gratings. *Electromagnetics* **5**, 17–36.
BRYAN-BROWN, G. P., SAMBLES, J. R. and HUTLEY, M. C. (1990) Polarization conversion through the excitation of surface plasmons on a metallic grating. *J. Mod. Opt.* **37**, 1227–1232.
CESCATO, L. H., GLUCH, E. and STREIBL, N. (1990) Holographic quarterwave plates. *Appl. Opt.* **29**, 3286–3290.
DEPINE, R. A., BRUDNY, V. L. and SIMON, J. M. (1987) Phase behavior near total absorption by a metallic grating. *Opt. Lett.* **12**, 143–145.
GAYLORD, T. K., GLYTSIS, E. N. and MOHARAM, M. G. (1987) Antireflection surface-relief gratings on dielectric and lossy substrate. *J. Opt. Soc. Am.* **4**, 92.
GUPTA, S. D. (1987) Theoretical study of plasma resonance absorption in conical diffraction. *J. Opt. Soc. Am. B* **4**, 1893–1898.
HAGGANS, C. W., LI, L. and KOSTUK, R. K. (1993a) Effective medium theory of zeroth order lamellar gratings in conical mountings. *J. Opt. Soc. Am. A* **10**, 2217–2225.
HAGGANS, C. W., LI, L., FUJITA, T. and KOSTUK, R. K. (1993b) Lamellar gratings as

polarization components for specularly reflected beams. *J. Mod. Opt.* **40**, 675–686.

INAGAKI, R., GOUDONNET, J. P. and ARAKAWA, E. T. (1986) Plasma resonance absorption in conical diffraction: effects of groove depth. *J. Opt. Soc.* B **3**, 992–995.

INAGAKI, T., MOTOSUGA, M., YAMAMORI, K. and ARAKAWA, E. T. (1983) Photoacoustic study of plasmon resonance absorption in a diffraction grating. *Phys. Rev.* B **28**, 1740–1744.

JAHNS, J. and BRUMBACK, B. A. (1990) Integrated-optical split and shift module based on planar optics. *Opt. Commun.* **76**, 318–320.

JAHNS, J. and HUANG, A. (1989) Planar integration of free-space optical components. *Appl. Opt.* **28**, 1602–1605.

KOK, Y.-L. and GALLAGHER, N. C. (1988) Relative phases of electromagnetic waves diffracted by a perfectly conducting rectangular-grooved grating. *J. Opt. Soc. Am.* A **5**, 65–73.

KOSTUK, R. K., KATO, M. and HUANG, Y.-T. (1990) Polarization properties of substrate-mode holographic interconnects. *Appl. Opt.* **29**, 3848–3854.

LI, L. (1993) A modal analysis of lamellar diffraction gratings in conical mountings. *J. Mod. Opt.* **40**, 553–573.

MANSURIPUR, M. (1990) Analysis of multilayer thin-film structures containing magneto-optic and anisotropic media at oblique incidence using 2×2 matrices. *J. Appl. Phys.* **67**, 6466–6475.

MARCHANT, A. B. (1990) *Optical Recording*. Reading, Massachusetts: Addison Wesley.

MAYSTRE, D. (1984) *Progress in Optics*, Vol. 21, Wolf, E. (ed.). Amsterdam: Elsevier North-Holland.

McPHEDRAN, R. C., BOTTEN, L. C., CRAIG, M. S., NEVIERE, M. and MAYSTRE, D. (1982) Lossy lamellar gratings in the quasistatic limit. *Optica Acta* **29**, 289–312.

MOHARAM, M. G. and GAILORD, T. K. (1982) Diffraction analysis of dielectric surface relief gratings. *J. Opt. Soc. Am.* A **72**, 1385–1392.

MOHARAM, M. G. and GAYLORD, T. K. (1983) Three-dimensional vector coupled-wave analysis of planar-grating diffraction. *J. Opt. Soc. Am.* **73**, 1105–1112.

NAQVI, S. S. H. and GALLAGHER, N. C. (1990) Analysis of a strop-grating twist reflector. *J. Opt. Soc. Am.* A **7**, 1723–1729.

NEVIERE, M. (1980) Electromagnetic theory of gratings. In Petit, R. (ed.) *Topics in Current Physics*, Vol. 22. Berlin: Springer-Verlag.

PALIK, E. D. (ed.) (1985) *Handbook of Optical Constants of Solids*, p. 294. New York: Academic Press.

PETIT, R. (ed.) (1980) Electromagnetic theory of gratings. In *Topics in Current Physics*, Vol. 22. Berlin: Springer-Verlag.

POPOV, E., TSONEV, L. and MAYSTRE, D. (1990) Gratings: – general properties of the Littrow mounting and energy flow distribution. *J. Mod. Optics* **37**, 367–377.

RAGUIN, D. H. and MORRIS, G. M. (1993) Antireflection structured surfaces for the infrared spectral region. *Appl. Opt.* **32**, 1151–1167.

RYTOV, S. M. (1956) Electromagnetic properties of a finely stratified medium. *Sov. Phys. JETP* **2**, 466–475.

SHARP CORPORATION (1988) *Laser Diode Users Manual*, Ref. No. HT509D.

SIMON, J. M. and SIMON, M. C. (1984) Diffractive gratings: a demonstration of the phase behavior in Wood anomalies. *Appl. Opt.* **23**, 970.

YARIV, A. and YEH, P. (1984) *Optical Waves in Crystals*, p. 207. New York: John Wiley.

YEH, P. (1979) Electromagnetic propagation in birefringent layered media. *J. Opt. Soc. Am.* **69**, 742–756.

Index